Underwater Intervention Conference 2016 (UI 2016)

New Orleans, Louisiana, USA
23 - 25 February 2016

ISBN: 978-1-5108-2297-9

Printed from e-media with permission by:

Curran Associates, Inc.
57 Morehouse Lane
Red Hook, NY 12571

Some format issues inherent in the e-media version may also appear in this print version.

Printed by Curran Associates, Inc. (2016)

For permission requests, please contact Underwater Intervention
at the address below.

Underwater Intervention
5206 Cypress Creek Parkway
Suite 202
Houston, Texas 77069
USA

Phone: 281-893-8539

www.underwaterintervention.com

Additional copies of this publication are available from:

Curran Associates, Inc.
57 Morehouse Lane
Red Hook, NY 12571 USA
Phone: 845-758-0400
Fax: 845-758-2633
Email: curran@proceedings.com
Web: www.proceedings.com

TABLE OF CONTENTS

*** Formatting Issue - Best Available Paper/Graphic ***

OCEANEERING®

Dynamic Mission Planning for AUVs

Sara Rayburn
Underwater Intervention
February 23, 2016

OCEANEERING®

1. Background/Motivation
2. System Design
3. Preliminary Results

*** Formatting Issue - Best Available Paper/Graphic ***

OCEANEERING®

Dynamic Mission Planning Background

Underwater Intervention 2014 to Now

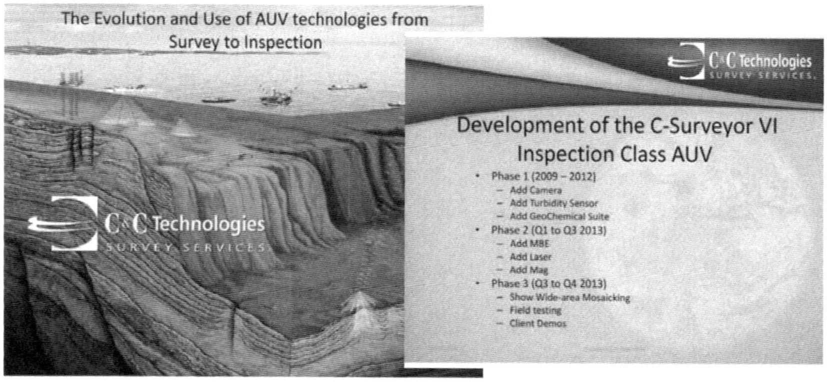

OCEANEERING®

*** Formatting Issue - Best Available Paper/Graphic ***

C-Surveyor VI Inspection Class AUV

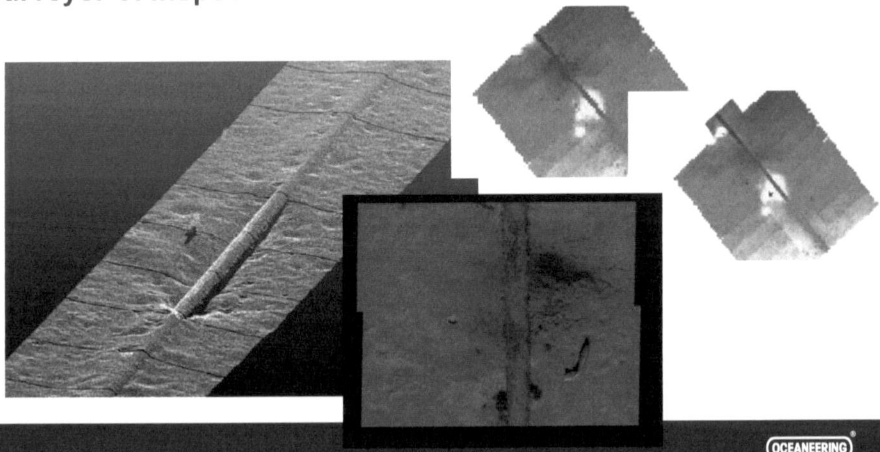

Inspection Class AUV Development 2016

- Pipeline tracking and following
- Dynamic mission planning
- Feature-based local navigation correction
- Robust object detection and obstacle avoidance
- Client Demos

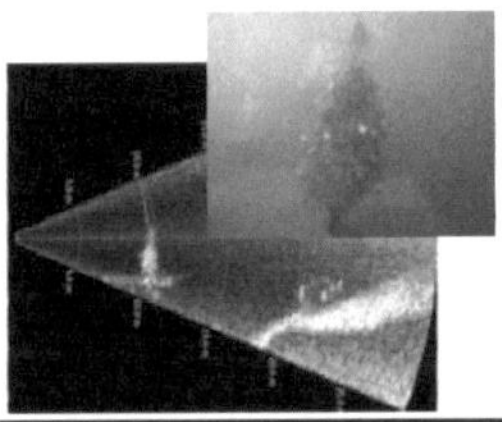

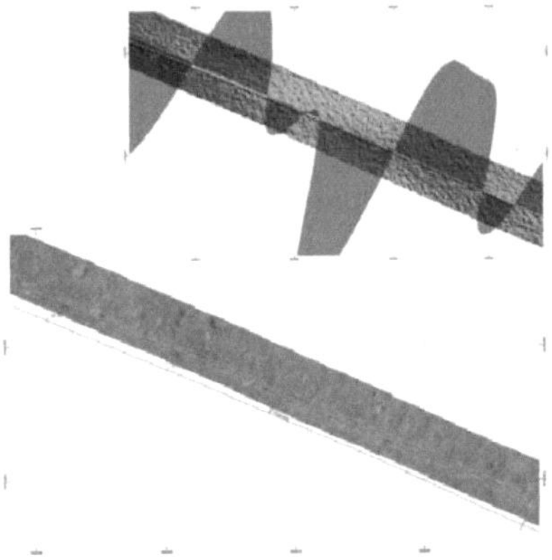

***** Formatting Issue - Best Available Paper/Graphic *****

Inspection Class AUV Development 2016

- Pipeline tracking and following
- Dynamic mission planning
- Feature-based local navigation correction
- Robust object detection and obstacle avoidance
- Client Demos

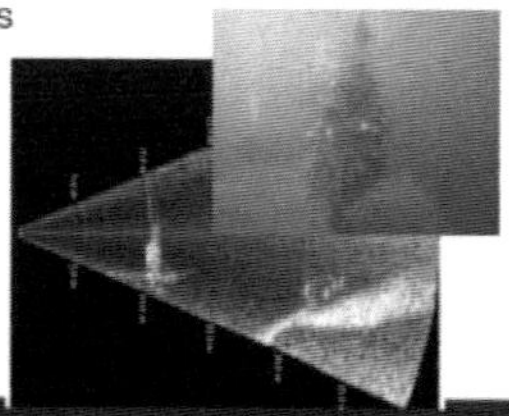

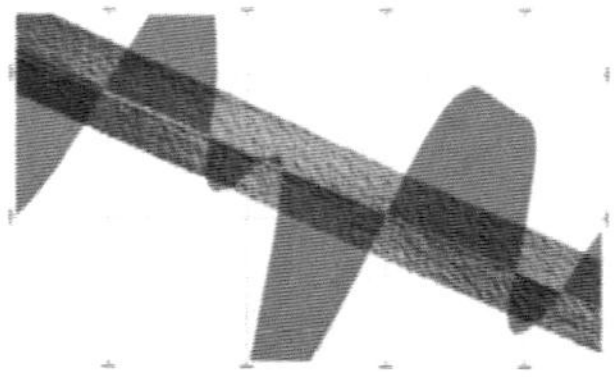

Dynamic Mission Planning for AUVs

Advantages of dynamic mission planning

- Efficient dive utilization (cost reduction)
  - Reduce need for multiple dives to capture relevant information.
  - Reduce need for multiple passes to ensure sensor coverage
  - Reduce need for preliminary hazard survey.
- Safety of operations
  - Object identification and avoidance in changing environment.
- Leverage operator insight for mission safety and high level planning, leverage real-time sensor data for reaction to changing environment.

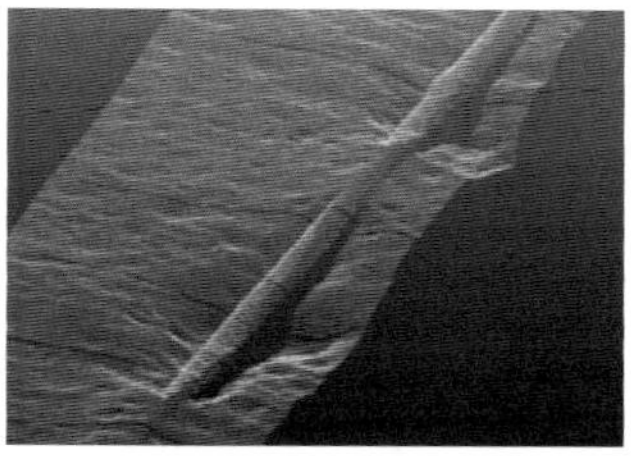

*** Formatting Issue - Best Available Paper/Graphic ***

Mission Plan

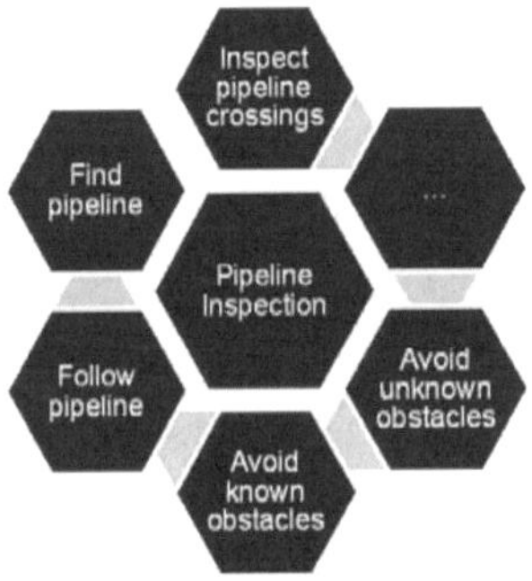

Dynamic Mission Planning for AUVs

Triggered Behaviors

- Real time sensor analysis allows autonomous behaviors to be triggered.
- Pre-dive mission plan enables behaviors of interest:
  - Leek and seep inspection
  - Pipeline following
  - Pipeline crossing inspection
  - Etc.
- AUV executes relevant behavior when trigger is detected in environment, reporting to operator and tagging data for post-processing.

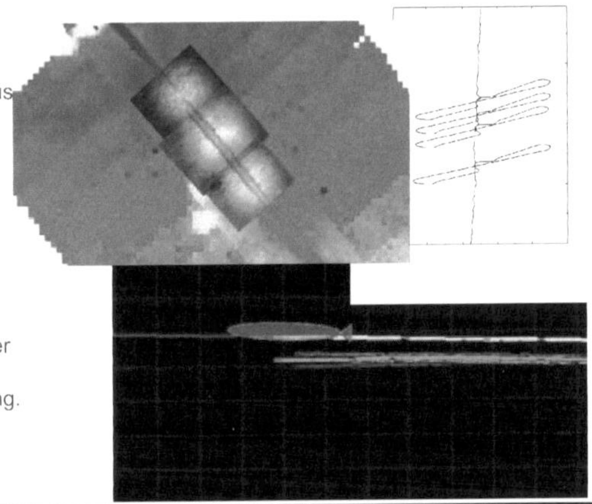

*** Formatting Issue - Best Available Paper/Graphic ***

Mission Plan

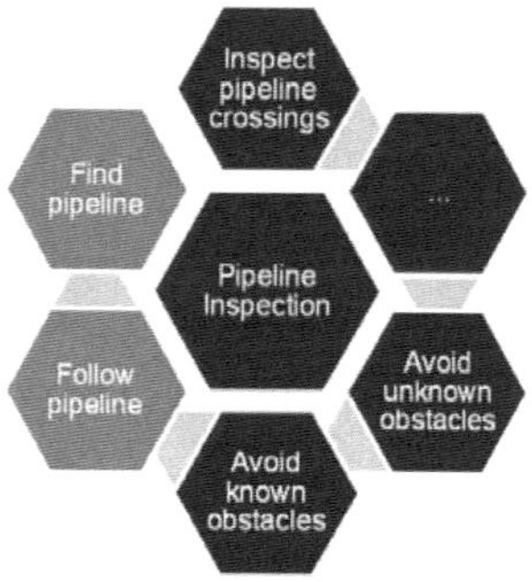

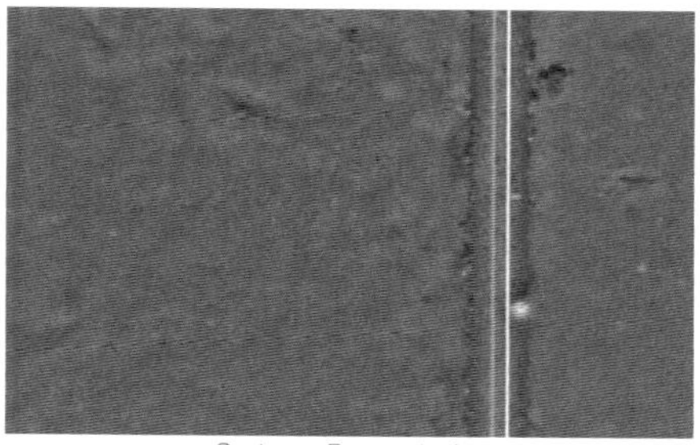

Customer Demonstration

Green line: Auto-tracked pipe

White line: Pipeline asbuilt

OCEANEERING®

Mission Plan

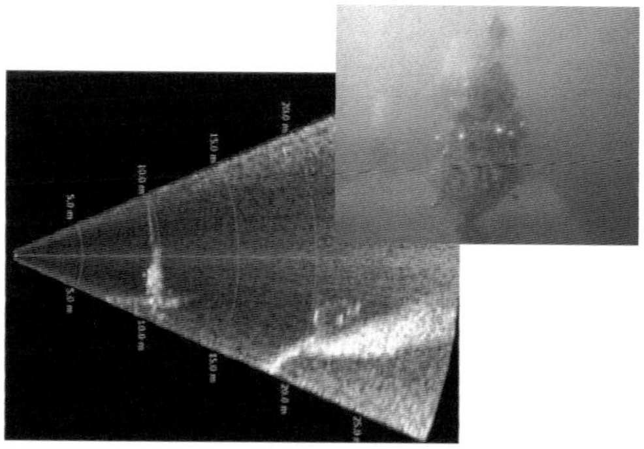

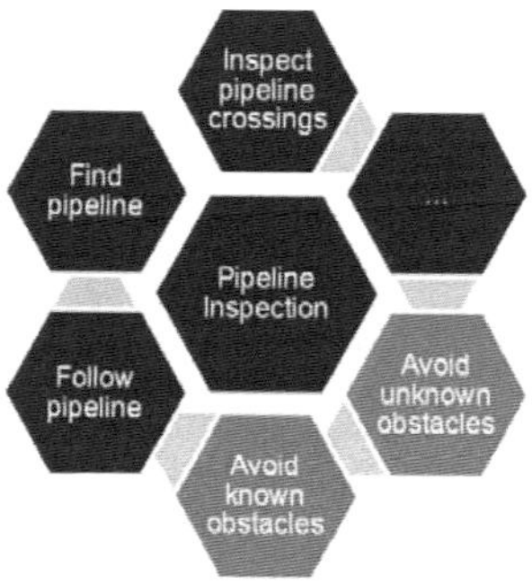

OCEANEERING®

*** Formatting Issue - Best Available Paper/Graphic ***

Mission Plan

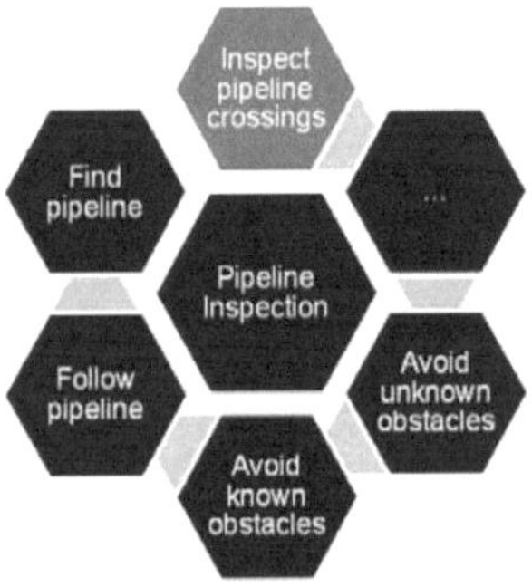

- Detect pipeline crossings
- Run 500m along the crossing pipeline
- Steer back to the primary pipeline prior to the detected crossing to ensure complete sensor coverage

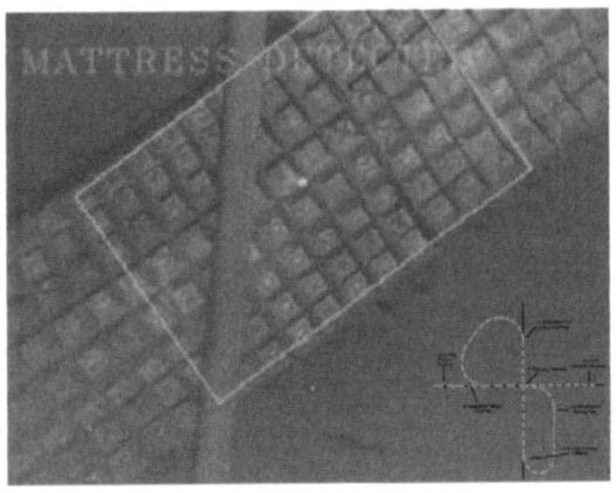

Dynamic Mission Planning for AUVs

Trials and Demonstrations

- 2015 – Autonomous pipeline inspection
  - Successes:
    - Pipeline detected and autonomously tracked for 20 km on several runs.
  - Challenges:
    - Water quality varied across length of pipeline.
- 2016 – Autonomous pipeline inspection
  - Successes:
    - Across track distance to tracked pipe reduced substantially across entire run.
- 2016 – Autonomous pipeline crossing inspection
  - Successes:
    - Inspected section of pipeline with 4 pipeline crossings. Each was detected and inspected from both directions.
  - Challenges:
    - Water quality
    - Reduced sensor set (camera, multibeam)
    - Pipe crossings are not visible in camera imagery, must be inferred.

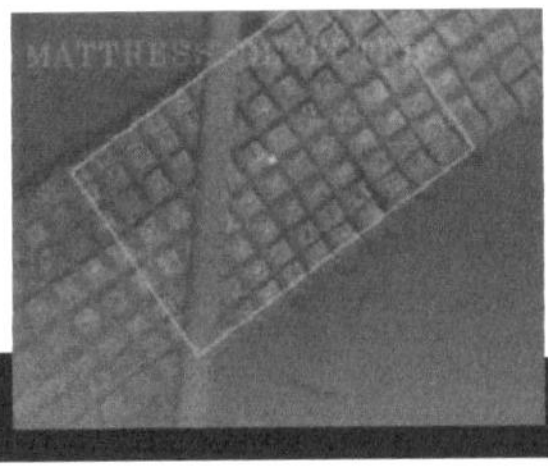

*** Formatting Issue - Best Available Paper/Graphic ***

OCEANEERING®

Connecting What's Needed with What's Next™

Robert Kraft, Vulcan Inc. Carl Barrett, 3U Technologies LLC.	Recovery of the Historical Bell from HMS Hood	Underwater Intervention February 2016

1. Abstract

On the 24th of May, 1941, a legendary WWII battle occurred when the British Battlecruiser HMS Hood engaged the German Battleship Bismarck in the Straits of Denmark, as the Bismarck attempted to break out into the North Atlantic. The sinking of HMS Hood during this engagement, resulting in 1,415 lives lost with only 3 survivors, has been a source of lore ever since.

On the 7th of August, 2015, a research team led by US philanthropist and entrepreneur Paul G. Allen successfully recovered the Bell of the battlecruiser HMS Hood, resting at a depth of 2,850 meters. Operations were performed aboard Mr. Allen's expedition yacht M/Y OCTOPUS, utilizing his 3000 meter rated ROV system and a suite of custom built recovery tooling. The operation was sanctioned and licensed by the UK Ministry of Defence and the Royal Navy. Bell restoration is currently in process at the dockyard facility in Portsmouth in preparation for display at the National Museum of the Royal Navy in Portsmouth England to respectfully serve as a tangible and fitting memorial for the 1,415 lives lost at sea. Bell unveiling is scheduled for 24 May 2016 on the 75th anniversary of HOOD's loss.

"This year marks the 70th anniversary of the conclusion of World War II, and this effort commemorates the hundreds of brave sailors who were lost at sea," said Allen. "It is a true honor to undertake the expedition to recover the Bell from 'The Mighty Hood'."

The Bell was first discovered and photographed during a July 2001 expedition by Bluewater Recoveries Ltd., where it was found wedged within significant steel debris, lying well away from the battlecruiser's main hull sections. In 2012 Mr. Allen led an expedition to the wreck site, but was unable to recover the Bell due to difficulties in extracting the Bell from surrounding debris and a very short weather window. However, video and photographic data from the 2012 expedition proved to be invaluable for planning the return mission and developing the appropriate tooling for the ultimately successful 2015 recovery operations.

This paper provides a brief history of the Hood Bell as provided by the written record on the Bell, followed by a review of the operations and recovery techniques utilized. The presentation will feature stunning photographs and video taken during recent recovery operations.

Robert Kraft is the Director of Subsea Operations with Vulcan Inc. Vulcan Inc., wholly owned by Mr. Allen, operates 3 private, non-commercial vessels internationally. Subsea assets include ROV's, AUV's and manned submersibles with capabilities to 6000 meters. Mr. Kraft has 23 years of experience in all aspects of subsea operations and intervention work.

Carl R. Barrett III (cbarrett@3utech.com) is a Project Manager and Senior Engineer with 3U Technologies LLC. (3U). 3U is an engineering and consulting organization specializing in projects Underground, Underwater and Under-Ice. Mr. Barrett holds both BSME and MSEE degrees and has 30 years of experience involving deep sea robotic systems, including specialized tooling for specific operations and complete robotic vehicles and systems.

2. Foreword

The following paper and presentation is not intended as a historical reference on the HMS Hood, nor the engagement with the Bismarck. The focus of this paper is the Bell recently recovered from the wreck site and the recovery operations occurring in August of 2015 in the Denmark Straits. The goal is to provide a record of the recovery operations as well as presenting observations and findings of the recovery team. But, this story is also intended to spin a new yarn, covering the last moments of the Bells service to the crew. After all, we know all sailors love a good yarn. The hope is that by adding this segment to the Bell's historical record, we are able to pay homage to the sailors lost 75 years ago, and honor those who made the ultimate sacrifice for their country.

3. The Bell

This story deserves to start at the end, as that is where the feature player comes in (see Figure 1). On initial examination immediately after recovery, the Bell was found to be a significant historic artifact, featuring a written record, covering 50 years of British Naval history. The Bell's history is literally written in Bronze on the Bell itself. Bronze that still shines after 125 years, including 74 years at the bottom of a cold ocean, wrapped in decaying steel.

According to very legible inscriptions along the sound rim of the Bell (lowest edge, shown in Figure 2), the Bell was first deployed on the HMS Battleship Hood, christened in 1891. In other words, the Bell had sailed on two warships over time, both named Hood.

Figure 1 - The Bell from HMS Hood

Figure 2 – Sound Rim Inscriptions

Two inscription rings along the sound rim read as follows:

THIS BELL WAS PRESERVED FROM HMS BATTLESHIP HOOD 1891-1914 BY THE LATE REAR ADMIRAL *

THE HONOURABLE HORACE HOOD K.C.B. D.S.O. M.V.O. KILLED AT JUTLAND ON 31ST MAY 1916 *

A second inscription was later added for the Christening of the HMS Battle Cruiser Hood. This inscription is located higher up, along the waist of the Bell. Figure 3 shows this inscription, which reads as follows:

IN ACCORDANCE WITH THE WISHES OF LADY HOOD IT WAS PRESENTED IN MEMORY OF HER HUSBAND TO HMS HOOD BATTLE CRUISER WHICH SHIP SHE LAUNCHED ON 22nd AUGUST 1918

Figure 3 – Second Inscription Commemorating the Christening of HMS Battle Cruiser Hood

These inscriptions show that this Bell was first deployed in 1891 on the HMS Battleship Hood. Admiral Horace Hood recovered this Bell after HMS Battleship Hood sank in 1914. Admiral Hood was later killed on the 31st of May, 1916, during the battle of Jutland, a keystone British naval battle of WWI near the Jutland peninsula of Denmark. His widow, Lady Hood was then presented with this Bell at some point after its first recovery.

Later, Lady Hood presented the Bell to the HMS Battle Cruiser Hood during commissioning, the 22nd of August, 1918. This same Bell then sailed on the HMS Battlecruiser Hood until sinking on the 24th of May, 1941. The inscriptions show that this Bell served through 50 years of British Naval history. As we now know, the Bell waited in the depths a further 74 years before returning, to tell its story again, after being salvaged from a vessel named Hood for the second time, another interesting point of this yarn!

4. Initial Wreck Site Investigations

4.1. Initial Discovery

HMS Hood was first located and documented in July 2001 during a Bluewater Recoveries Ltd. expedition. This operation included a strict "Look but Do Not Touch" policy. During this expedition, the Bell was located and photographed, but was not disturbed. While this initial investigation provided location and general site information for the Bell's final resting place, the difficulty of potential recovery was not apparent.

4.2. 2012 Recovery Attempt

In the summer of 2012, under license from the UK Ministry of Defence and the Royal Navy, M/Y OCTOPUS sailed to location on an expedition to recover this Bell, to be used as a memorial to the sailors lost. With limited information regarding extraction requirements, the crew of the OCTOPUS utilized their ROV to perform an initial dive to the wreck site at a depth of 2,850 meters. During this dive, the crew located the Bell and attempted to attach a customized snap hook through the Bell's crown. The bell location was found to be deep inside of an overhanging metal structure, resulting in significant impediments to working access. Due to many factors, the hook did not fit properly and the Bell was not initially captured or recovered. However, during these operations, the Bell was moved sufficiently to show the true extent of entrapment by various pieces of steel debris. In short, at this point it became obvious that the Bell was not going to come easily!

Figure 4 – Manipulator with Hook in Foreground. Bell in the Background. From 2012 Expedition.

4.3. Diving Conditions

It is worth noting some unique issues associated with the wreck site. The HMS Hood sits in 2,850 meters (9,350 feet) of water, at the bottom of the Denmark Strait. While depth alone prevented access to the site for years, depth alone is not an impediment to modern robotic diving systems. In general, Iceland and the Arctic Ocean are to the east, Greenland is to the north-west and the Atlantic Ocean is to the south-west. The sloping seabed and water density differences combine to create a 3,500 meter "Cataract" that has been termed the "World's Largest Waterfall."

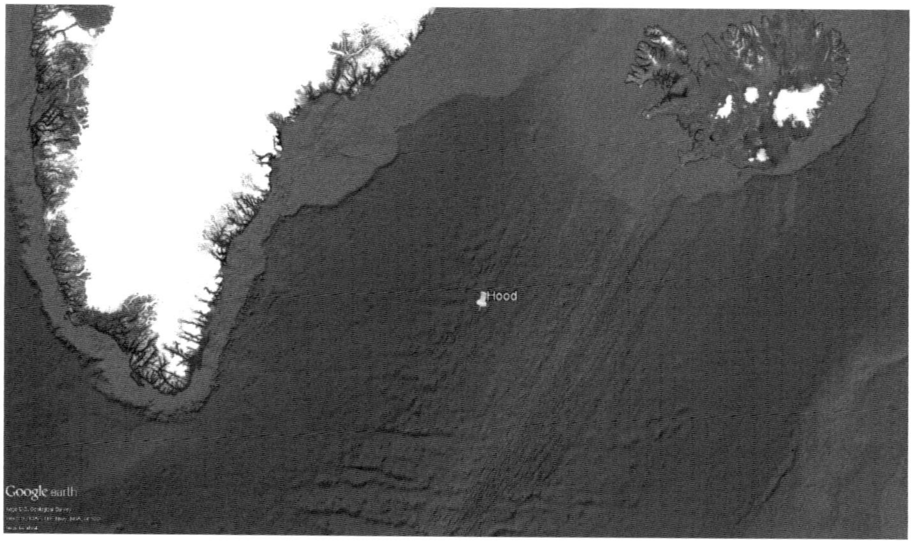

Figure 5 – General Location of HMS Hood in the Denmark Straits

The eastern, arctic reaches of the Strait are relatively shallow, with depth dropping sharply by 3,500 meters moving towards the south-west. The Strait is open to the cold waters of the Arctic Ocean to the north-east with the warmer Atlantic and Gulf Stream water to the south-west. The Hood and Bismarck battled here because the Bismarck was looking for a route to the open Atlantic and the Hood was sent to prevent that from occurring.

The unique oceanographic conditions result in a flow of dense, cold water sinking below lower density, warmer surface water. This cold water then runs along the seabed slope, generally north-east to south-west. Warmer surface currents in the area run generally towards the west, but shift north to south, leading to variable sea conditions. The wreck site is subject to significant and constant bottom current, with varying velocities of up to approximately 3 knots.

From the standpoint of ROV operations, the worksite can be extreme. High bottom currents directly impede ROV progress, while surface countercurrents play havoc with the umbilical. As a free swimming, live boat system, OCTOPUS's ROV has no TMS to stabilize the umbilical. On the positive side, the constant current insures excellent visibility. As for the Hood herself, a constant flow of oxygen rich arctic water provides an environment conducive to high oxidation rates and supports significant marine flora and fauna.

4.4. 2012 Dive #2

Alternative snap hooks had been configured for the 2012 operations, so the ROV was recovered from the first dive, re-configured with alternative tooling and sent back to site. Bottom currents were running strong during this expedition and flying to the specific location took significant time. The Bell worksite is protected by the wreckage, which provides shelter from bottom currents, but not from current and wave effects on the 3,000 meter umbilical. By the time the ROV was settled back into location for a second extraction attempt, weather conditions had started to deteriorate and the dive was terminated. Unfortunately, the fickle weather conditions of the Denmark Strait continued to worsen and the expedition was aborted after a single extraction attempt, to the disappointment of all.

5. 2015 Mission Planning

While viewed as a failure at the time, the 2012 expedition was actually critical to the ultimate success of the 2015 expedition. The first response was significant power and propulsion upgrades to the ROV in 2013. The system deployed in 2012 was 30 Hp rated. Upgrades in 2013 brought this to 67 Hp, with larger thrusters, double the thrust and a new Kevlar umbilical. An auxiliary tooling system was added to drive high powered tooling. Video and photographic records of the 2012 expedition also provided a much improved understanding of the obstacles to recovery plus a better understanding of the Bell features and size.

Figure 6 – Zoomed View from 2012 Expedition Video Archives

5.1. <u>Other Expeditions</u>

After completion of upgrades to the ROV, and acquisition of a 1500 meter rated AUV, M/Y OCTOPUS set off on a series of very successful expeditions in the Pacific. In late 2014, M/Y OCTOPUS performed a detailed survey of Iron Bottom Sound in the Solomon Islands, the site of multiple WWII Naval battles, including The Battle of Guadalcanal. During March of 2015, M/Y OCTOPUS successfully located and documented the wreck of the Japanese Yamato Class Battleship Musashi, the pride of the WWII Japanese navy, sunk on the 24th of October, 1944 in the Sibuyan Sea, Philippines.

Throughout these new expeditions, the crew tuned in the upgraded ROV and operations in preparation for a return to HMS Hood. After completion of these successful expeditions, planning commenced for a second recovery attempt of the Bell from HMS Hood. However, for this trip, the team intended to arrive fully prepared for success. Recovery planning started with a detailed review of video and photographic documentation from the 2012 trip. HD quality video and lighting by the ROV provided an excellent record with which to study the Bell in-situ. With no records from 1891 to provide details, comparison measurements were made between features on the Bell and known references included in the video content. From these measurements, a facsimile 3D software model was generated. This model and knowledge of the impediments to recovery (gained from the video record) provided the basis for detailed tooling design.

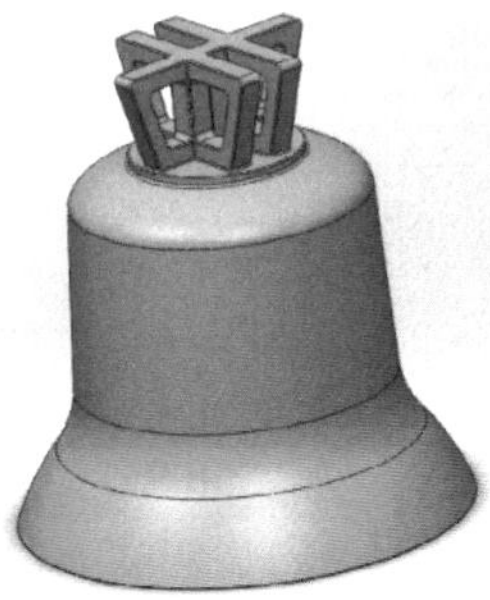

Figure 7 – Bell Model

5.2. New Tooling

As a starting point, it was determined that a suction cup system would be the primary recovery tool, however, to maximize the probability of success it was decided that several different types of tools should be developed and built. The snap hooks utilized previously were deemed sufficient for hooking a crown eye. To round out the tool kit, two different hydraulic powered C-clamp tools were designed and fabricated to provide a range of alternative recovery capabilities.

5.2.1. Suction Cups

A suction cup system allows for a wide range of surface conditions and dimensions, without concern of tool damage to the Bell. A 150 mm, bellows style suction cup was selected as the optimum match to the Bell dimensions, according to the drawing developed. In order to allow for all possibilities, a range of different size suction cups, as shown in Figure 8, were procured from the vacuum handling industry.

Figure 8 – Suction Cup Selection

Power for the suction cup system would require a pump solution. While something along the lines of a Tritech AnchorZip pump system would be optimum for such an operation, the ZipJet system also provides an eductor style dredge system which was deemed a plus for future operations. A ZipJet pump system was therefore selected and modified for purpose as shown in Figure 9. Modifications involved bypassing the eductor to generate 4 ATM of suction directly from the pump inlet. 4 ATM of pressure applied to a 150 mm suction cup produces over 300 Kgf of holding force, well in excess of the ROV manipulator capacity. Ultimately, the suction cup holding capacity was impressive, including holding tight while the 19 mm SST round bar "T" handle was bent during the recovery operation!

Figure 9 - ZipPump

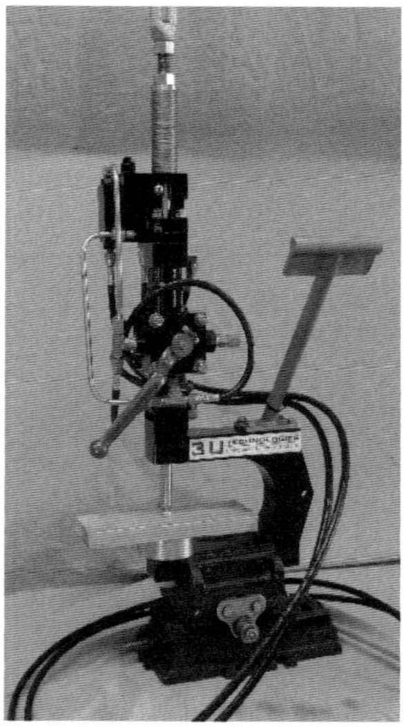

Figure 10 – Large Clamp Tool

5.2.2. C-Clamp Tools

The hydraulic C-clamp tool kit included two size clamps, shown in Figures 10 and 11. A larger C-clamp tool (110 mm throat) was configured with soft jaws to grip the Bell lip, as well as assist in debris removal as required. A smaller (38 mm throat) unit was designed specifically to grip a crown ear or capture the ear by passing the cylinder rod through a crown eye. Potential configurations of each tool included different grippers and hydraulic configurations (auto release or manipulator valve release) to allow for a range of potential scenarios.

After completion of shop testing, the tooling was delivered to M/Y OCTOPUS in early summer, 2015 and tested with the ROV. M/Y OCTOPUS and her ROV were ready for an early August expedition to the HMS Hood!

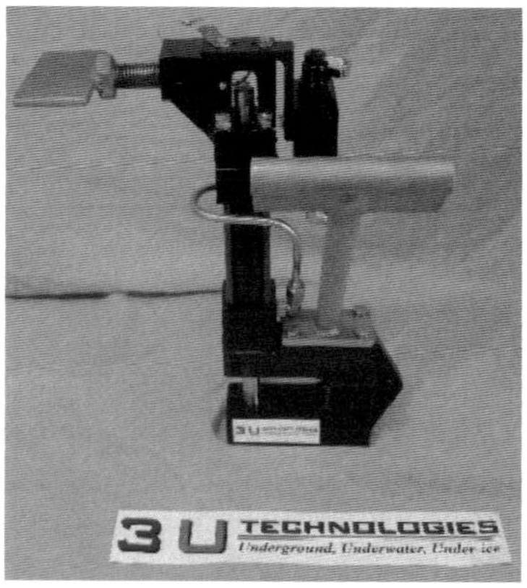

Figure 11 – Small Clamp Tool

6. Recovery Operations

6.1. Dive #1

MY OCTOPUS reached the dive site early the 7th of August, 2015. The first dive to the site was underway by 2 am. Conditions were more amenable to diving operations than encountered in 2012. Specifically, the Cataract driven seabed current was lower than experienced in 2012, but still more than sufficient to maintain visibility. With the upgraded ROV performance capability, currents did not prove to be a significant issue. However, Murphy's Law was in evidence during the first dive, with failure of both manipulators prior to any recovery. By noon, August 7th, after 10 hours on station, the ROV had been to the site and was back on deck for repair. After completion of manipulator repairs, the ROV was back on location 7 hours later to commence operations.

Figure 12 – ROV Being Deployed from MY OCTOPUS

6.2. Dive #2

Figure 13 shows commencement of final recovery operations using a suction cup system.

Figure 13 – Suction Cup Approach to Bell

With the recovery suction cup tool positioned inches away from the Bell, the pump was switched on and the tool was instantly attached to the side of the Bell! Unfortunately, working the Bell free was going to prove difficult. For what seemed like an hour (but actually only 15 minutes according to the log) the ROV operators worked to free the Bell, as shown in Figures 14 and 15. Ultimately, it was decided that the suction cup was holding just fine and the debris was showing signs of movement. It was time to give it a go! After a few hard yanks and some maneuvering teamwork, the Bell was free and moved to a seabed location away from the wreck.

Figure 14 – Bell Extraction

The suction cup tooling performed to expectations and provided a key advantage in the ability of the ROV to maneuver the Bell during removal. The suction cup provided an essentially rigid attachment of the ROV manipulator to the Bell. Had a snap hook been utilized for this operation, removal from the debris entrapment would have proved much more difficult because the ROV would have had much less control over the Bell.

Figure 15 – Finally Free!

Since a vacuum is not considered a reliable solution for transport to the surface, the Bell was rigged with a snap hook through a crown eye as a recovery safety line. One of the eyes appeared to be filled with hard steel, but was ultimately found to contain loose rust. The open eye was chosen for attachment of the recovery line. As would ultimately be noted, this turned out to be a lucky choice!

Figure 16 - Clearing Wreck with the Bell

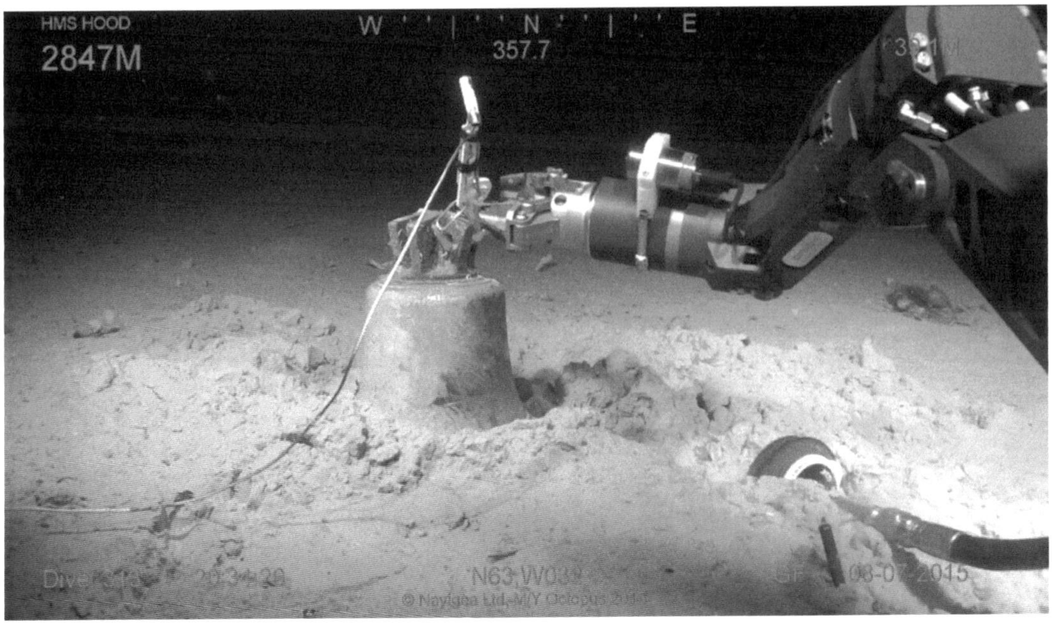

Figure 17 – Bell Rigged for Recovery

With a safety line successfully attached, the Bell was brought onto a platform fitted on the front of the ROV, and recovery to the surface commenced. By 22:30 of day one on site, the Bell from the HMS Hood finally broke the surface, after 74 years of deep submersion.

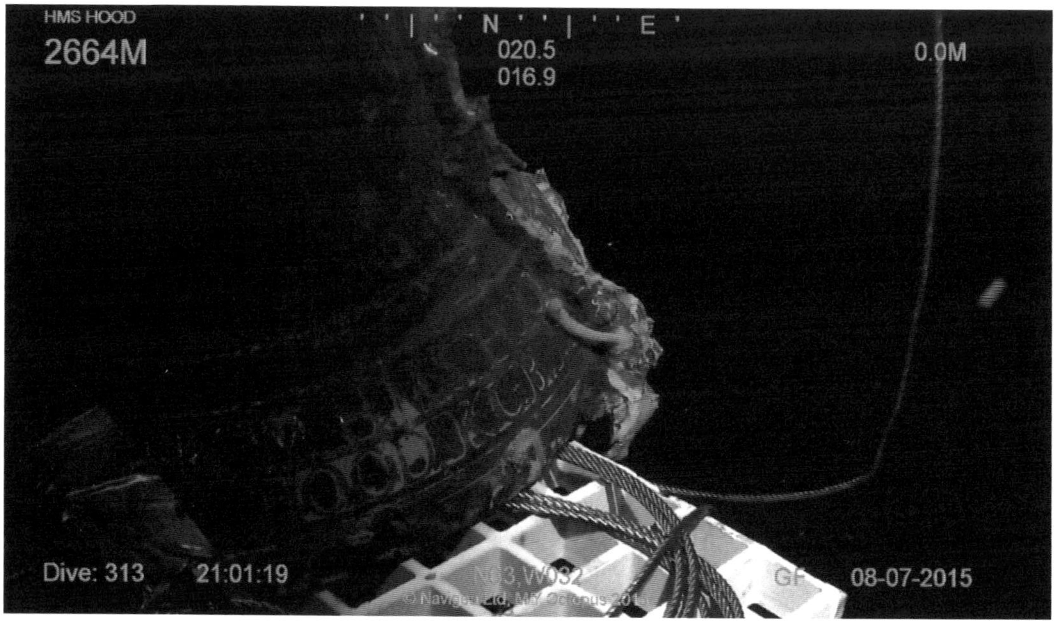

Figure 18 – The Long Trip to the Surface

7. A Subsea Memorial

While recovery of the Bell was the priority goal for the expedition, addition of a memorial at the wreck site was also requested by the family members of the crew and the H.M.S. HOOD Association. The 2001 expedition placed a memorial plaque dedicated to the sailors lost. For this expedition, the plan was to unfurl a British White Ensign flag at the Hood's bow, near the previously placed plaque. In preparation, the Ensign flag was outfitted with flotation at the top and weighted lines at the bottom to achieve a "Flying Flag" effect.

Figure 19 – Memorial Plaque Left in 2001

After standing by for 20 hours to wait out changing weather conditions, a memorial dive to the site was performed and then work to unfurl the flag began. During unfurling, the flag and weight lines became entangled, a less than desired result. After several attempts and a delicate touch, the entanglement was resolved and the flag unfurled to full effect. At this point, the dive was immediately terminated due to building weather conditions. After a relatively rough recovery, the ROV was brought on deck, having completed the last dive to the 'Mighty Hood'.

Figure 20 – British White Ensign Fully Deployed

8. Epilogue

On initial removal of the Bell from the ROV, a single crown ear immediately broke off of the Bell in the hands of one of the crew. While damaging such an artifact was an obvious concern, the broken ear could be repaired and it was considered fortunate that the ear had been recovered to the surface with the Bell versus being lost forever. It was also fortunate that the safety hook had been attached to the other ear! On further inspection it was found that a majority of the cracked section showed weathering from long term exposure (not corrosion), while a roughly 12 mm area of shiny metal, shown in Figure 21, revealed the small section of casting that had remained intact through the years. Realizing that cast bronze is brittle, it was surmised that the ear was damaged during the sinking and must have endured a significant impact, which almost resulted in loss of this segment.

Figure 21 – Broken Crown, Showing Fresh Metal

After return to port in Reykjavík, Iceland, the HMS Hood Bell was flown back home to the UK National Museum of the Royal Navy in Portsmouth, England to undergo restoration. On the 24th of May, 2016, on the 75th anniversary of HOOD's loss, the Museum and the H.M.S. Hood Association will host a coming out celebration as the Bell goes on display at the museum, forming part of the Jutland exhibition in the "Aftermath of the Battle" section of the Museum through 2018.

Speaking on behalf of the ROV operators and the entire M/Y OCTOPUS crew, it was an honor to be involved in this project, bringing back this important artifact to serve as a memorial to the 1,415 crew lost in the sinking of the HMS Hood. It was a further honor to assist with arranging the subsea memorials to the lost sailors of HMS Hood.

Figure 22 – MY OCTOPUS, Recovery Operations Crew

9. Further Information

For further information on upcoming Bell displays and History of H.M.S. Hood, please visit the following websites:

- www.hmshood.com
- http://www.nmrn-portsmouth.org.uk/
- http://www.historicdockyard.co.uk/site-attractions/attractions/national-museum-of-the-royal-navy-portsmouth

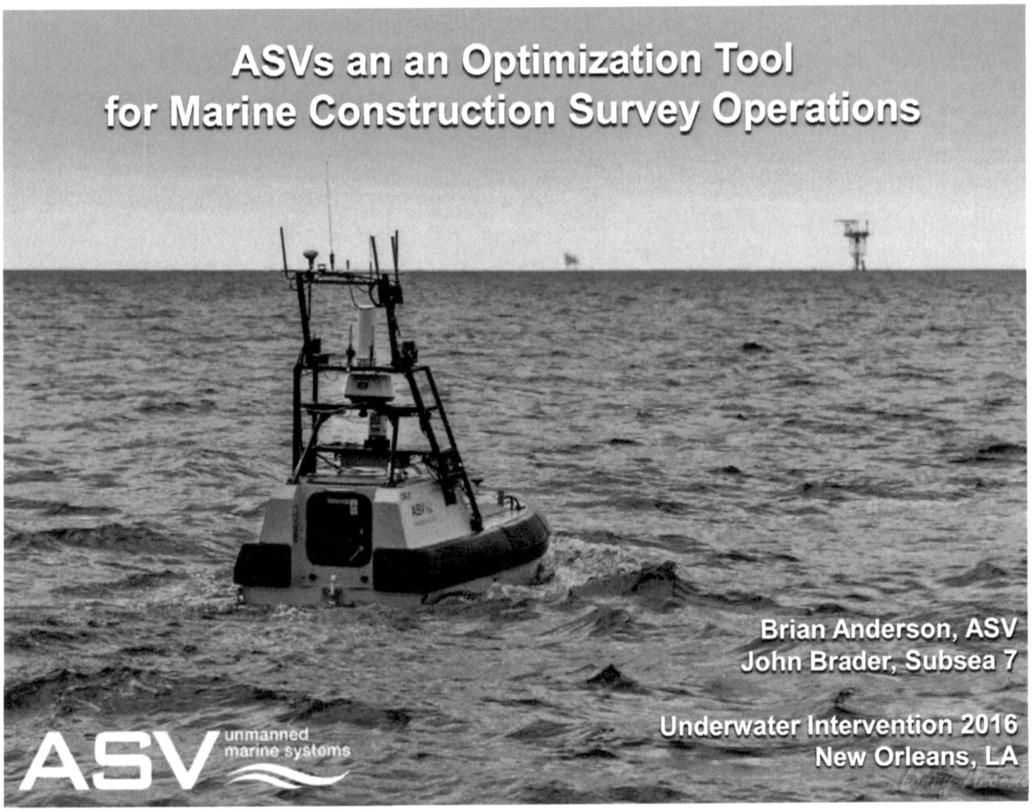
ASVs an an Optimization Tool
for Marine Construction Survey Operations
Brian Anderson, ASV
John Brader, Subsea 7
Underwater Intervention 2016
New Orleans, LA
ASV unmanned marine systems

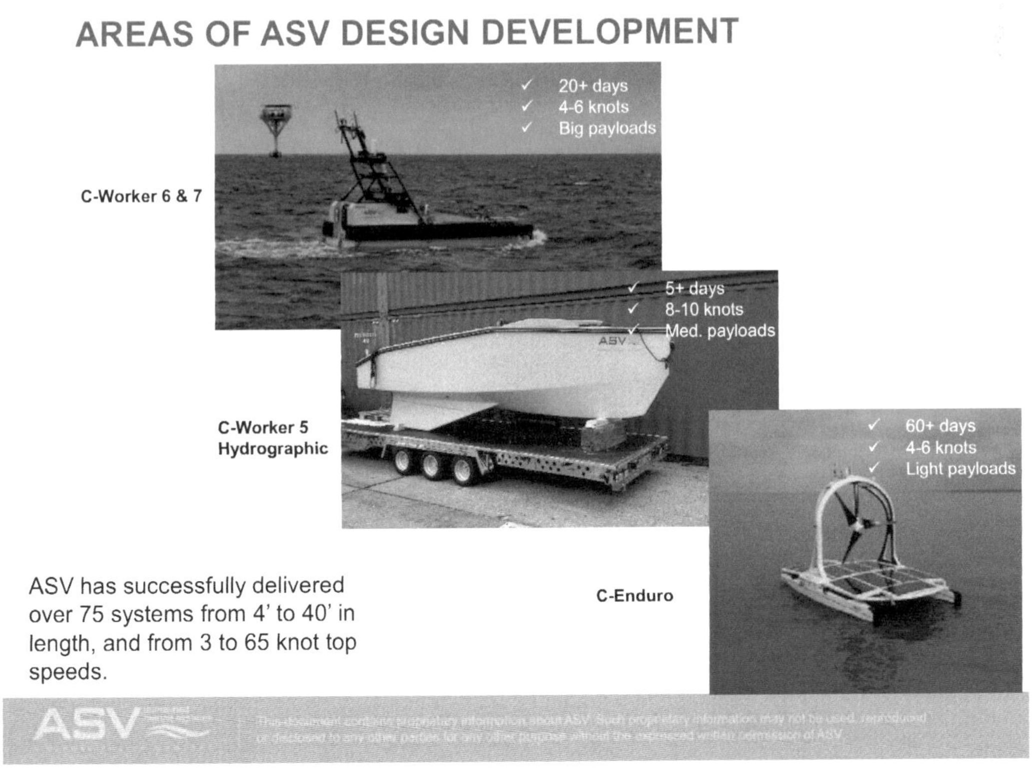
AREAS OF ASV DESIGN DEVELOPMENT
✓ 20+ days
✓ 4-6 knots
✓ Big payloads
C-Worker 6 & 7
✓ 5+ days
✓ 8-10 knots
✓ Med. payloads
C-Worker 5
Hydrographic
✓ 60+ days
✓ 4-6 knots
✓ Light payloads
C-Enduro
ASV has successfully delivered over 75 systems from 4' to 40' in length, and from 3 to 65 knot top speeds.
ASV

AREAS OF DEVELOPMENT AND OPERATIONS

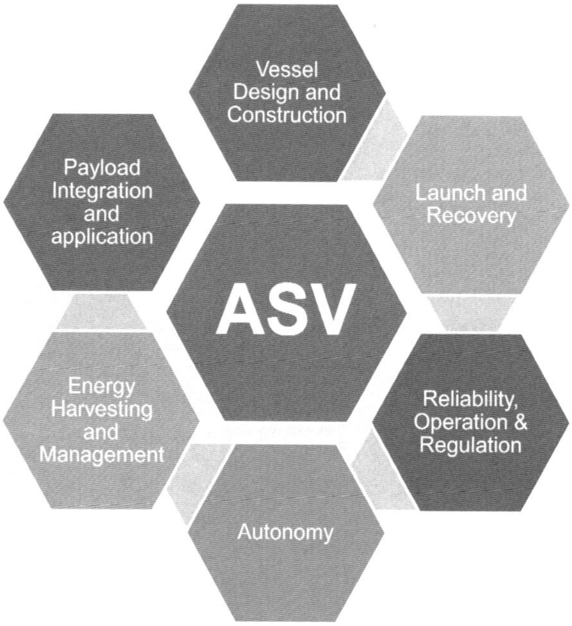

C-Worker 6

- 6m Vessel
- Industrial Grade
- 20 Day Endurance
- 4 Knot Cruise
- Dual Redundancy
- Radar, AIS, Cameras
- 1.5x1x1.5m Payload
- LOS or OTH Supervised
- Ship by Air, Sea, Road
- Robust LARS

ASV unmanned marine systems | This document contains proprietary information about ASV. Such propriety information may not be used, reproduced or disclosed to any other parties for any other purpose without the expressed written permission of ASV.

C-Worker video

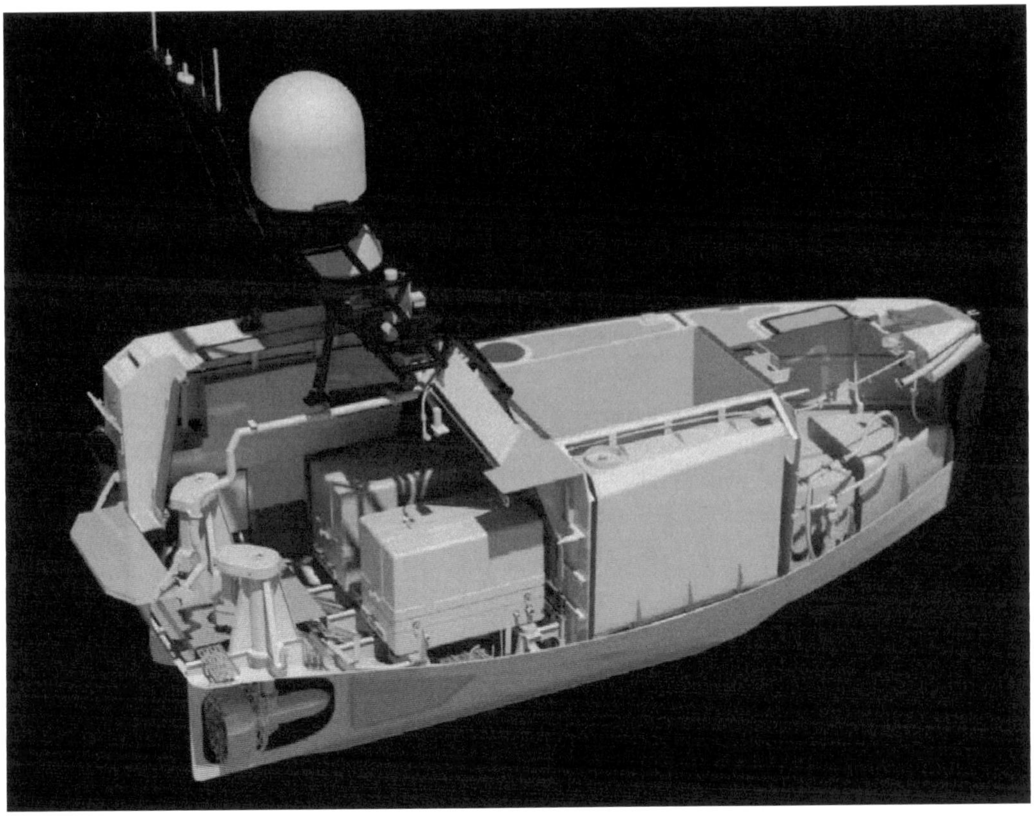

Current Developments: C-Worker 7

- Increased Endurance
- Larger Payload Bay (increased length to 2.5m)
- Design Approved July 2015
- Fabrication commenced Aug 2015
- Sea Trials scheduled to be completed May 2016
- **Entry into Service May 2016**

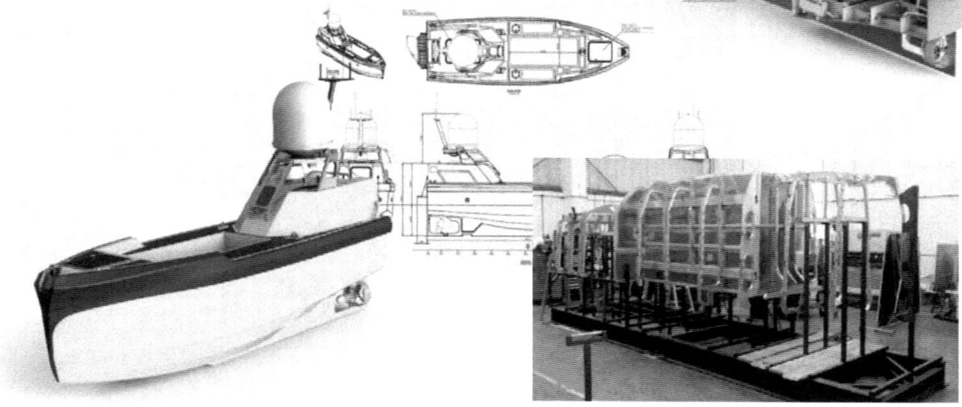

Supervised Autonomy

- **ASV Operations Supervised from:**
  - **Mothership via 30km Radio Link or**
  - **Shore via VSAT Satellite Link**

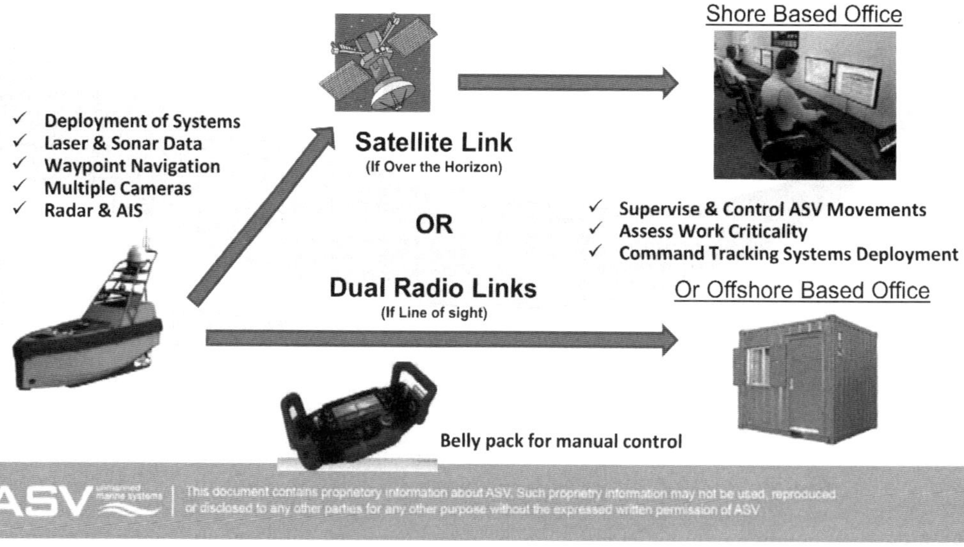

C-Worker 6 Design Overview – Control System

User Interface

View controls

Setup controls

Key Data

Selectable Tabs

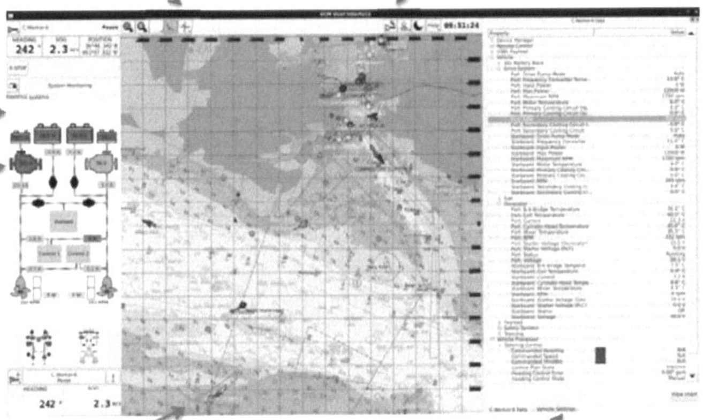

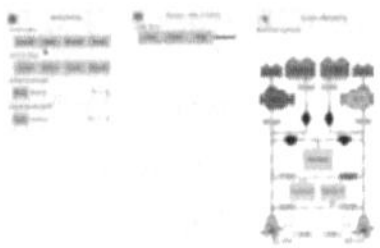

Top down displaying;
Full S57 ENC's
AIS targets from receiver on USV
Vessel track
Vessel mission plans

Data window displays all vehicle telemetry data. All data can be viewed in a trend plot

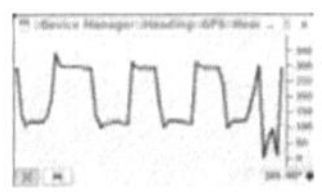

C-Worker 6 Design Overview – Control System

User Interface

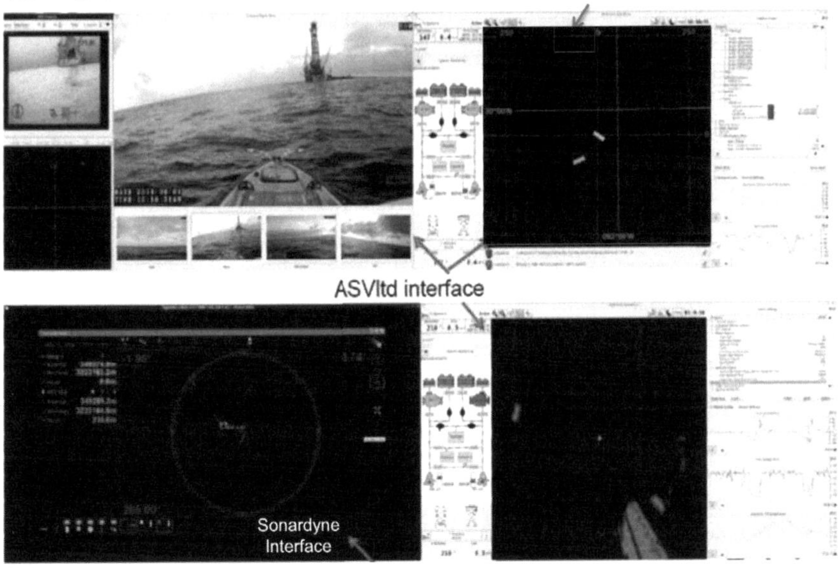

Payloads Operated to Date

- Single Beam Echo Sounder (Odom CV100)
- Multi-beam Echo Sounder (Kongsberg 2040, R2sonic, Teledyne MB1, Edgetech 6205)
- Ultra Short Baseline Positioning System (Sonardyne Lodestar GyroUSBL)
- Sidescan (Edgetech 4200, Edgetech 4125, Tritech Starfish)
- Acoustic Doppler Current Profiler (Teledyne RDI and Nortek)
- Passive Acoustic Monitor arrays (various)
- Conductivity, Temperature, & Depth (various)
- Acoustic Fish Tracking (Vemco)
- Wetlabs Triplet Puck (Chlorophyll, Dissolved Oxygen, Fluro.)
- Inspection ROV (Saab Seaeye Falcon)
- Acoustic Modem (various)
- Helikite w/ Camera
- Meteorological (wind speed, pressure, temperature, humidity)
- Laser (Renishaw Dynascan, Velodyne)

Interest in…

- Small Seismic Compressor
- Oil Spill Dispersant Application
- Oil Boom Towing
- Hull Mounted & Towed Subbottom Profiler
- Magnetometer & Gradiometer Surveys
- UAV Deployment & Recharge
- Remote Tide Modelling
- AUV Deployment / Comms / Positioning
- Drifter Buoy Deployment
- XBT Deployment
- Sonardyne Blue-comm LED Modem
- Fire Suppression Foam Application

CW6 Payload Developments

Hydrographic surveying EM 2040C

The EM 2040C multibeam echosounder for high resolution mapping and inspection applications.

- High resolution
- Wide frequency range
- Short pulse lengths and large bandwidth
- Extended range due to Frequency Modulated (FM) chirp
- Complete roll and pitch stabilization
- Water column data (standard)
- Seabed image (standard)

Subsea Positioning USBL

Lodestar GyroUSBL combines a Sonardyne 6th (6G) generation high performance HPT USBL transceiver and a Lodestar Attitude and Heading Reference System (AHRS) / Inertial Navigation System (INS) in the same mechanical assembly.

With the AHRS / INS in fixed mechanical alignment to the USBL's acoustic array, the Lodestar GyroUSBL can be quickly deployed without need for a USBL calibration to determine the alignment of the ship's motion sensors to the acoustic transceiver.

This document contains proprietary information about ASV. Such proprietry information may not be used, reproduced or disclosed to any other parties for any other purpose without the expressed written permission of ASV.

C-Worker 6 Payload Developments

ROV & Innovatum Pipeline & Cable Survey Tool

Obs Class ROV to ~150m

"Force Multiplier" Examples

ASV unmanned marine systems | This document contains proprietory information about ASV. Such proprietry information may not be used, reproduced or disclosed to any other parties for any other purpose without the expressed written permission of ASV.

MBES Results – Mississippi Oyster Bed Test

- "Force Multiplier" program in parallel with DEA hydrographic survey vessel
- Excellent feedback from NOAA, US Army Corps of Engineers, others.
- EdgeTech 6205 MPE & Sidescan

Narrow Beam Survey – Bering Sea

- "Force Multiplier" program in parallel with Terrasond hydrographic survey vessel
- SMBB200-3. 200kHz, 3 degrees. Used with an Odom CV100. "eChart" software was used to control the singlebeam and Hypack 2014 was used to collect all the data. Post-processed kinematic (PPK rather than RTK) GPS positioning with a Trimble system. Hemisphere V113 for heading, heave, pitch, and roll.

Image Courtesy of Terrasond

ASV unmanned marine systems | This document contains proprietory information about ASV. Such proprietry information may not be used, reproduced or disclosed to any other parties for any other purpose without the expressed written permission of ASV.

Multi Beam & Sidescan Sonar Survey – GoM

- "Force Multiplier" program in parallel with C&C Technologies hydrographic survey vessel
- Kongsberg 2040 Multi Beam Bathymetry, EdgeTech 4125 Sidescan with Tow Fish & winch
- IXSEA Octans 3000 TI IMU

Sidescan towfish holder and winch

Field Data Examples

ASV unmanned marine systems | This document contains proprietary information about ASV. Such proprietry information may not be used, reproduced or disclosed to any other parties for any other purpose without the expressed written permission of ASV

Operational Video – Heavy Weather

Casius Report

Casius Report Results (Oct. 2015)

October 2015, 500m Central GoM – better than 0.5% of water depth

Statistics:

	Before CASIUS (distance)	After CASIUS (distance)	Before CASIUS (% depth)	After CASIUS (% depth)
39.4% Beacon Positions (1 sigma)	2.5m	1.0m	0.50	0.20
50.0% Beacon Positions (CEP)	2.7m	1.2m	0.55	0.24
63.2% Beacon Positions (1 Drms)	3.0m	1.4m	0.60	0.29
86.5% Beacon Positions (2 sigma)	3.6m	2.2m	0.73	0.46
98.2% Beacon Positions (2 Drms)	4.7m	3.9m	0.96	0.80

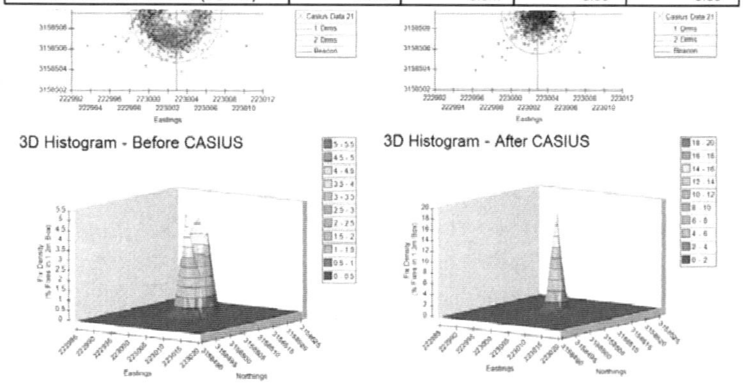

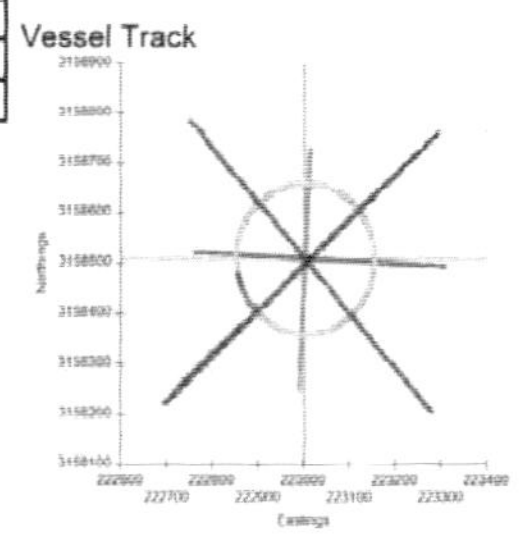

C-Worker 6 Payloads

USBL Field Data Comparison – CW6 vs. Stinger Mounted Gyro USBL Data

- Intentional offset of data for display purposes
- C-Worker 6 data is quiet, indicates ROV stoppages at pipe joints
- Over 100m Horizontal and vertical offsets of stinger USBL from GPS antenna; less than 2m offsets for C-Worker 6.
- Data recorded during pipe lay operations in the GoM, 2014.
- A dramatic increase in signal to noise ratio with the C-Worker 6 USBL

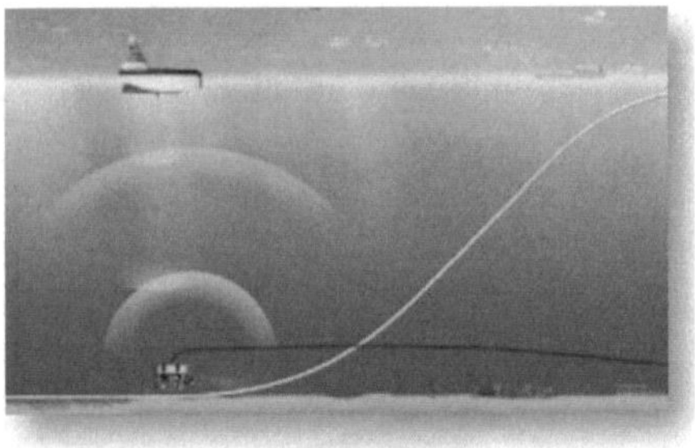

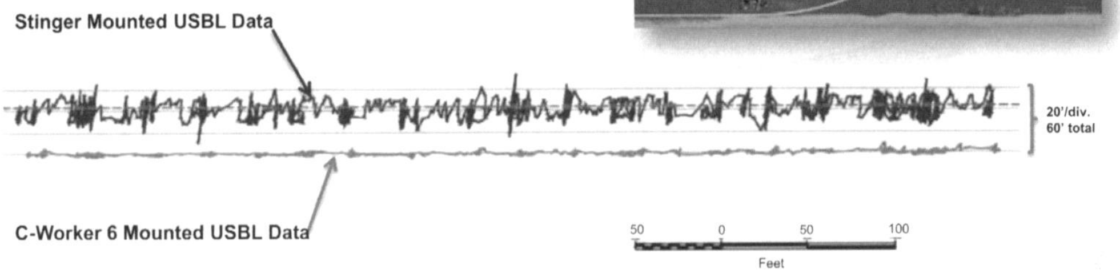

Water Depth = 1,300' (~400m)

C-Worker 6 Acoustic Signature of the ASV

Passive Acoustics –GoM Marine Mammal Monitoring

Single Hydrophone Channel, C-Worker 6
Horizontal Scale = 5 minutes, Vertical Scale 0 to 24 kHz

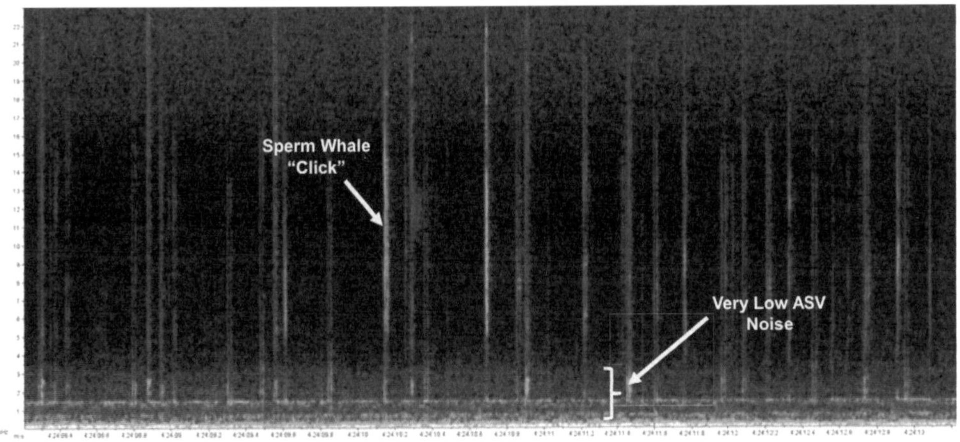

"The low acoustic noise level of the C-Worker 6 makes it an excellent platform for subsea acoustics" – Seiche Measurements

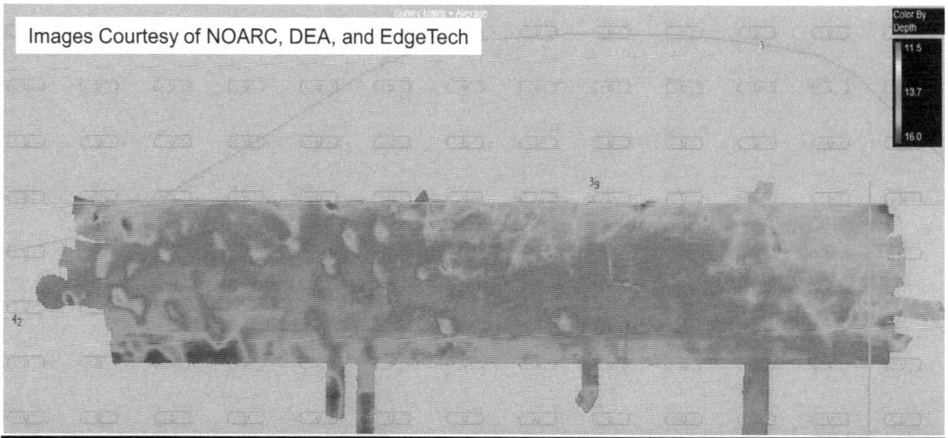

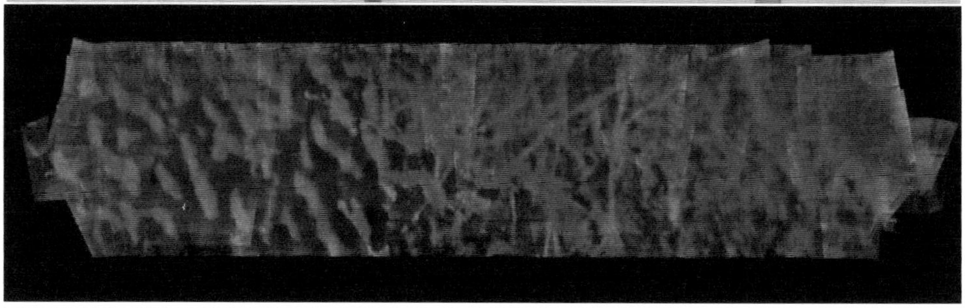

MBES initial results – Stokes bay demo

Data Examples: Lidar & Meteorological Data

- Renishaw Dynascan M250 system
- Enables ability to provide above and below the waterline surveys simultaneously
- Evaluated range and resolution
- Potential future use for collision avoidance
- System provided excellent range (~250m)

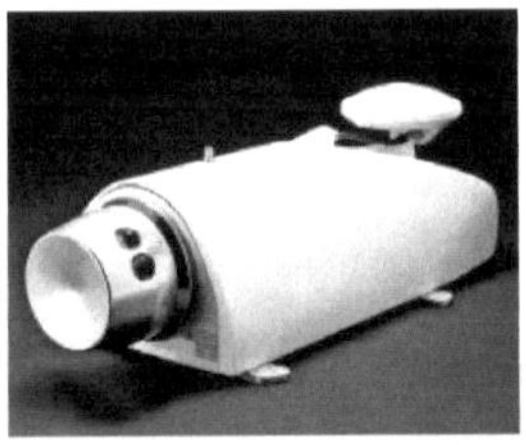

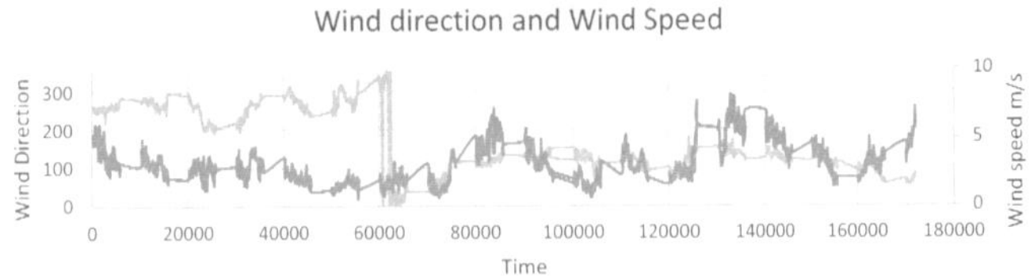

ASV unmanned marine systems | This document contains proprietary information about ASV. Such propriety information may not be used, reproduced or disclosed to any other parties for any other purpose without the expressed written permission of ASV.

Loop Current Profiling - GoM

- GOM work with Horizon Marine monitoring Loop Current
- Also working with Marine Science Scotland for Tidal Power site

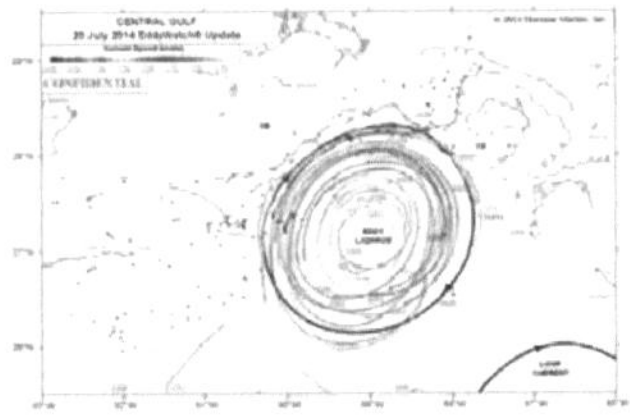

C-Worker ADCP vs. Horizon Marine FAST Eddy Transects

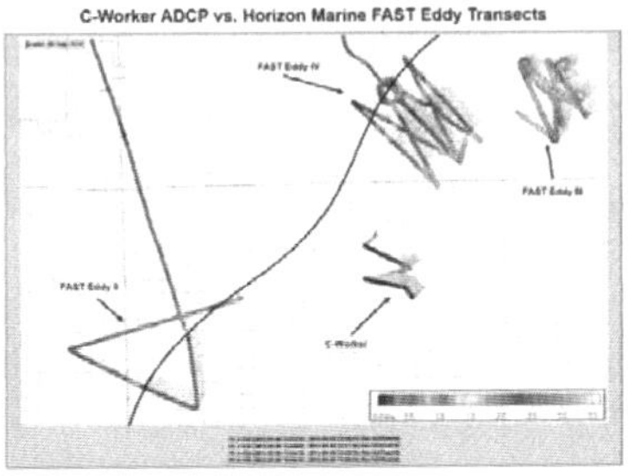

C-Worker 6 ADCP Payload, C-Enduro Below

ASV unmanned marine systems | This document contains proprietary information about ASV. Such proprietary information may not be used, reproduced or disclosed to any other parties for any other purpose without the expressed written permission of ASV.

Data Upload via Acoustic Modem

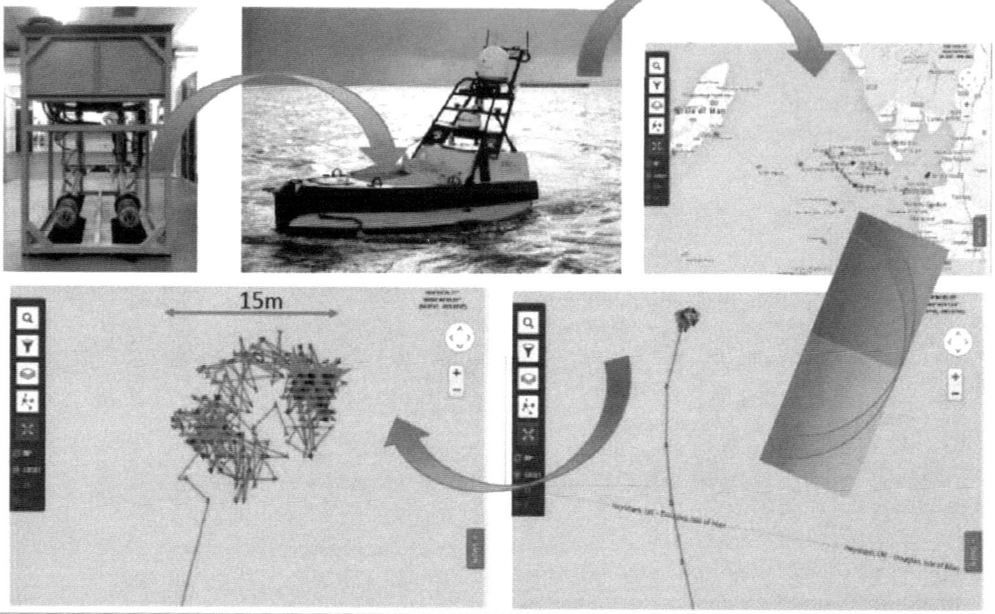

ASV unmanned marine systems | This document contains proprietory information about ASV. Such proprietry information may not be used, reproduced or disclosed to any other parties for any other purpose without the expressed written permission of ASV

C-Worker 6 - Summary

Benefits

- Easily mobilized by sea or air
- Decreases man hours on the water, improves HSE risks
- Robust design, heavy duty
- Flexible payload for multiple parallel applications
- No dedicated LARS gear – only a deck crane is required
- Significant cost savings over surface ship operations

Currently available & build plan

- Two units, one in GoM, one in Europe
- Eight total in fleet by mid 2016

Special Thanks to BIOSwimmer Sponsor: Homeland Security
Science and Technology

BIOSwimmer Overview Discussion Presenter: **Mike Rufo, Director ASG**

Date: **23 FEBRUARY 2016**

Who are we? BOSTON ENGINEERING™

Company **OVERVIEW**

- Founded in 1995 by Two Engineers (MSEE and MSME)
- Service Firm That Helped 100's of Others Commercialize
- Electromechanical Products and Systems
- Expertise in DFx (Design for Manufacturability, Assembly, Service, Reliability, etc.)
- Small business - 70 people / ~50 Technical Staff
- Diversified Across Many Markets
- Internal verticals include medical, industrial and…
    - ….the Advanced Systems Group (ASG)…

 2

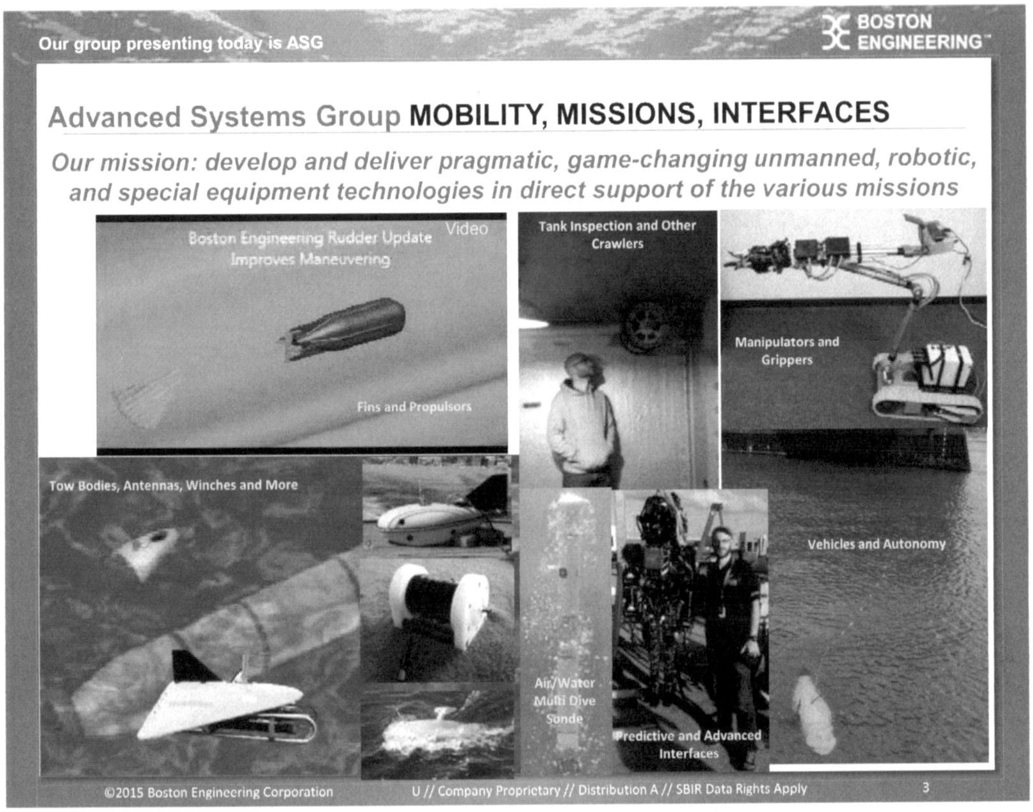
Our group presenting today is ASG
BOSTON ENGINEERING
Advanced Systems Group MOBILITY, MISSIONS, INTERFACES
Our mission: develop and deliver pragmatic, game-changing unmanned, robotic, and special equipment technologies in direct support of the various missions
Boston Engineering Rudder Update Improves Maneuvering
Video
Fins and Propulsors
Tank Inspection and Other Crawlers
Manipulators and Grippers
Tow Bodies, Antennas, Winches and More
Vehicles and Autonomy
Air/Water Multi Dive Sonde
Predictive and Advanced Interfaces
©2015 Boston Engineering Corporation
U // Company Proprietary // Distribution A // SBIR Data Rights Apply
3

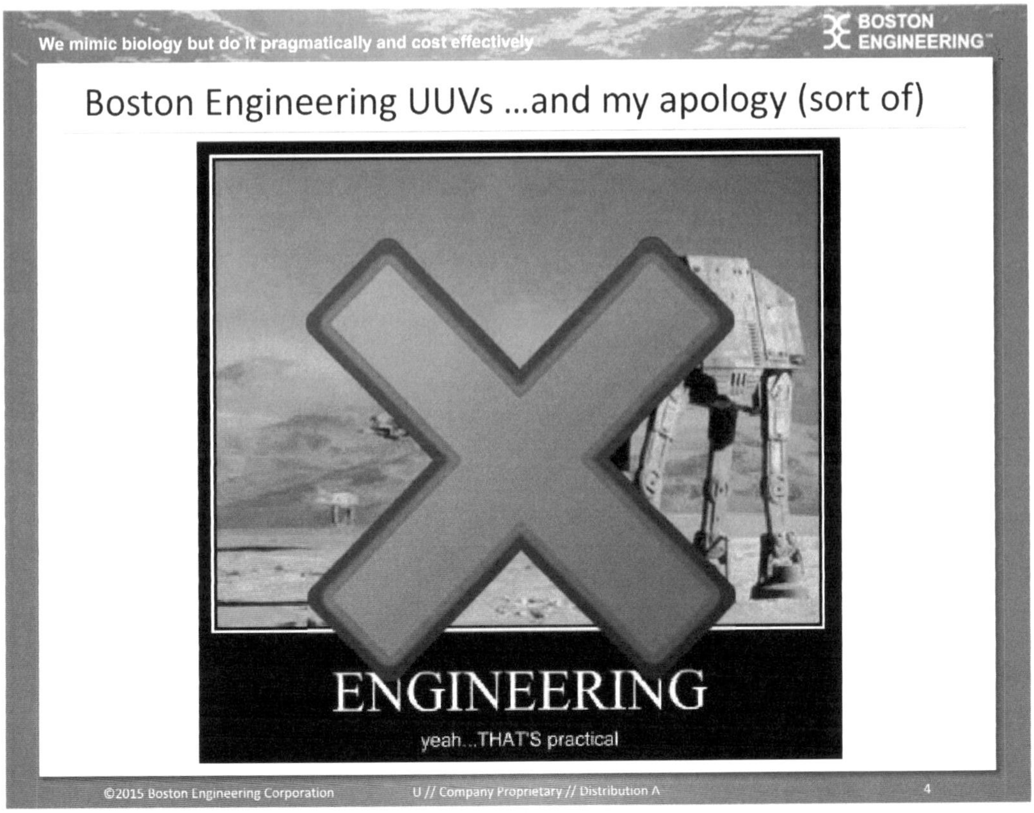
We mimic biology but do it pragmatically and cost effectively
BOSTON ENGINEERING
Boston Engineering UUVs ...and my apology (sort of)
ENGINEERING
yeah...THAT'S practical
©2015 Boston Engineering Corporation
U // Company Proprietary // Distribution A
4

Introduction to our technology

BOSTON ENGINEERING™

Homeland Security
Science and Technology

Our Technology **HIGHLY MANEUVERABLE AUVs**

BIOSwimmer AUV – Flexible body + thruster; highly maneuverable; near real time data; & lower cost

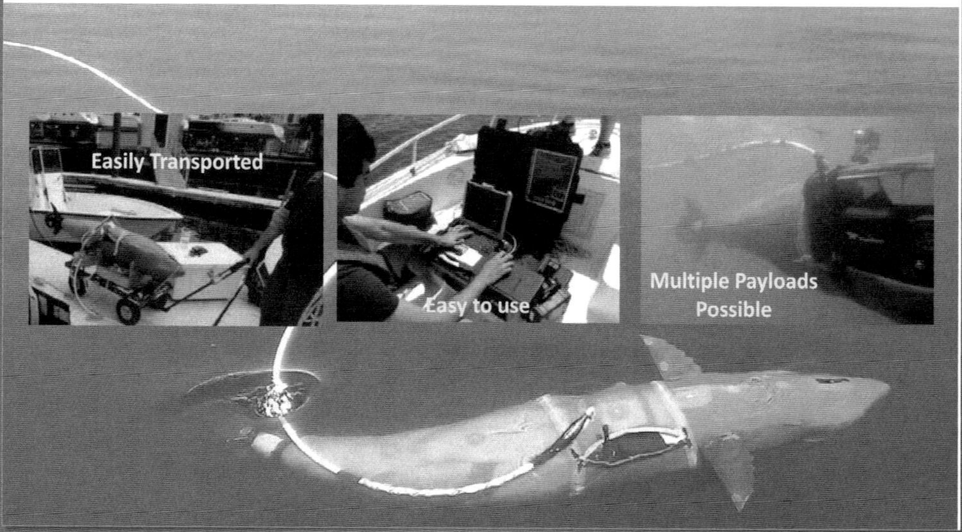

Our biological model

BOSTON ENGINEERING™

Our model **WHY A TUNA?**

- Conducive to replication:
- Streamlined, "rigid" forebody, propulsive motion in rear 1/3; small tail amplitude

The GhostSwimmer AUV for ONR that started it all

- The intersection of speed, maneuverability & endurance! (Scombridae family)
- ½ - 2.2 body lengths per second (BL/s). Max to 10-27 BL/s (~50 knots!)
- High continuous cruise speed with little body motion
- Turn 180° & re-attain 10 m/s in ~2-3 BL

Where is the gap?
BOSTON ENGINEERING™
Homeland Security
Science and Technology
How BIOSwimmer fits MISSION
How to rapidly check infrastructure and perform search and recovery across large areas? Houston 2015
How to rapidly find the small vessel smuggling the nuclear material?
How to inspect sensitive assets in densely populated areas?
How to help reduce costs and help divers do their job?
©2015 Boston Engineering Corporation
U // Company Proprietary // Distribution A
7

How big is the gap?
BOSTON ENGINEERING™
Homeland Security
Science and Technology
How BIOSwimmer addresses many tasks MISSION
Current AUV systems can't work here
... or here
... or here
... but BIOSwimmer can and does
Video
©2015 Boston Engineering Corporation
U // Company Proprietary // Distribution A // SBIR Data Rights Apply
8

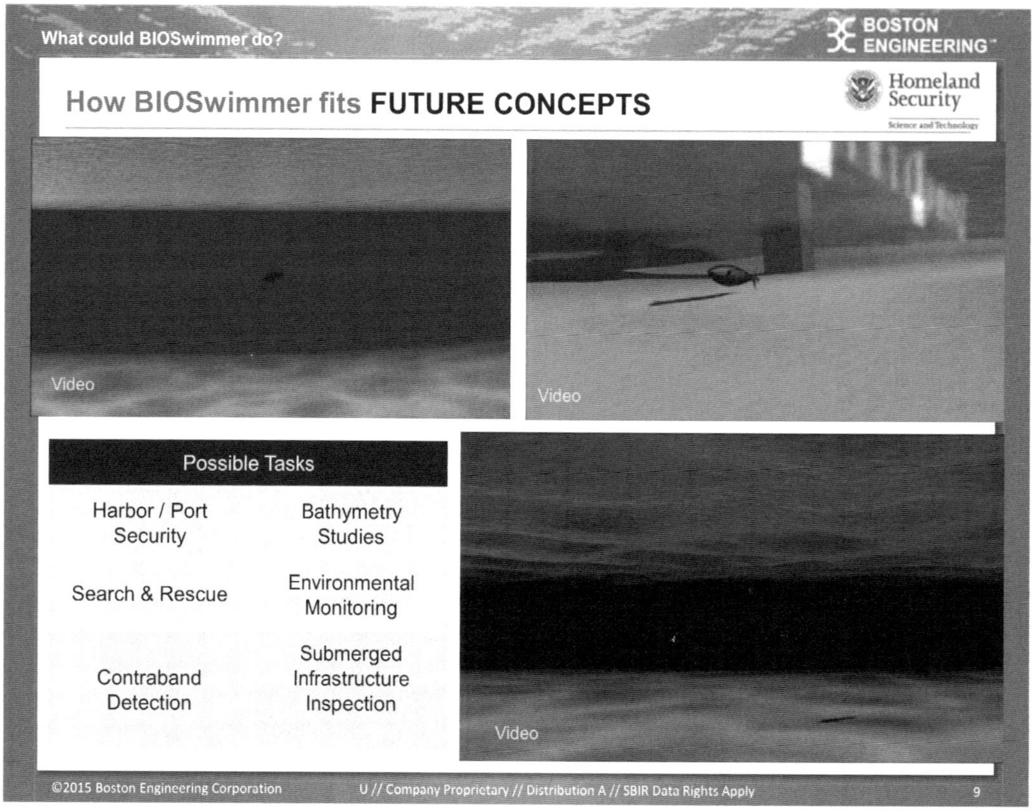
What could BIOSwimmer do?
BOSTON ENGINEERING
Homeland Security
Science and Technology
How BIOSwimmer fits FUTURE CONCEPTS
Video
Video
Possible Tasks
Harbor / Port Security
Bathymetry Studies
Search & Rescue
Environmental Monitoring
Contraband Detection
Submerged Infrastructure Inspection
Video
©2015 Boston Engineering Corporation
U // Company Proprietary // Distribution A // SBIR Data Rights Apply
9

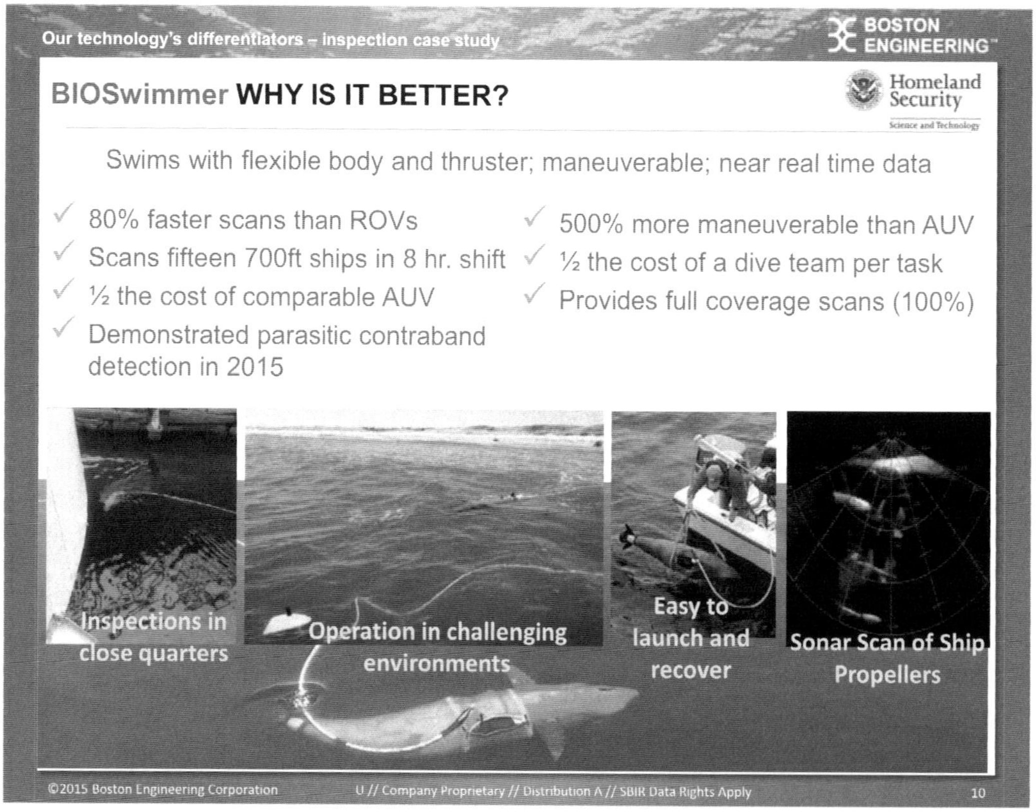
Our technology's differentiators – inspection case study
BOSTON ENGINEERING
Homeland Security
Science and Technology
BIOSwimmer WHY IS IT BETTER?
Swims with flexible body and thruster; maneuverable; near real time data
✓ 80% faster scans than ROVs
✓ Scans fifteen 700ft ships in 8 hr. shift
✓ ½ the cost of comparable AUV
✓ Demonstrated parasitic contraband detection in 2015
✓ 500% more maneuverable than AUV
✓ ½ the cost of a dive team per task
✓ Provides full coverage scans (100%)
Inspections in close quarters
Operation in challenging environments
Easy to launch and recover
Sonar Scan of Ship Propellers
©2015 Boston Engineering Corporation
U // Company Proprietary // Distribution A // SBIR Data Rights Apply
10

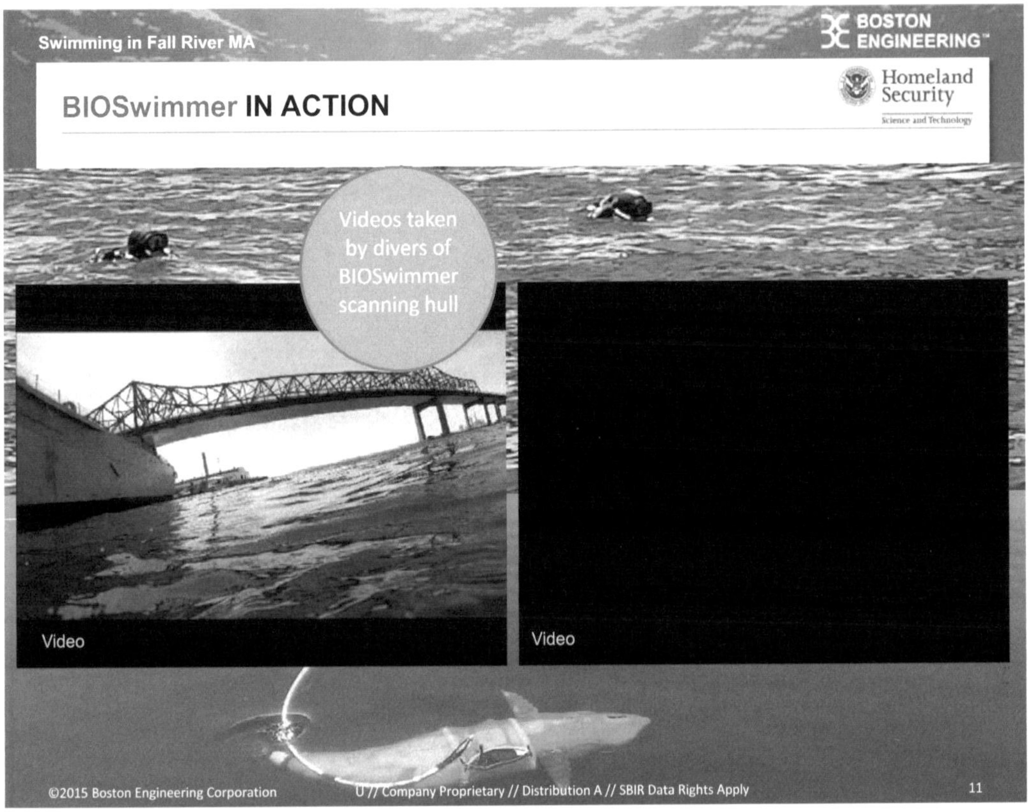
Swimming in Fall River MA
BOSTON ENGINEERING
Homeland Security
Science and Technology
BIOSwimmer IN ACTION
Videos taken by divers of BIOSwimmer scanning hull
Video
Video
©2015 Boston Engineering Corporation
U // Company Proprietary // Distribution A // SBIR Data Rights Apply
11

Demonstrated Application
BOSTON ENGINEERING
Homeland Security
Science and Technology
Find Contraband and Inspect Running Gear CASE STUDY
©2015 Boston Engineering Corporation
U // Company Proprietary // Distribution A // SBIR Data Rights Apply
12

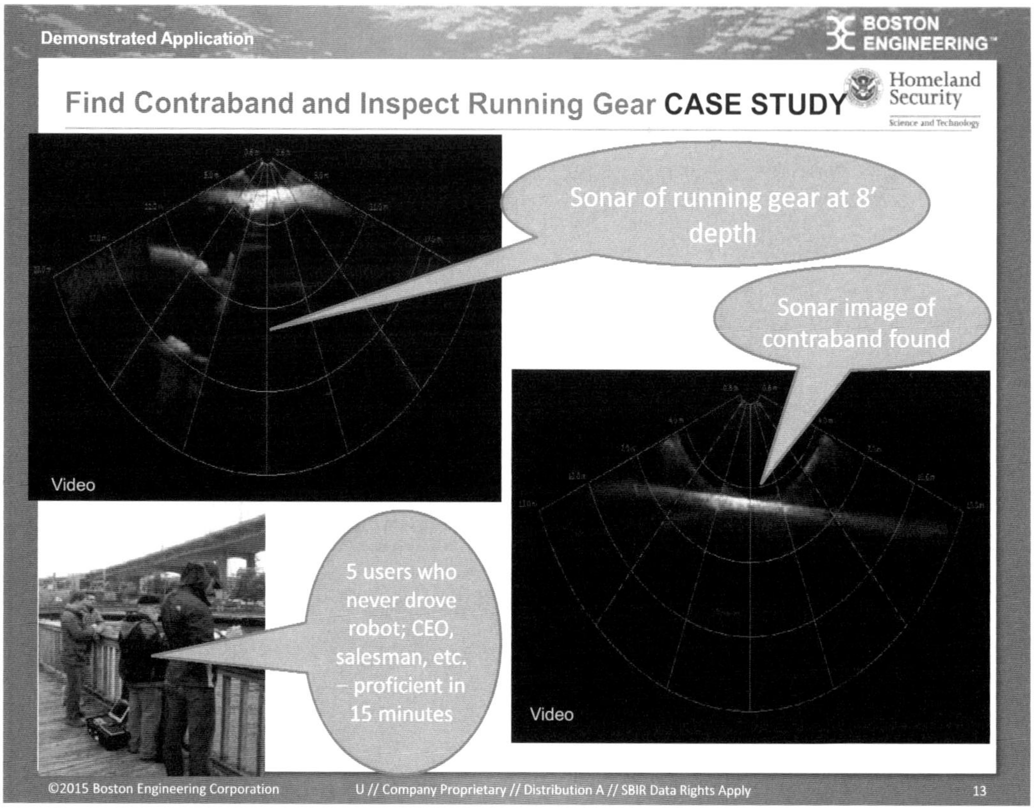
Demonstrated Application
BOSTON ENGINEERING
Homeland Security
Science and Technology
Find Contraband and Inspect Running Gear CASE STUDY
Sonar of running gear at 8' depth
Sonar image of contraband found
Video
5 users who never drove robot; CEO, salesman, etc. – proficient in 15 minutes
Video
©2015 Boston Engineering Corporation
U // Company Proprietary // Distribution A // SBIR Data Rights Apply
13

What about topside rig and FSPO?
BOSTON ENGINEERING
Straightforward L&R CRANE LAUNCH
What about here?
http://www.ogj.com/articles/shell/perdido/2010/04/h-ps-rig-205-ready-for-perdido-spar-drilling-completions-interventions.html
We will work with you: suggest launching as conventional ROVs do with a custom cradle
... or here?
https://en.wikipedia.org/wiki/Floating_production_storage_and_offloading
Deck checks
Custom cradle that straddles belly
Lowered via crane
Swims out of cradle
Wireless comms from OCU on deck to towed antenna
Using onboard camera perhaps supported by acoustic beacon, swim back into cradle
Lift via crane
Post mission checks
http://www.offshore-technology.com/contractors/rovs/cns/cns3.html
http://www.oceaneering.com/rovs/rov-media/
©2015 Boston Engineering Corporation
U // Company Proprietary // Distribution A // SBIR Data Rights Apply
14

What did we talk about?

BOSTON ENGINEERING™

Take - Away **SUMMARY**

- BIOSwimmer's 'hybrid' design provides high access, fast scanning, near real time data at lower cost than AUVs
- Standard payloads have been proven and the ability to include many more is in place in an easy-to-use package
- Broad application space with many opportunities to work with us for your application short term

Thank you to UI for the opportunity to speak here

Special thanks to our supporters at DHS S&T

U // Company Proprietary // Distribution A

15

UUV Business Commercialization Team / Mike Rufo

781-314-0723

mrufo@boston-engineering.com

Boston Engineering Corporation: 300 Bear Hill Road, Waltham, MA 02451
www.boston-engineering.com

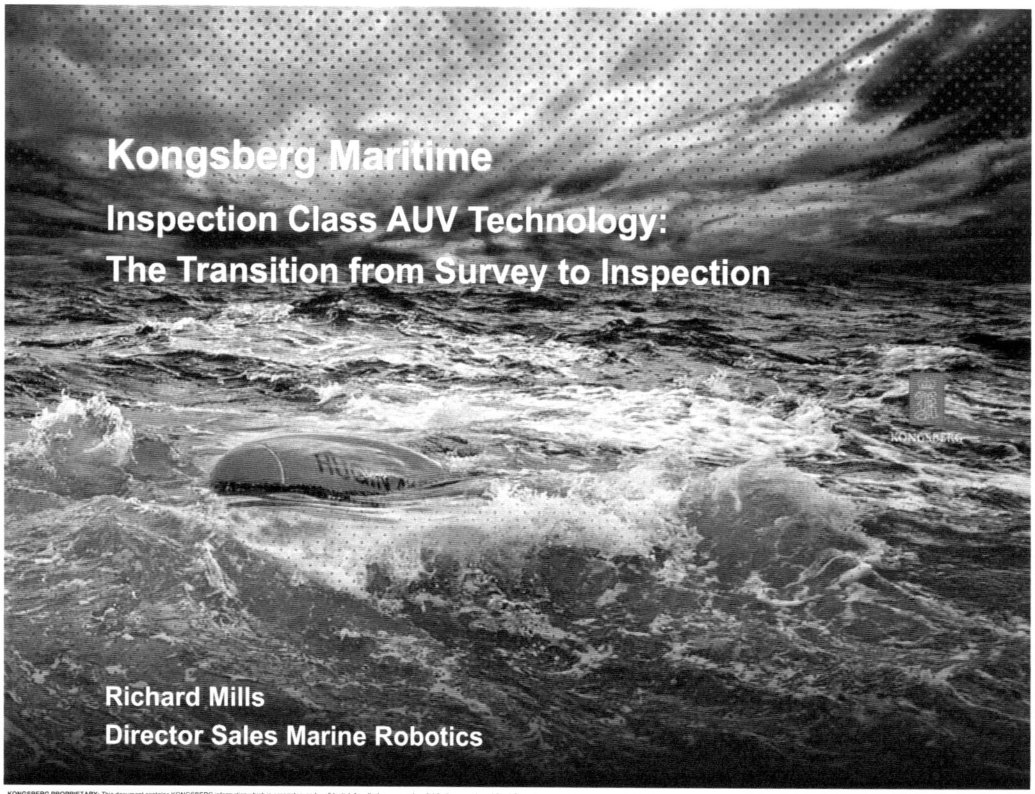

Disclaimer:

KONGSBERG

The following presentation is designed to identify some of the technology gaps to enable survey class cruising AUVs to transition to inspection class autonomous vehicles.

The information presented is not exhaustive and does not necessarily indicate any particular product or development by Kongsberg Maritime. There may be other technologies under development that are not included.

Thanks for their input:
Memorial University Newfoundland
Fugro
AXA Engineering
KM Marine Robotics and Underwater Mapping Departments
University of Southern Mississippi

What are AUVs Used for Today?

KONGSBERG

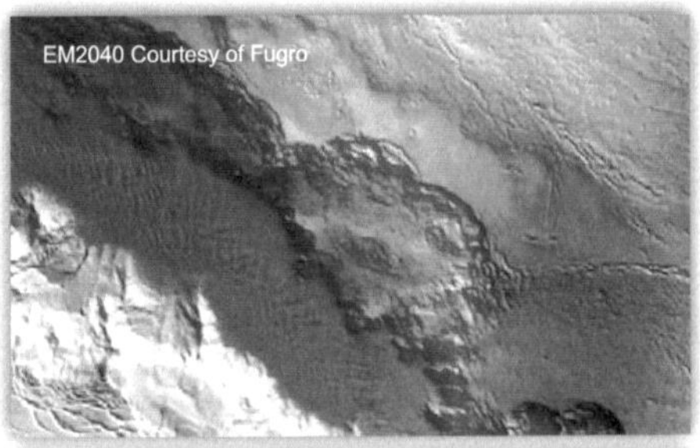

- Hydrography
- Oceanography
- Geohazard Surveys
- Pipeline Inspections
- Route Surveys
- As-Built Surveys
- Environmental Monitoring
- Benthic Habitat Mapping
- Marine Archeology
- EEZ/UNCLOS Surveys
- Search and Salvage
- And more....

The State-of-the-Art Survey Class AUV?

KONGSBERG

Accurate

Flexible

Low Downtime

IHO Grade

Varied Payload

Reliable

Repeatable

Endurance

State-of-the-Art Survey Class AUV: HUGIN AUV System

KONGSBERG

Dimensions:
- Length: 5.2 to 6.5 m
- Diameter: 75 cm

Depth Ratings:
- 3000, 4500 and 6000 m

Power Supply:
- Rechargeable and swappable Lithium Polymer batteries

Endurance:
- 24-70+ hours with payload sensors running

Typical Payload Sensors:
- HISAS Synthetic Aperture Sonar
- EM2040 Multibeam Echosounder
- EdgeTech SSS & Sub-Bottom Profiler
- CathX Ocean Colour Still Image Camera
- CathX Ocean Laser Profiler
- OFG Magnetometer
- Contros HydroFlash CH4
- Franatech METS
- WetLabs Turbidity

What is the Challenge?

KONGSBERG

How do we transition from Survey to Inspection?

1. *Technology*

- *Integration of new sensors (laser profiler, color camera etc.)*
- *Greater adoption of SAS*

TECHNOLOGY GAPS

- *CP Assessment*
- *Mass spectrometry for locating and identifying leaks for carbon fingerprinting*
- *Ultrasonic/Non-Destructive Testing for structural integrity*
- *Integration of sensor technology on subsea infrastructure with data harvested by the AUV*

New Sensor Integrations

KONGSBERG

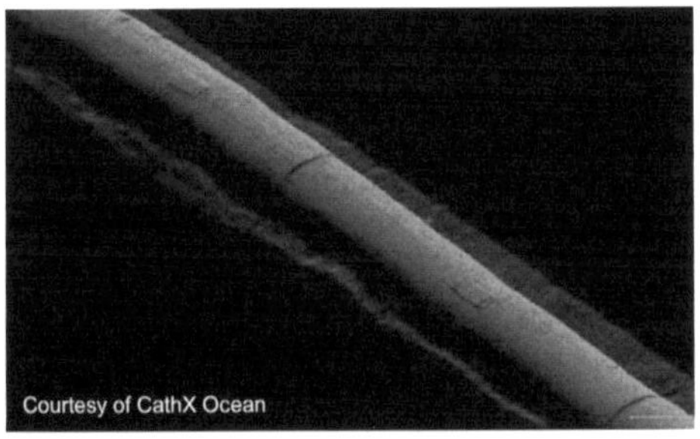

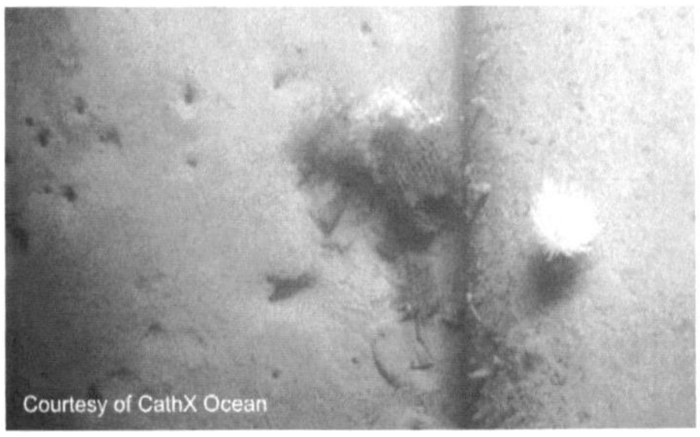

- *Cathx Ocean color camera with laser profiler is being integrated for delivery 2016*
- *Ocean Floor Geophysics magnetometer is being integrated for delivery in 2016*
- *Environmental sensors being integrated for delivery in 2016 include:*
  - *WetLabs ECO Puck CDOM fluorometer*
  - *WetLabs ECO Puck Chlorophyll A*
  - *OCR-504 Radiometer*
  - *Biospherical Instr. QSP-2150 PAR*
  - *Satlantic DeepSuva SUNA V2 Nitrate*
  - *Contros HydroFlash DO*

KONGSBERG PROPRIETARY - See Statement of Proprietary Information 15.01.2016 WORLD CLASS - through people, technology and dedication Page 7

Greater Adoption of SAS

KONGSBERG

- *Synthetic Aperture Sonar*
- *HISAS 1032 theoretical resolution 2 x 2 cm*
- *HISAS 2040 theoretical resolution 1.2 x 1.3 cm*
- *Practical resolution 4 x 4 cm*
- *Range independent resolution*
- *HISAS 1032 range: 300 m either side*
- *HISAS 2040 range: 150 m either side*
- *In-Mission SAS and Bathymetry processing*

KONGSBERG PROPRIETARY - See Statement of Proprietary Information 15.01.2016 WORLD CLASS - through people, technology and dedication Page 8

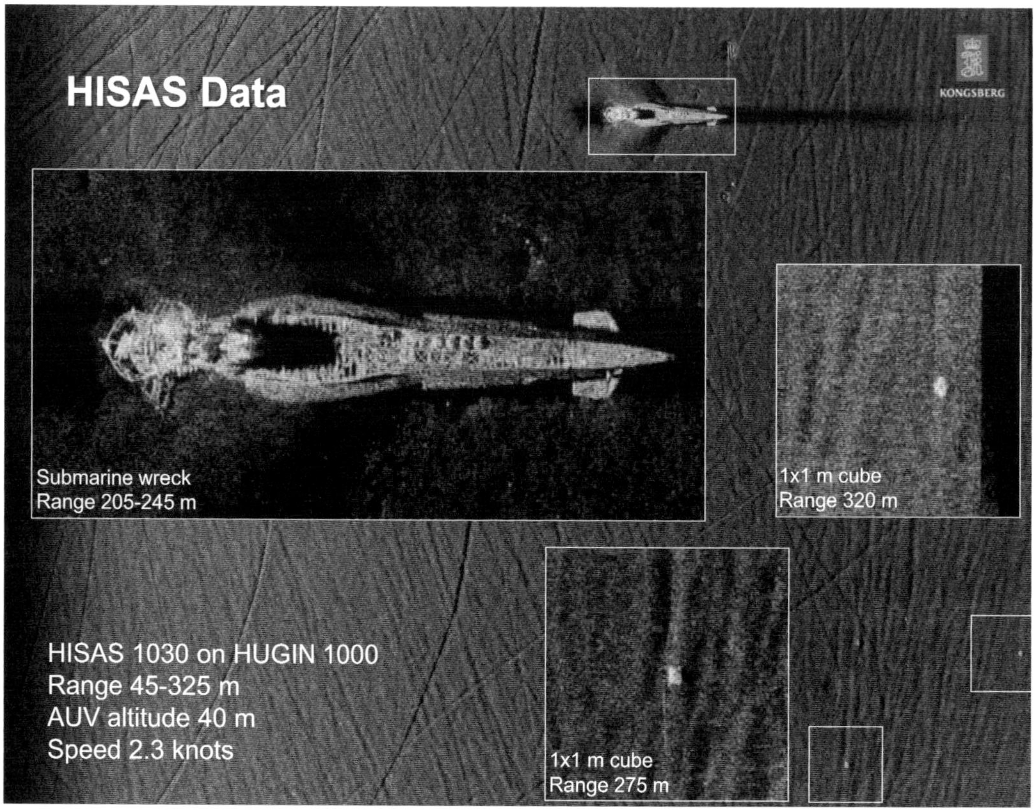

What is the Challenge?

KONGSBERG

How do we transition from Survey to Inspection?

1. *Technology*
2. *Vehicle Behaviours*

CAPABILITY GAPS

- *Efficient hovering/slow speed manoeuvring*
- *In-mission adaptive control*
- *Relative navigation*
- *Subsea residency*
- *Data harvesting from in-situ sensors*

What is the Challenge?

KONGSBERG

How do we transition from Survey to Inspection?

1. *Technology*
2. *Vehicle Behaviours*

Pipe Tracking

KONGSBERG

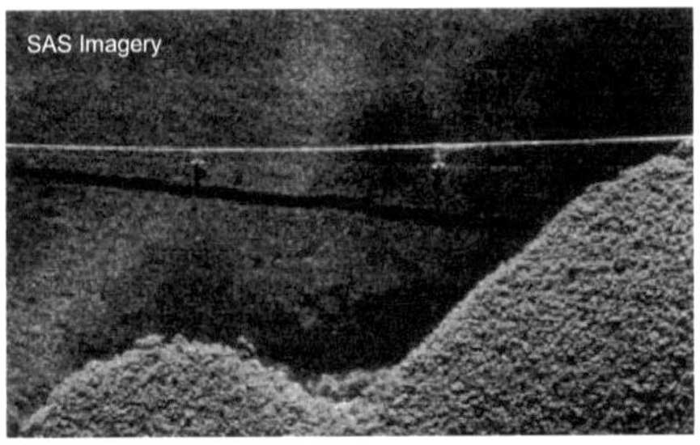

- *Autonomous tracking using HISAS/SSS and EM2040*
- *Pass 1: alongside pipe to collect SAS/SSS imagery, bathymetry and create a-priori map*
- *Pass 2: overhead collecting 400 kHz bathymetry and photo mosaic*
- *No interaction with operator required*
- *Track points sent to surface in real-time*

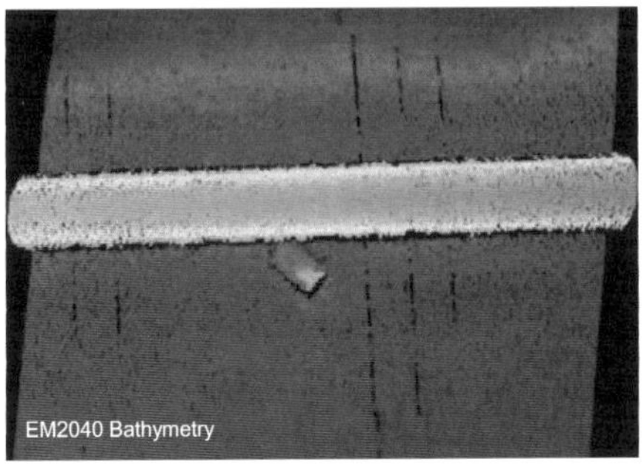

Adaptive Control

KONGSBERG

For MCM Operations the latest generation HUGIN MR is equipped with HISAS 1032

1. HISAS processing occurs on-board in real time

2. In-Mission automatic target recognition algorithms classify mine-like objects

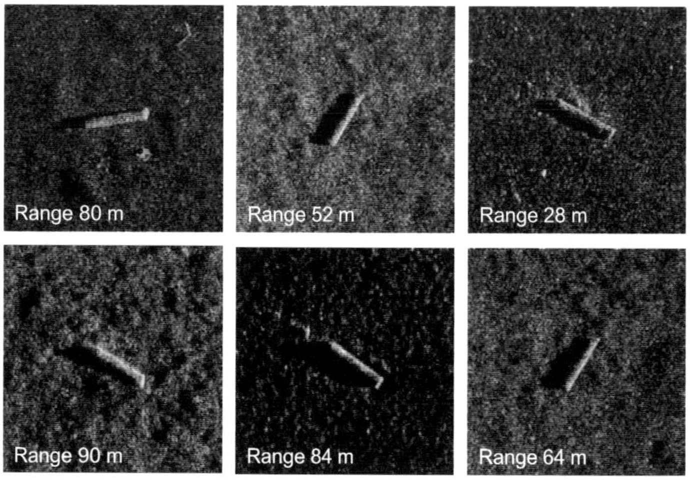

Adaptive Control

KONGSBERG

3. On-board adaptive control re-plans the mission to photograph high-priority targets

4. Revised mission plan transmitted to operator for approval

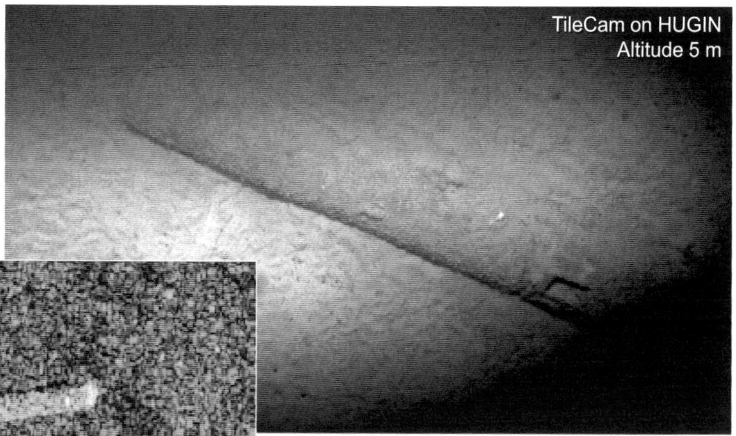

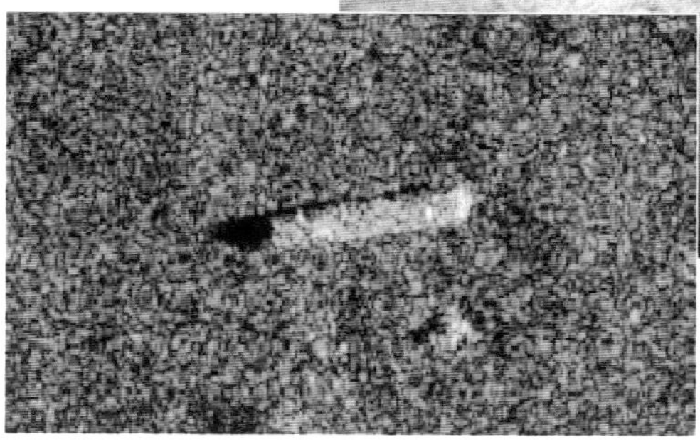

5. AUV completes **detection**, **classification** and **identification** in a single dive

Adaptive Control

KONGSBERG

Example: **LEAK/SEEP DETECTION during Pipe Survey**

1. Analyse EM2040 Water Column data in mission
2. Analyse METS/PAH sensor data in mission
3. If bubbles in water or hydrocarbons are detected then HUGIN stops current mission objective
4. Report to operator
5. Autonomously circles back to detection area
6. Conducts multiple slow speed passes to record high resolution data

Example: **CHANGES IN BATHYMETRY or SEARCH AND RECOVERY**

1. Analyse EM2040 and SSS/HISAS imagery in real-time
2. Apply known bathymetry/imagery filters
3. Run autonomous change detection algorithms
4. Identify discrepancies
5. Report to operator
6. Autonomously return to identified discrepancies
7. Conducts multiple slow speed passes to record high resolution data

What is the Challenge?

KONGSBERG

How do we transition from Survey to Inspection?

1. *Technology*
2. *Vehicle Behaviours*
3. *Improve the benefits of using AUVs*

BENEFITS

- *Higher survey speeds*
- *Better quality, more consistent data with better geo-referencing*
- *Lower life cycle costs*
- *Flexibility of operations*

What is the Challenge?

KONGSBERG

How do we transition from Survey to Inspection?

1. *Technology*
2. *Vehicle Behaviours*
3. *Improve the benefits of using AUVs*
4. *Commercial challenges*

CHALLENGES...?

- *Acceptance that AUVs work differently from ROVs and produce a different data set*
- *Acceptance that AUVs will augment, not replace ROVs*
- *Lack of real-time visual feedback for the majority of operations*
  - *Could be overcome by the introduction of real-time 3D visualisation of AUV operations*

kongsberg.com
KONGSBERG
KONGSBERG PROPRIETARY - See Statement of Proprietary Information

Development of a Low-Cost Expandable AUV Platform for Engineering Education and Research

CDR Andy Gish, PhD, PE
US Naval Academy

Underwater Intervention '16
New Orleans, LA

1

Presentation Outline

- Background and Motivation
  - Ocean Engineering at USNA
- Similar projects (USNA and elsewhere)
  - USNA-1 (Wick and Stilwell, 2002)
  - Autonomous SeaPerch (Laskey, Pendleton, and Wongissares, 2015)
  - Other academic projects
  - Commercial products
- Latest iteration: SmartPerch
  - Vehicle design and specifications
  - Educational applications
- Future work

2

Background and Motivation

- **PROBLEM:** Complexity and cost of existing commercial AUVs prohibit widespread use in engineering education and research
- Common commercial AUVs:
  - Very useful tools for many applications, but……
  - Not conducive to modification and experimentation
  - Cost prohibits purely educational uses, discourages risk
  - Complexity requires knowledgeable operators and maintainers (Time + $$$)
- **SOLUTION:** Develop an inexpensive (<$1000) AUV platform that can be built, maintained, and operated by faculty and undergraduate students
  - Use the platform to teach underwater vehicle design principles, experiment with new vehicle technologies, and support other research

REMUS 100

Iver 3-580

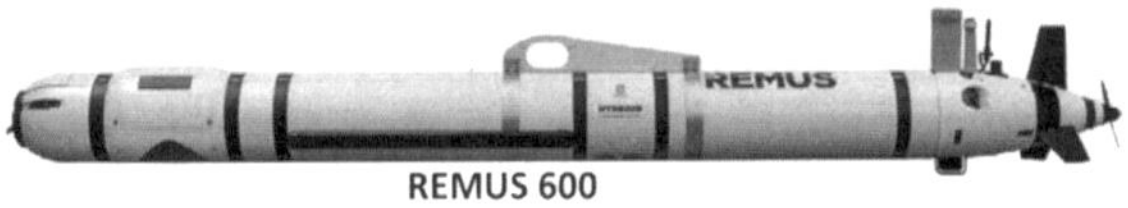

REMUS 600

Gavia Offshore Surveyor

3

United States Naval Academy Ocean Engineering Program

- 4000 undergraduate students, 0 graduate students
- 25 different majors (10 engineering)
- Faculty: 50% tenure-track civilian, 50% military officers
- Very rigorous academic/military/physical regimen
- 60-70 Ocean Engineering majors per class
  - 14 required core engineering courses
  - 4 engineering electives (senior year)
    - Underwater track: Underwater Work Systems and Life Support Systems
  - Senior Capstone Design Project (2 semesters)
  - Opportunities for independent research projects (1 or 2 semesters)

4

Similar Projects (USNA and Elsewhere)

- Claus, Brian, et al. "Development of a low cost instructional platform for submersible design." *Electrical and Computer Engineering, 2009. CCECE'09. Canadian Conference on*. IEEE, 2009.
  - Based on LEGO Mindstorms kit
- Busquets, Javier, et al. "Low-cost AUV Alba13 as multi-sensor platform with water sampler capabilities, for application in multi-agent ocean research applications." *Oceans-St. John's, 2014*. IEEE, 2014.

- Wick, Carl E., and Daniel J. Stilwell. "USNA-1: A MINIATURE, LOW-COST AUTONOMOUS UNDERWATER VEHICLE." *Sea Technology,* June 2002.
- Autonomous SeaPerch (USNA senior design project, spring 2015).

USNA-1 miniature AUV

- Commercial examples: Riptide Micro-UUV

Source: www.RiptideAS.com

Autonomous SeaPerch

5

Latest Iteration: SmartPerch

- **Design Objectives**:
  - Operating depth: 15 ft (max. depth of largest test tank at USNA)
  - Neutrally buoyant in freshwater, statically stable
  - Maneuverable in longitudinal, transverse, and vertical directions
  - Wi-fi comms with laptop (at or near surface)
  - Command depth, course, and duration of maneuvers
  - Total cost <$1000
- Start with basic SeaPerch kit
  - Readily available and affordable
  - Open frame conducive to modification and add-ons
  - No control surfaces – all control via thrusters

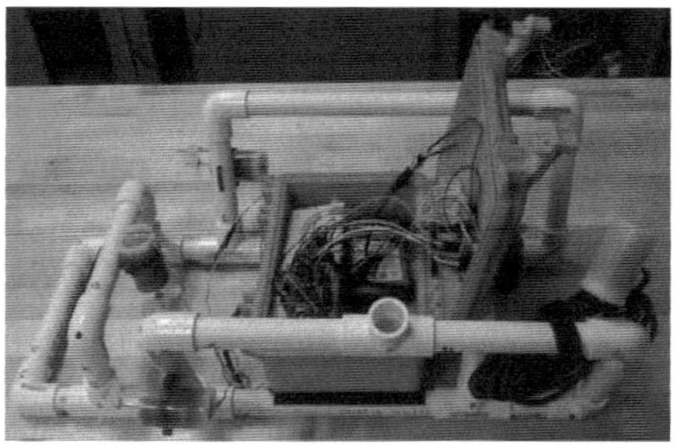

- Pressure hull: Pelican 1120 case
- Waterproof cable glands for hull penetrations (3)
- Added 2 hose clamps around box to improve seal
- Tested watertight to 15 ft
- Extra vertical thruster to counter added weight and volume

6

Processing and Data Logging

- **mbed LPC1768 microcontroller**
  - Large number and variety of I/O pins
  - Readily-available open-source code for components
  - C++ programming

- **microSD card data logging**
  - Record pressure (depth), temperature, heading, orientation
  - User defined logging frequency
  - Can be expanded with additional sensors

Sensors and Power

- **DS18B20 Temperature sensor**
  - Waterproof
  - Range: -55C to +125C
  - Accuracy: +/- 0.5C
  - Requires 5 Vdc from mbed

- **MPX5050DP Pressure sensor**
  - Range: 0-50 kPa (0-7.25 psi, 0-16.7 ft H_2O)
  - Temperature compensated
  - Requires 5 Vdc from mbed
  - Sensing port connected to pressure hull fitting with tygon tubing

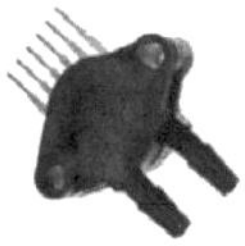

- **MC33926 Dual Motor Drivers**
  - Full H-bridge driver
  - 3 amp continuous, 5 amp peak current
  - PWM input for variable speed control

- **12.6V rechargeable Li-ion battery (for motors)**
  - 4800 mAh capacity

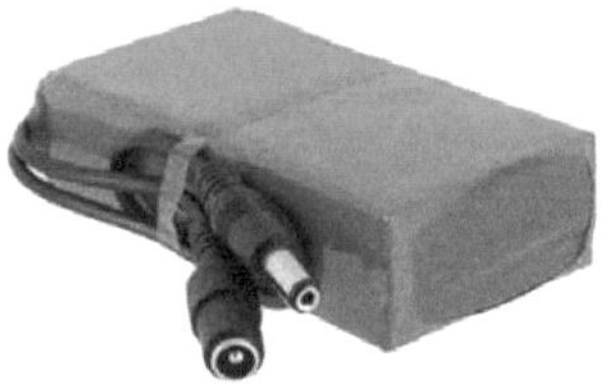

- **5.5V rechargeable battery pack (for mbed+sensors)**
  - 1000 mAh capacity

Navigation and Communications

- **InvenSense MPU-9150: 9-DOF compass/gyro/accelerometer**
  - 3.3Vdc from mbed
  - Onboard Digital Motion Processor™ integrates all sensor data
  - Draws 4.25 mA at full power with all sensors
  - Max. 400 kHz sampling
- **XBee 1mW wi-fi modules**

– Wireless comms between AUV and control laptop
– 300 ft range
– Max data rate: 250 kbps

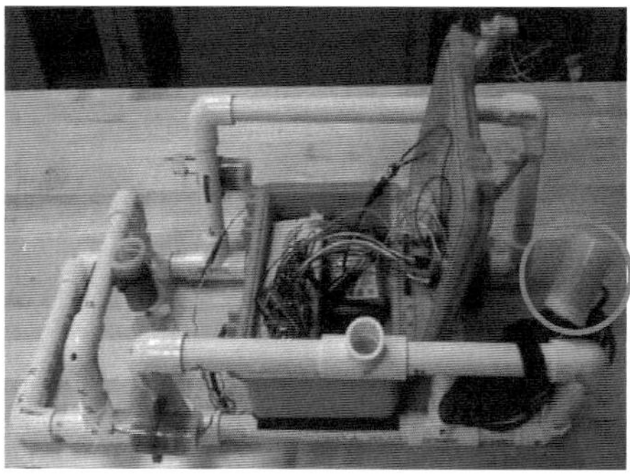

Onboard unit in waterproof enclosure outside pressure hull. Can be released as a tethered floating antenna to provide continuous comms while AUV is submerged.

9

Video – Tank Testing

10

SmartPerch Cost Breakdown

Component	Cost
Basic SeaPerch Kit (w/out battery)	$142
Pelican case	$25
Mbed LPC1768 microcontroller	$55
MPX5050DP pressure sensor	$16
MPU-9150 9-DOF sensor	$35
DS18B20 temperature sensor	$10
MicroSD card and board	$25
XBee wi-fi units	$50
MC33926 Dual Motor Drivers	$54
Batteries (12.6v and 5.5v)	$28
Waterproof cable glands	$5
Misc (jumper wires, epoxy, protoboards, etc)	$15
TOTAL	**$460**

11

Applications for SmartPerch

- Educational platform for underwater technology and vehicle design:
  - Pressure hull design and testing
  - Hydrostatics and stability
  - Hydrodynamics and maneuvering performance modeling
    - Experimentally measure hydrodynamic coefficients
  - Resistance and powering
    - With/without floating antenna to demonstrate tether drag effect
  - Power budget / endurance calculations
  - Control systems
  - Underwater navigation

12

Future Development Possibilities

- Improved vehicle control
  - Control vertical thrusters independently, to give pitch control
  - Incorporate more advanced control algorithms
    - Use as test platform for different algorithms
  - Incorporate sensor-driven decisions and simple autonomy (possibly elements of MOOS-IvP)
  - Develop basic inertial navigation software (limited by MPU-9150 capability)
- Additional sensors and communications
  - Hydrophone(s)
  - Camera
  - Acoustic comms
  - Other sensors???
- Platform for experimentation with other technologies
  - Inductive battery recharging
  - Optical (laser) comms
  - Biomimetic propulsion methods (flapping foils, synthetic jet, etc.)

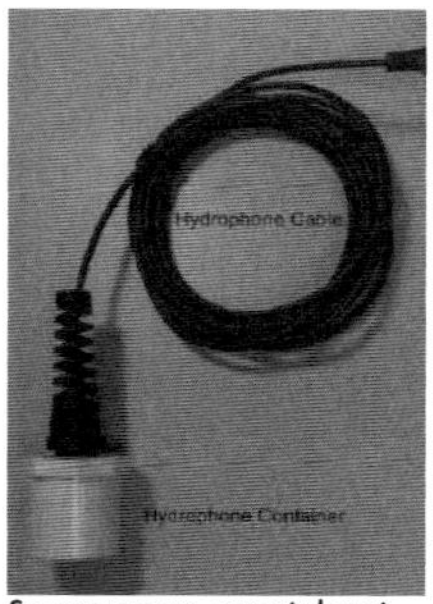

Source: www.coseetek.net

13

Questions?

<u>Contact information:</u>
CDR Andy Gish
U.S. Naval Academy
410-293-6585
gish@usna.edu

14

Backup Slides

15

United States Naval Academy Lab Facilities

- **380 foot towing tank:**

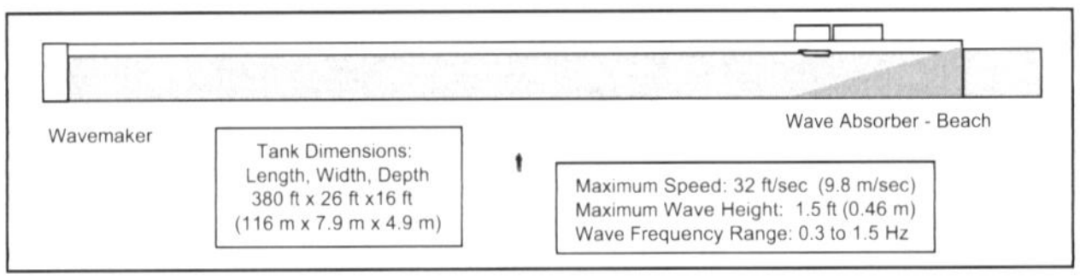

- **120 foot towing tank:**

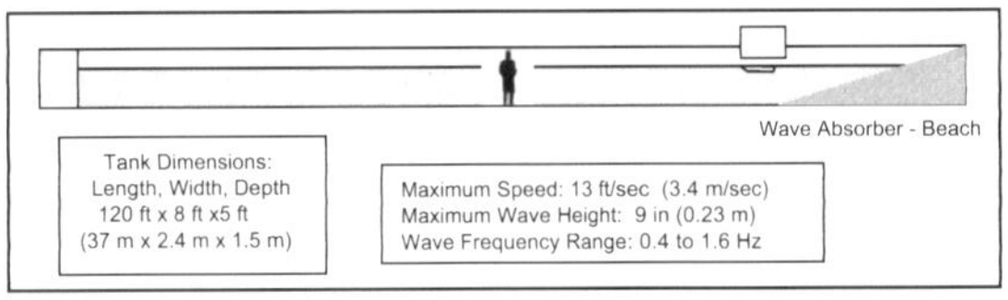

- **Circulating water channel:** max. speed 25 ft/sec
  - test section 16"x16"x5'
- **Coastal engineering basin:** max. depth 1 foot
  - used to study wave effects on beaches and structures
- **Ballast tank:** depth=4 feet
  - used for static tests and ballasting vehicles

16

United States Naval Academy
Field Facilities

- Pier/seawall access to Severn River and College Creek
- Small boat/Yard Patrol (YP) craft access to river and Chesapeake Bay

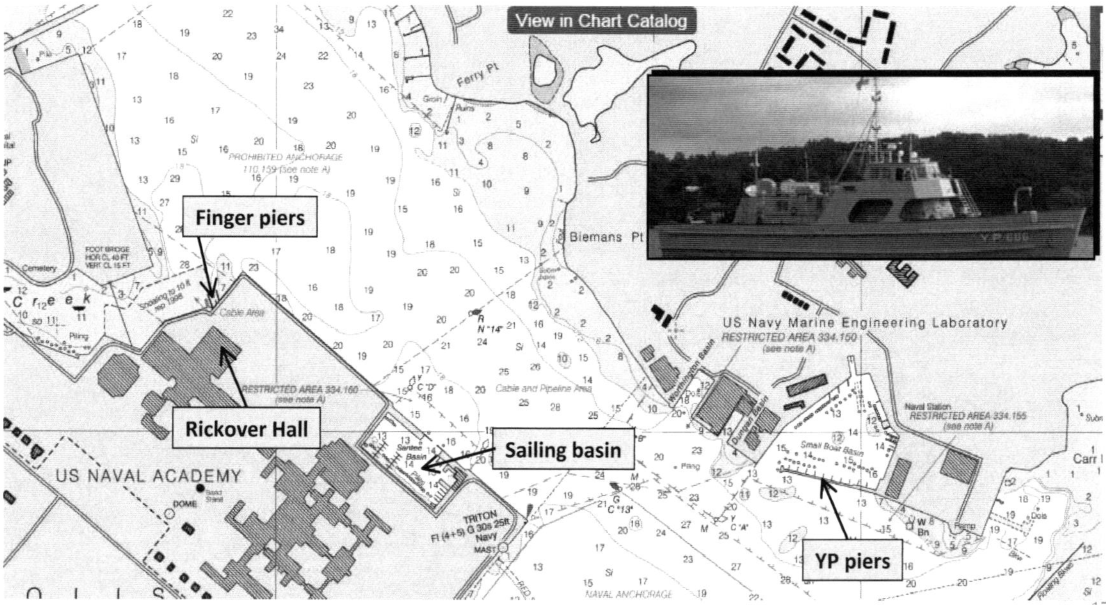

17

LAUNCH OF THE SUBMARAN S10

KEN CHILDRESS
VICE PRESIDENT, CO-FOUNDER

CHANGING OCEAN EXPLORATION

Submaran: A highly ruggedized, autonomous surface and sub-surface vessel.

•Wind and solar powered
•Autonomous surface and underwater capability
•Long Range & Extended missions
•Easily deployable
•Sensor and payload agnostic

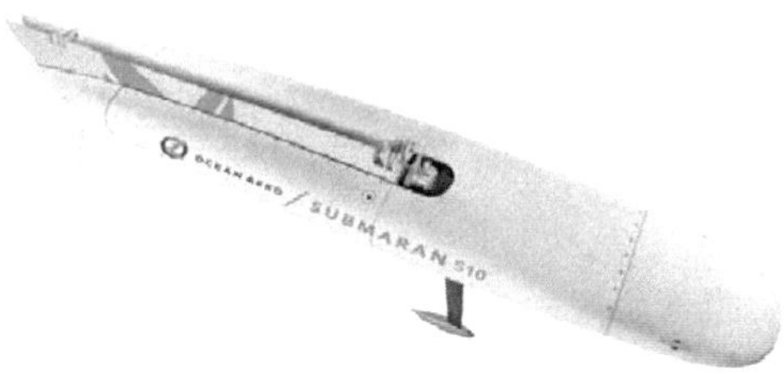

SUBMARAN OVERVIEW:

Submaran S10: UUSV (Unmanned, Underwater, Surface Vessel)

Wind and Solar-Powered Freedom

- Surface speed 5+ knots
- Submerges up to 10m
- Payload: 23kgs
- 300w peak/50w
- Stability adjusted
- Auxiliary thruster
- Multi-mission capabilities
- Scalable design

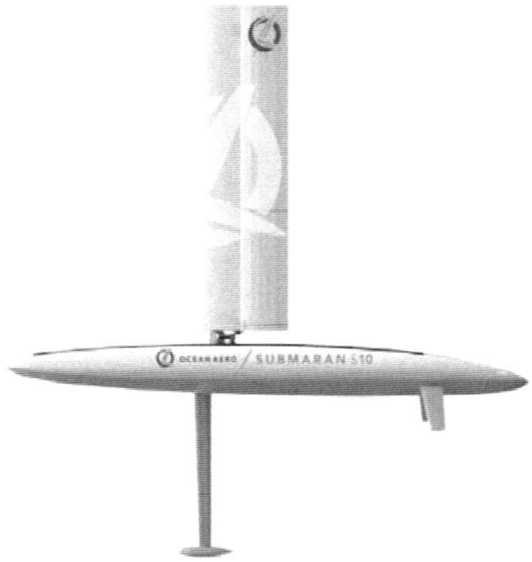

- Length: 4.14M
- Height: 2.45M Wingsail up
- Draft: 1.2M

SUBMARAN DIFFERENTIATORS

- Hybrid design: Sail, electric and buoyancy gliding
- Faster: Speeds greater than other competitive unmanned platforms
- Autonomy & Ease of Deployment: Helps reduce required manpower
- Submerging capability: Provides stealth and survivability

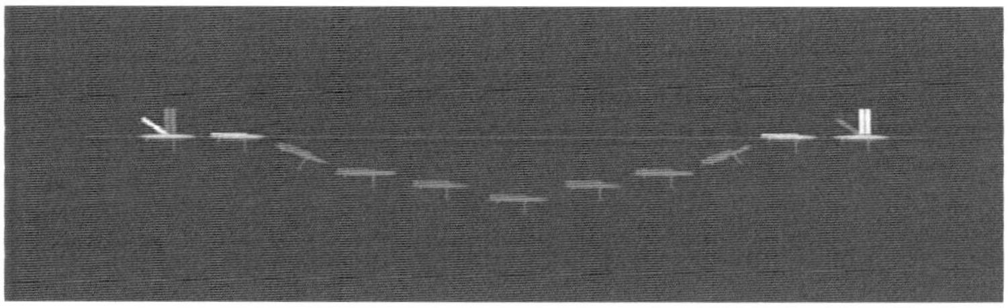

STANDARD CONFIGURATION DETAILS

NAVAGATION:
GPS
Masthead Anemometer
Boat Speed Indicatior
IMU/Compass

COMMUNICATIONS:
Iridium Satellite
Celular data
Wi-Fi

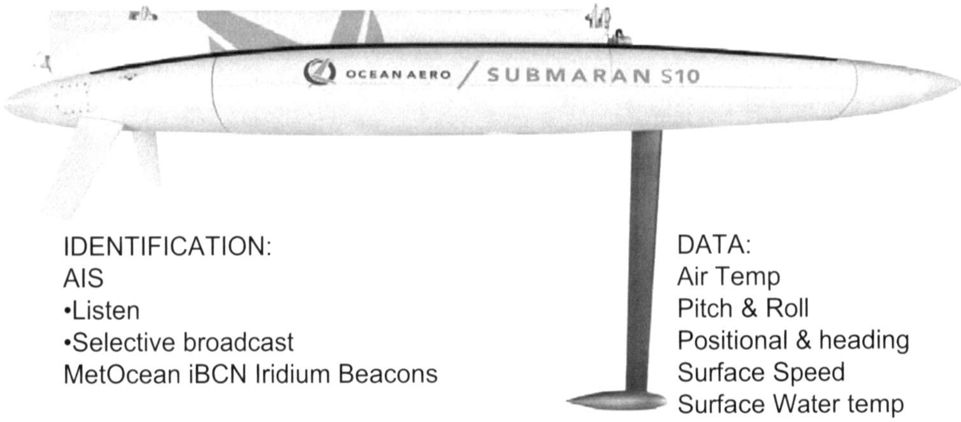

IDENTIFICATION:
AIS
•Listen
•Selective broadcast
MetOcean iBCN Iridium Beacons

DATA:
Air Temp
Pitch & Roll
Positional & heading
Surface Speed
Surface Water temp

MISSION OVERVIEW

Commercial

- Communications Gateway
- Hydrography
- Oceanography
- Meteorology
- Fisheries Production
- Environmental Mgmt.
- Telecommunications
- Asset Mgmt.
- Security
- Oil and Gas Production

Government

- Communications Gateway
- Hydrography
- Oceanography
- Meteorology
- Fisheries Mgmt.
- Environmental Mgmt.
- Telecommunications
- Security
- Oil and Gas Mgmt.
- Defense*
- Intelligence*

Scientific

- Communications Gateway
- Hydrography
- Oceanography
- Meteorology
- Fisheries Research
- Environmental Research
- Debris field Mgmt.

PRODUCT OVERVIEW

	S-10	S-200	S2-200
Speed			
Surf	5kts	5kts	10+kts
Sub Surf	N/A*	1-2	1-2
Depth	10M*	200M	200M
Payload	23kgs	23kgs	795kgs
Total Weight	127kgs	127kgs	2265kgs
Length	4.14M	4.14M	12M
Height (wing up)	2.45m	2.45m	5.0m
Power	50w	50w	>100w
Peak	300w	300w	>300w
Release Date	Q1 2016	Q4 2016	Q2 2017

* The S-10 will be able to submerge but not operate underwater (i.e. navigate)

SP1 V2

SP1 V2.2

The Mule

SP2 Hull 1
OCEAN AERO

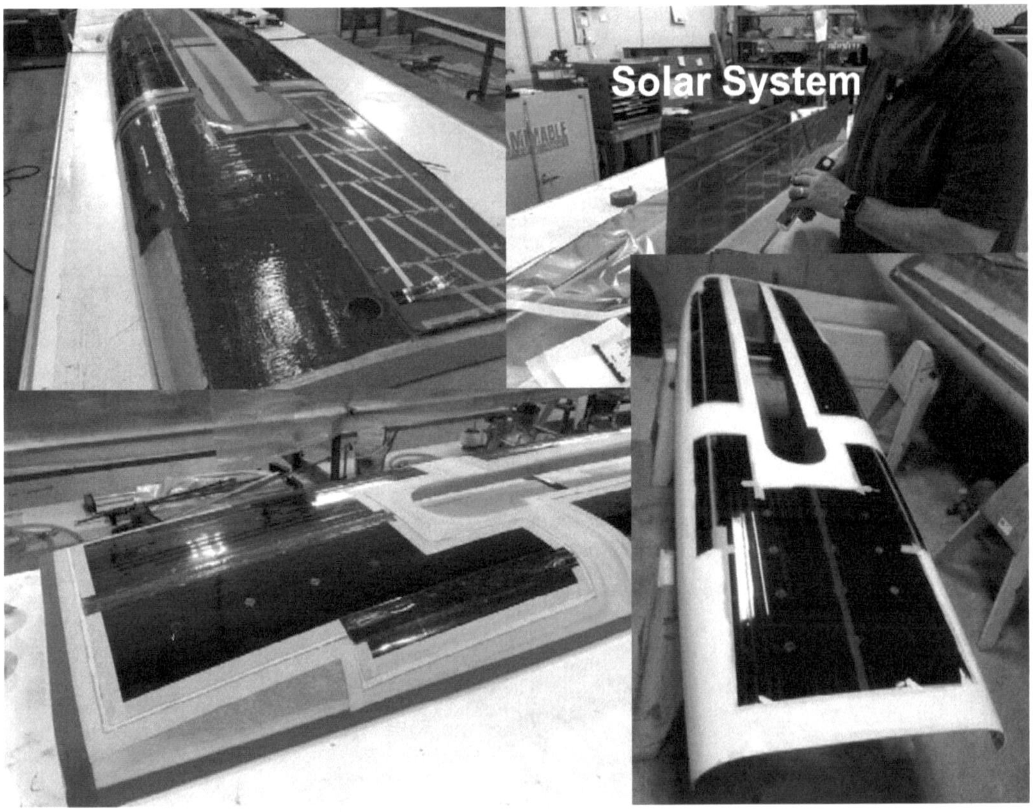
Solar System

Submerging

SP2 Hull 2

Weather Testing

SUBMARAN S10 VESSEL

Final pre-production version

- Open ocean waypoint navigation
- Autonomous tack and jibe
- Cross track 3 meters
- Autonomous “go to” & “loiter”
- Tested with towed bodies
- Mast mounted payloads
- Submersible to 10 meters

THANK YOU

Ken Childress
Vice President/COO
10755 Scripps Poway Pkwy
Suite 367
San Diego CA 92131
kenc@oceanaero.us
858-223-0142

The Riptide UUV: A New Modular, Open, Low-Cost Micro-UUV

Riptide Autonomous Solutions

Presented by Justin Manley

February 2016

Riptide Founders / Leads

Riptide Autonomous Solutions

Jeffrey M. Smith
President & CEO
BS - Mech Eng (WPI)
MS - Mech Eng (RPI)
MBA - (RPI)
20+ Years in technical management, program management, executive management, Navy Program Development, UUV Subject Matter Expert

Dr. Dani Goldberg
Software Principal
BS -Comp Sci (Brandeis)
MS - Comp Sci (USC)
PhD - Comp Sci (USC)
20+ Years in technical management and software development for autonomous systems (space and undersea), architecture development, hybrid systems

Dr. Stefano Brizzolara
Naval Architecture Principal / Founder
PhD – Numerical Hydrodynamics for Ship Design (Univ of Genoa)
~20 Years in naval architecture for advanced surface craft, architecture development, hydrodynamic design

Lenny Baker
Sr. Lead Systems Engineer / EE / SW
BS - Elec Eng (WPI)
MS – Elec Eng (Umass)
10+ Years in electrical design, shipboard electronics, power and controls

Sam Godin
Lead Mechanical Engineer
BS - Mech Eng (WPI)
20+ Years in mechanical design, SolidWorks Super User

SCPO (ret) Dan Lawrence
Lead Analyst
Retired ONI ACINT Specialist, Operational Test Director, 30+ years experience in at-sea technical assessment for operational utility

Company Background

Riptide Autonomous Solutions

- Small business focused on Unmanned Maritime Vehicle (UMV) development and production
  - Autonomous Undersea Vehicles (AUVs) / Unmanned Undersea Vehicles (UUVs)
  - Unmanned Surface Vehicles (USVs)
- Customer Focus
  - US Military
  - Research & Development
  - Commercial
- Product Offering
  - System Development
  - Transition to Production
  - Production

Initial Vehicle Offerings

Riptide Autonomous Solutions

- Micro-UUV (μUUV)
  - Small yet capable platform suited for a host of applications from development to full mission.
  - Low Cost – Could be expendable
  - Short duration, low risk to highly efficient, persistent operations

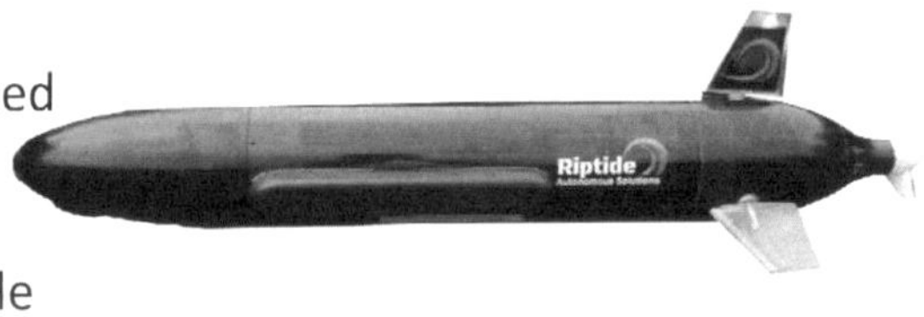

- USV Mothership
  - High stability hull form ideally suited to launch and recover numerous unmanned systems
  - Highly efficient hull form
  - Exploring multiple funding opportunities

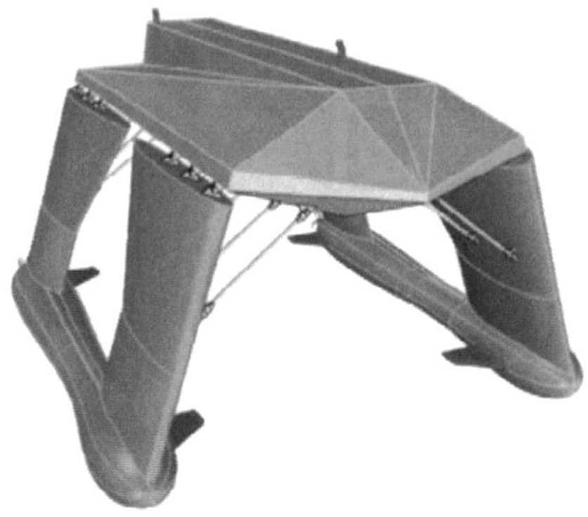

μUUV Overview

Riptide
Autonomous Solutions

Initial Demonstration Vehicle

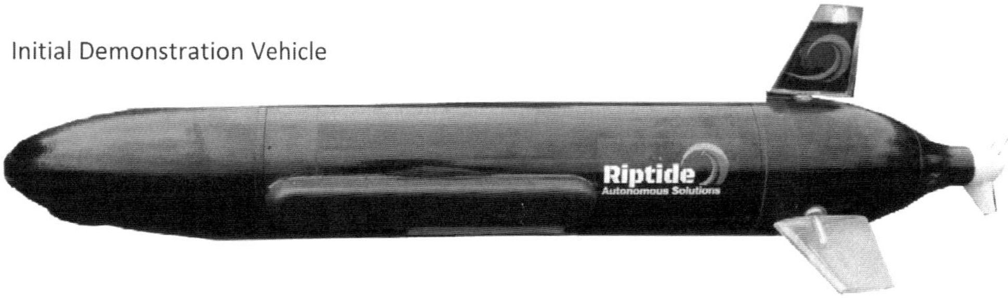

- 4 7/8 inches in diameter
- 33.5 Inches in length
- 19 lbs
- 200 m rated

μUUV Overview

Riptide
Autonomous Solutions

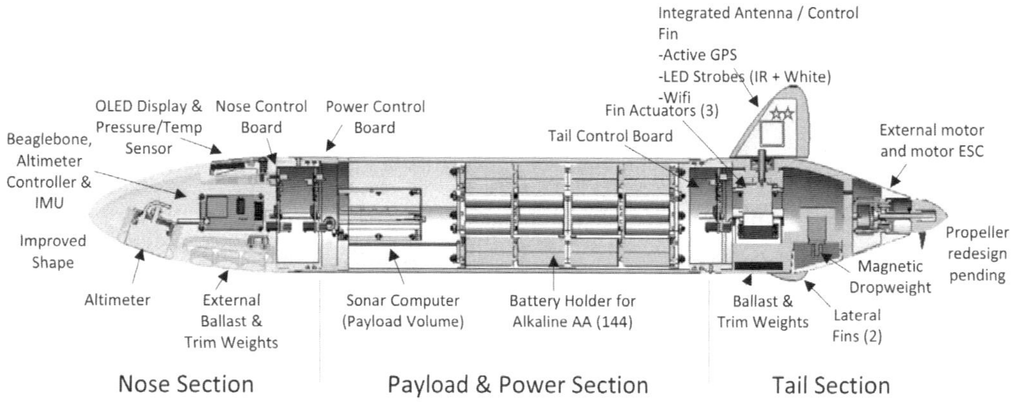

Sonar Imagery

Riptide Autonomous Solutions

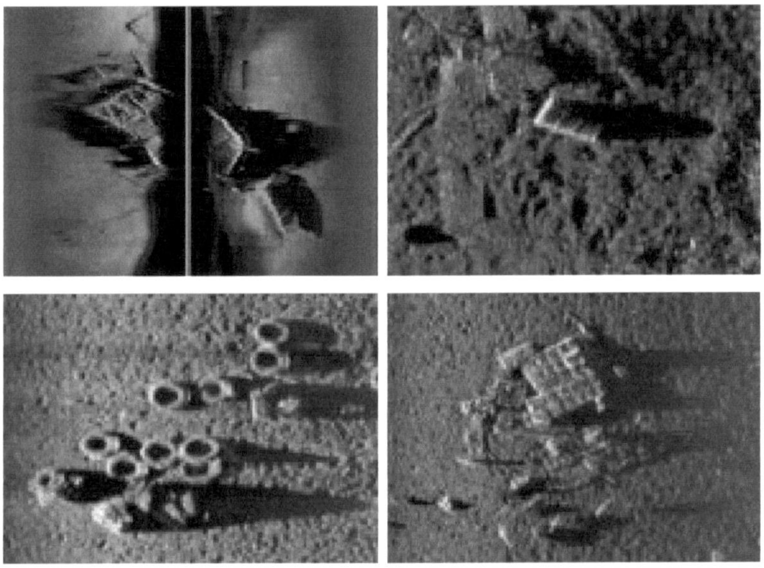

μUUV Energy Options

Riptide Autonomous Solutions

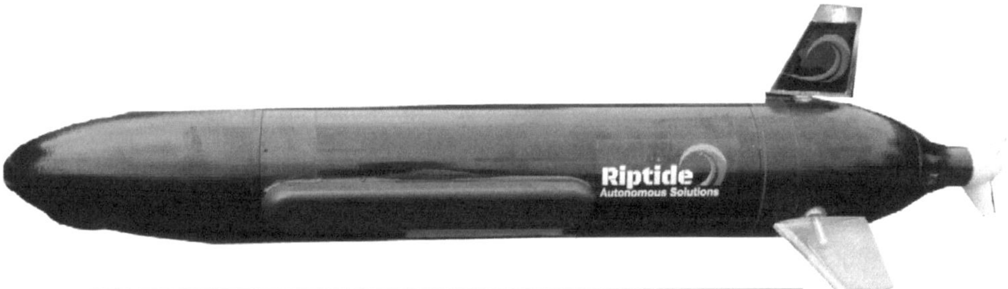

Energy Option	Predicted Endurance (per foot of battery)
Alkaline	30 hours
Lithium Ion (Rechargeable)	36 hours
Lithium Ion (Primary)	100 hours
Open Water Power	500 hours

Assumes 36" μUUV with 10W payload @ 2 knots

μUUV and MOOS

Riptide
Autonomous Solutions

- MOOS-IvP running on a Beaglebone Black
  - Provides a well supported, cost effective, power conscious hardware environment for MOOS-IvP and MOOS applications
  - Interface to top-side is via Wi-Fi connected to the Beaglebone Black USB Host port
- GPS and Wi-Fi have direct interfaces (via PWB's) to the Beaglebone Black
- MOOS applications interface to rest of the hardware through the Nose, Power, or Tail Controllers via I2C
  - Low power Arduino Pro Micros handle unique interface needs to hardware located in the respective compartments
- Flexible payload configuration with a range of standard interfaces as supported by the Beaglebone Black such as RS-232, USB, I2C, SPI, and Ethernet

μUUV Features

Riptide
Autonomous Solutions

- Superior Hydrodynamics
- Highly Efficient
- Externally Swappable Motor
- Integrated Display
- Active Roll Control
- Flexible Payloads
- Highly Flexible HW Architecture
- Low Cost
- Flexible Energy Options
- Highly Extensible
- Highly Flexible SW Architecture

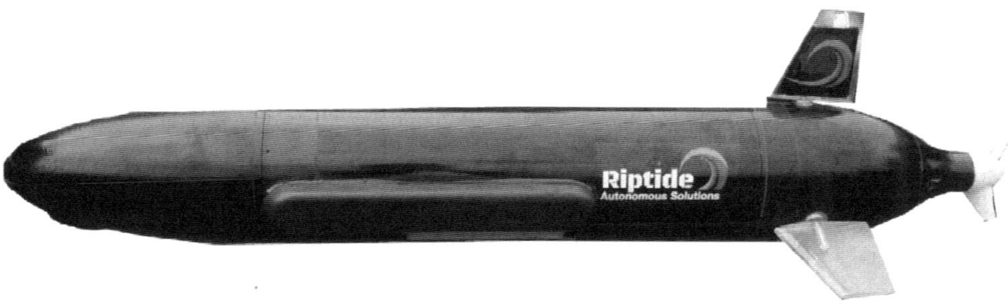

Comparison to Other Aluminum Technologies

Aluminum as a Heat Source

MIT / Lincoln Lab "Beaverworks" class 2011-2012

Evolved to

Similar Work:

PENNSTATE ARL

ARL @ Penn State's aluminum-seawater combustor[1]

Pros: high power

Cons: dangerous powder, hard to control and operate continuously, limited Carnot efficiency

Prognosis: infeasible for UUV & sensor use.

Aluminum as an H_2 Source

MIT / Lincoln Lab "Beaverworks" class 2012-2013

Evolved to

Similar Work:

GENERAL ATOMICS

General Atomics' "ALPS" System for the LDUUV Program[2]

Pros: 1-3x endurance vs. li-ion batteries[3]

Cons: O_2 storage limits energy density and creates safety hazards

Prognosis: Feasible for large UVVs, but not for small ones.

Al-H_2O Direct Fuel Cell

OpenWater Power

Foundational Work:

1) Al-O_2 fuel cells[4]
2) Al-H_2O corrosion batteries[4]

Pros: 10x endurance vs. li-ion batteries. Inherently safe, non-toxic & air-shippable.

Cons: Primary cell: can refuel, can't recharge.

Prognosis: Superior power system for UUVs and ocean sensors of all sizes.

[1] Miller, T.F., Walter, J.L., Kiely, D.H., "A next generation AUV energy system based on aluminum-seawater combustion," *Workshop on Underwater Vehicles*, June 20-21, 2002

[2] http://fuelcellseminar.com/wp-content/uploads/b2b32-2-1.pdf Accessed April 5th 2015

[3] Mendez, Alejandro, Teresa J. Leo, and Miguel A. Herreros. "Current State of Technology of Fuel Cell Power Systems for Autonomous Underwater Vehicles." *Energies* 7.7 (2014): 4676-4693.

[4] Li, Qingfeng, and Niels J. Bjerrum. "Aluminum as anode for energy storage and conversion: a review." *Journal of Power Sources* 110.1 (2002): 1-10.

www.OpenWaterPower.com

OpenWater Power

Comparison to Other Batteries and Fuel Cells

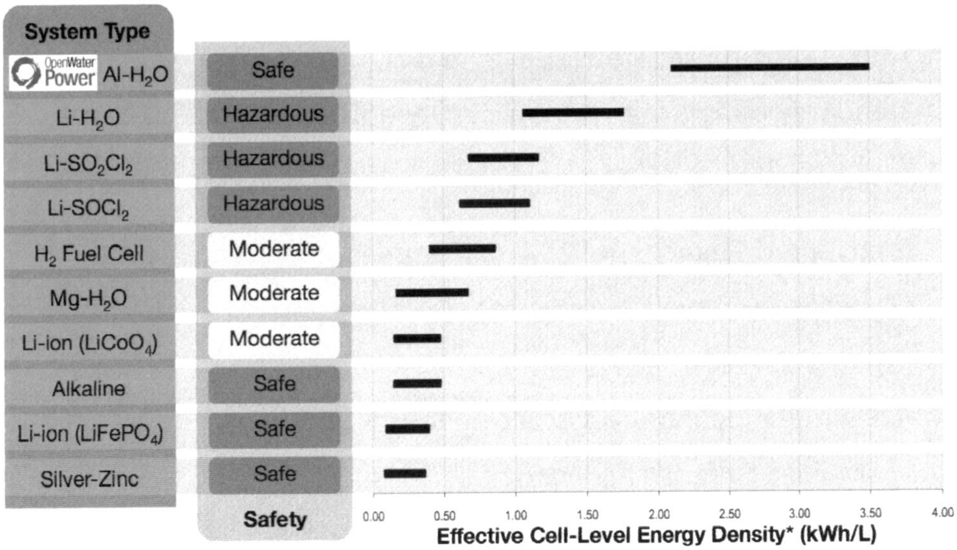

*The ranges of energy densities represent various formulations of each chemistry. Typically, for a given chemistry, power density and energy density can be traded off against each other, with higher energy densities being achievable when power requirements are low.

www.OpenWaterPower.com

OpenWater Power

Riptide µUUV

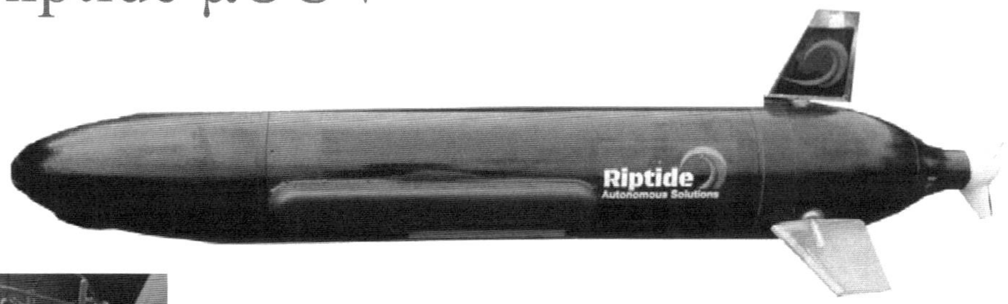

144 Alkaline AA batteries yields ~400-500 Wh → Open Water Aluminum-Seawater yields 5 kWh

www.OpenWaterPower.com

OpenWater Power

Missions / Applications

Riptide Autonomous Solutions

Short Duration Missions (Base µUUV)

Standard configuration at low cost enables wider use of UUVs in known applications

Anticipated Science, Defense or Commercial Roles:

- Rapid Environmental Assessment (REA)
- Marine Survey
- Localized pipeline/infrastructure inspection

Specialized Defense applications:

- Confined Area Search
- Neutralization

Missions / Applications

Riptide Autonomous Solutions

Long Endurance / High Persistence (Advanced μUUV)

Low cost and long endurance allows for consideration of expendable applications (HSE permitting)

Anticipated Science, Defense or Commercial Roles:

•Environmental Monitoring (persistent presence offshore pollution sites, or similar features)

•Oceanographic Data Collection (enhance floats & gliders)

•Extended pipeline/infrastructure inspection

•Anti-Submarine Warfare

Recent Field Results

Riptide Autonomous Solutions

UI Briefing will include recent photos and videos from field tests planned in February 2016

Summary

Riptide
Autonomous Solutions

- Riptide is contracted to deliver multiple micro-UUVs to the US Navy
- All specifications are being validated in test and in-water demonstrations
    - Alternate Energy Options will be validated when available
- Riptide has an exclusive partnership with Open Water Power for the micro-UUV vehicle class and is supporting their ongoing power system development
- Riptide is actively seeking development partners to explore new applications for this new, state-of-the-art μUUV

Riptide
Autonomous Solutions

For further information, please contact: Jeff Smith
617-820-4586
Jeff@RiptideAS.com

Sabertooth - Hybrid ROV/AUV capable of performing Survey and Intervention Tasks

By

Chris Roper and Bjorn Borring

Chris Roper is the sales manager of Saab Seaeye North America. Bjorn Borring is a Project Manager at Saab.

This paper describes a November 2015 demonstration of the Sabertooth hybrid vehicle at the NASA Neutral Buoyancy Laboratory (NBL) in Houston, TX. The Sabertooth is a hybrid electric vehicle with both Remotely Operated Underwater Vehicle (ROV) and Autonomous Underwater Vehicle (AUV) capabilities. This event was organized by OneSubsea to demonstrate existing commercially available technologies to the O&G community. Saab demonstrated the following the Sabertooth capabilities:

1) Exit the subsea docking station in autonomous mode and transit to a pre-set point.
2) Hover above a fixed location and harvest data from a subsea sensor using Free Space Optics (FSO).
3) Perform a detailed inspection of a 40 foot long subsea jumper in fully automated mode.
4) Transit to a valve location in AUV mode, switch to ROV mode and manually pilot the un-tethered ROV to engage a torque tool into a valve bucket, open and close the valve. Depart valve location and switch back into AUV mode.
5) Transit to a subsea docking station in AUV mode, then manually pilot the un-tethered ROV into its subsea docking station using Free Space Optics.
6) Detect an obstacle, evaluate avoidance options and take the best path to next way point.
7) Use the obstacle avoidance sonar to detect a structure, then move forward step-by-step and position in the centre of the structure.
8) Demonstrate the Sabertooth 6 Degrees of Freedom (DOF) capability.

Contents

Background: Saab

The Sabertooth Hybrid ROV/AUV traces its history back to the military MCM (Mine Counter Measure) ROV system Double Eagle SAROV. (SAROV refers to "Saab Autonomous Remotely Operated Vehicles"). Currently there are over three dozen Double Eagle & Double Eagle SAROV systems operated by 5 NATO Navies.

The Sabertooth couples existing commercial components from Saab Seaeye's range of commercial ROV systems with defense technology which significantly increase the functionality of this hybrid AUV/ROV system.

Error! Reference source not found.

Figure 1 Saab components in Sabertooth

Sabertooth takes advantage of mature technologies from a number of Saab Seaeye products:

1) SubROV, a compact 6 Degree of Freedom ROV launched from a submarine torpedo tube. The winch and launch assembly remain inside the flooded torpedo tube.
2) Work class ROV systems (Seaeye Leopard and Seaeye Jaguar) provide the majority of the sub-assemblies. Both ROV systems are supported by a seasoned development team.

Sabertooth was introduced to the US Market in 2012 with the sale of a Single Hull Sabertooth to Hibbard Inshore. The Single Hull Sabertooth is similar to the battery-powered SubROV, but has greater onboard power and payload capabilities. In ROV mode the Single Hull Sabertooth is controlled via a reinforced fiber optic tether up to 13,000 meters long. The tether is managed using a Saab constant tension winch, shown below.

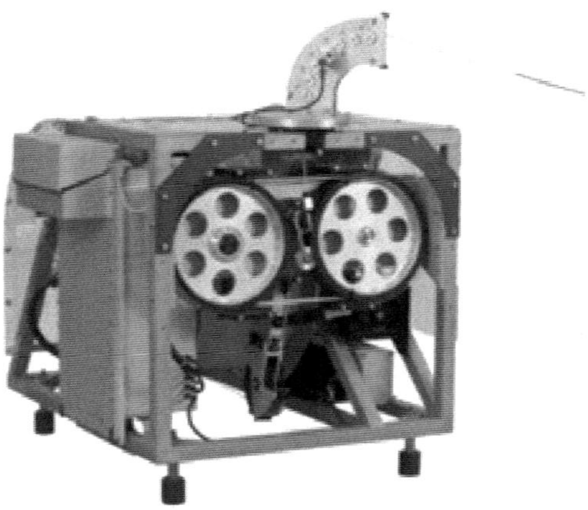

Figure 2 Saab constant tension winch

Technology Demonstration at NBL

In Spring 2015 OneSubsea invited Saab to bring its Sabertooth Hybrid ROV/AUV system to Houston as part of a technology demonstration. The main reason for this demo was to show how several mature technologies could be combined to provide connectivity with assets on the seafloor. By using smart hardware and Free Space Optical (FSO) communication, it becomes possible to monitor seafloor assets continuously and transmit data in real time from the seafloor to operational center.

Saab agreed to mobilize a Dual Sabertooth system and personnel to support this demonstration.

Additional Objectives

Later the initial setup (which included a subsea docking cassette with Bluecomm FSO) was expanded to include deploying an electric torque tool. The torque tool was configured to activate a Ringo subsea valve. OneSubsea provided the subsea valve and the jumper that would be inspected by the Sabertooth. Saab developed a 30 minute demonstration which would demonstrate the eight objectives described above.

Set up and logistics

Saab air-freighted Sabertooth, the docking station, torque tool, battery management system and a full set of support spares and tools from Sweden to Houston. They arrived on Nov. 4, 2016 along with a three man technical support team which installed the Sabertooth's rechargeable lithium batteries and prepared the Sabertooth for operation.

Figure 3 Arriving at NASA NBL.

Figure 4 Sabertooth is being lowered into the NBL tank.

NASA NBL is well-equipped to handle lifting the hardware from the receiving dock directly to the tank. In the photos above, the onsite Oceaneering ROV is shown directly below Sabertooth as she heads to the tank. The Swedish flag hangs prominently on the wall as Sabertooth makes her maiden plunge into the NASA NBL on November 7th.

Figure 5 Electric Torque Tool. The blue and white torque tool is rigidly mounted on the bow of the Sabertooth.

The electric Torque tool was shipped from Seanic Ocean Systems in Houston directly to Saab in Linkoping, Sweden where it was installed and tested on Sabertooth. Sabertooth has a full 6 Degrees of Freedom (DOF) which allows the tool to be engaged with a valve in any orientation. Control of the tool is managed by the Sabertooth Icon control software and displayed on the Seanic GUI.

Test before the big Day

The November 12th-13th OneSubsea demonstration included a number of subsea equipment manufacturers. Equipment from different manufacturers operate at different frequencies in a warm water tank. That can cause communication difficulties; the warm water itself presented a challenge. The NBL water is held at a very balmy 96 degrees F (35 C) which is intended to keep divers comfortable during their four hour exposures. That can cause internal temperature of some electronic housings to climb into the 3 digits. On November 11, Saab technicians encountered issues with the navigational sonar that caused scanning sectors to shift arbitrarily. This kind of issue had never been experienced in actual ocean water.

Problem and Workaround

Most of the tasks in the demo would require fully operational navigation sonar. Because of the heat-related sonar issues, the engineers devised a workaround which involved pre-programmed autonomous missions using local coordinates. The Doppler Velocity Log (DVL) was fully operational so the DVL/Internal Navigation System (INS) was used for navigation within the tank. During the Jumper inspection and the operations around the subsea structure, the Sabertooth worked under full control of its internal pre-programmed autonomous mission controller. Docking and valve operations were performed (as planned) under human control via Free Space Optics (FSO).

Demonstration Schedule

Four Sabertooth demonstrations were performed over a four day period.

On the first two days (Nov. 12th and 13th), OneSubsea hosted the event, arranging for industry and academic leaders to address the attending oil and gas clients. Speakers presented technologies from WHOI, ONR, Sonardyne and Saab.

On Nov. 16th and 17th Saab did additional demonstrations for 85 attendees from Oil & Gas, Military and Academia.

Demonstration Tasks

The rest of this report highlights the Sabertooth's performance on five key tasks.

Task 1: Inspect the 6 inch diameter jumper

All inspection data is stored onboard the Sabertooth and uploaded to the surface when Sabertooth returns to the docking station. The inspection was performed as a pre-programmed autonomous mission. All video inspection data was captured and stored onboard Sabertooth.

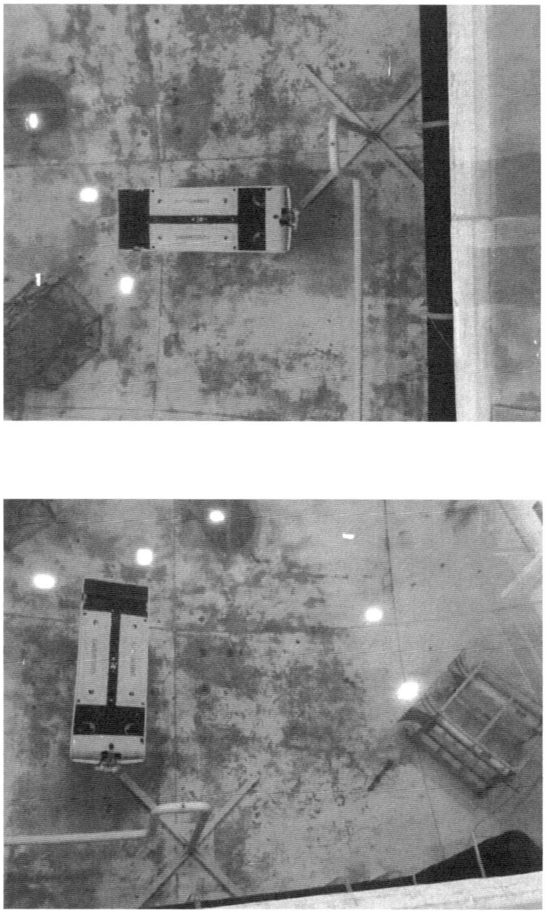

Figure 6 These two photos show Sabertooth performing a detailed inspection of the jumper.

Task 2: Operating a Valve

For this task, Sabertooth approaches a subsea valve in AUV mode and switches into ROV mode. In ROV mode the pilot_ can fly the Sabertooth forward and engage the torque tool with the valve bucket. The torque tool is controlled by the pilot through the Sabertooth surface controller. Sabertooth and the torque tool were controlled through the Bluecomm Free Space Optical (FSO) system. In fact, the task of opening and closing a subsea valve by an untethered vehicle has never been performed before; during the demo it was performed successfully multiple times.

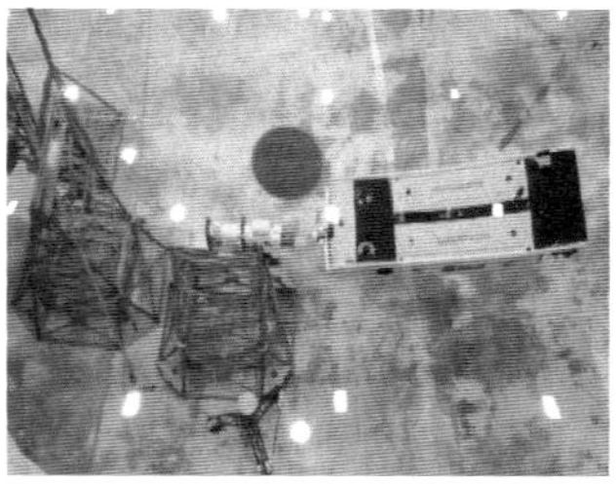

Figure 7 Sabertooth is flown in an untethered ROV to engage the torque tool in the Onesubsea Ringo valve bucket.

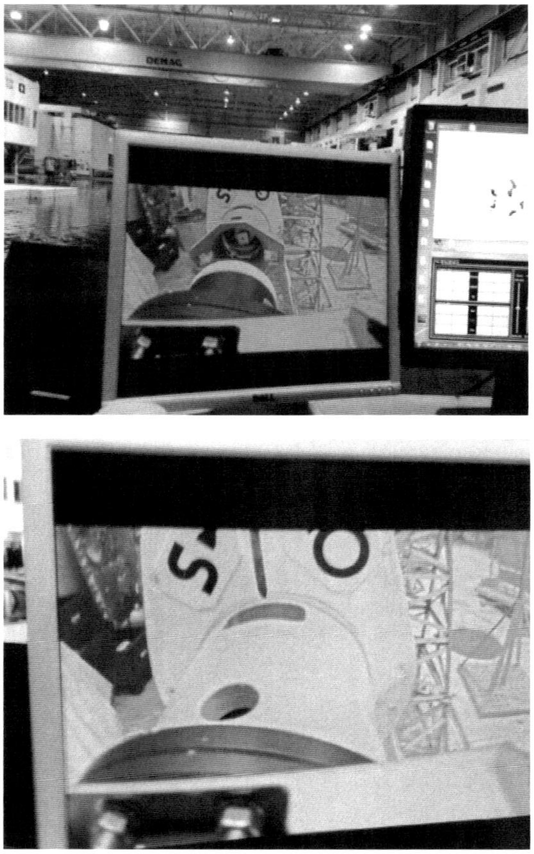

Figure 8 Two photos showing the operator's view as the torque tool is maneuvered into the bucket.

In the photo above, once the electric torque tool is engaged, it is activated to stroke the valve. The indicator is shown mid-position between Open and Shut.

After this task was complete, the pilot using FSO downloaded the next pre-programmed autonomous mission for the Sabertooth: Obstacle Avoidance & Go to Structure.

Task 3: Obstacle Avoidance & Go to Structure

To accomplish this, Sabertooth is programmed to use its navigation sonar to 1)Avoid an obstacle or 2) "Go to Structure." When running the "Go to Structure" function, the Sabertooth locates the object and then positions itself in the center of that object.

Figure 9 Demonstration of "Go To Structure" using a structure on the tank floor. The Bluecomm FSO appears here as blue light reflecting off the floor in the photo.

Task 4: Perform a diagonal dive

Most ROVs & AUVs operate in 3 or 4 degrees of freedom (DOF). But the Sabertooth operates in all 6 DOF. Having 6 DOF allows Sabertooth to position itself to optimize sensor, camera, and tooling orientation.

Figure 10 Sabertooth rolled to starboard. Bluecomm is clearly visible mounted to the underside.

Task 5: Docking and Data Uploading

Sabertooth has a "Go Home" function that retraces the original path to return to the docking station or launch point. Go Home is used in the event the Sabertooth gets lost, encounters significant onboard problems or receives the abort command. The Sabertooth locates the docking station and auto docks using its onboard sonar. During the demonstration docking and undocking functions were performed by the pilot flying the Sabertooth in un-tethered ROV mode. Once Sabertooth was docked, data from the Jumper inspection was uploaded to and displayed at the surface station via Free Space Optics.

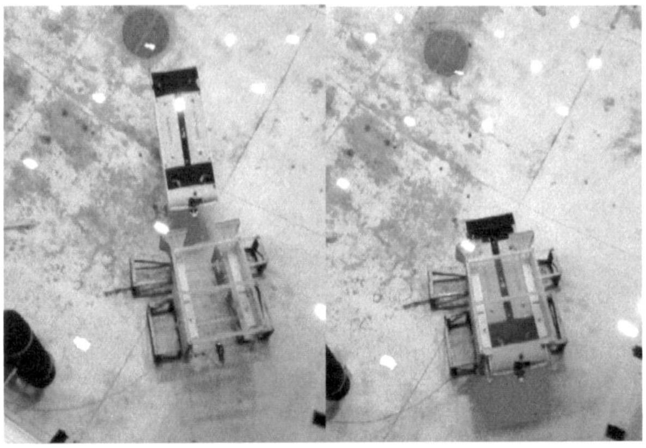

Figure 11 Docking Station. This station recharges the Sabertooth batteries and provides a high speed data link to the operator.

The docking station serves as a battery recharger and also as a high speed data link to the surface. The 6Mb/sec data link rate opens up the capacity to perform real time intervention tasks in untethered ROV mode.

Using Free Space Optics, the following tasks have been accomplished:

1. Connect and disconnect an ODI subsea connector (WHOI Nerious AUV / ROV fitted with a Kraft manipulator)
2. Valve operation – Saab Sabertooth fitted with a Seanic Ocean Systems electric torque tool.
3. Data uploaded from subsea sensors to Sabertooth.
4. Data upload for Sabertooth to surface.
5. Data downloaded from Surface to Sabertooth
6. Manually fly the Sabertooth into its docking station.

•

Figure 12 Photo showing how portable tables were set up poolside to initiate operations.

As can be seen in the photos, NBL has a fantastic facility holding 6.5 million gallons of the cleanest water imaginable. A full mockup of the International Space Station (minus the solar panels) sits in the tank, an impressive feat of engineering to behold.

Figure 13 Sabertooth (with the International Space Station in the background).

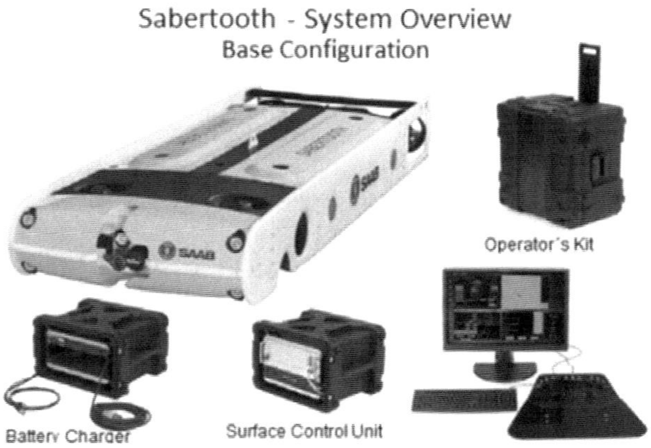

Figure 14 Control Station. Sabertooth is recharged between missions using standard 110 or 220 VAC facility power.

Figure 15 Two photos of attendees viewing the control room monitors. Different monitor show Sabertooth data and images from different camera angles.

Conclusion

The demonstration was very successful and all attendees left with a much clearer understanding of the state of the Saab Sabertooth seafloor resident development program. The attendees witnessed a number of firsts during the demonstration.

1) The first time a Double Hull Sabertooth has been demonstrated in North America
2) The first time a subsea valve has been activated by an untethered ROV.
3) The first time a Saab underwater system has been operated in close proximity to the International Space Station.

Figure 16 (Left) Project Manager Bjorn Borring; (Right) Sabertooth Pilot Per Eriksson

Figure 17 (Left) Saab NA Sales Manager Chris Roper; (Right) Software Engineer Thomas Andreasson

Acknowledgements

- Dr. Diana Grauer, Tobas Voelkel and their team at Onesubsea
- Staff at the NASA NBL
- Saab Management: Agneta Kammeby Bert Johansson, Sven-Inge Svensson & Jan Siesjo
- The Saab Technical Support team

Special thanks to the NASA Astronauts that put up with our photo opportunity requests.

Richard Jenkins, CEO
UNDERWATER INTERVENTION
SAILDRONE

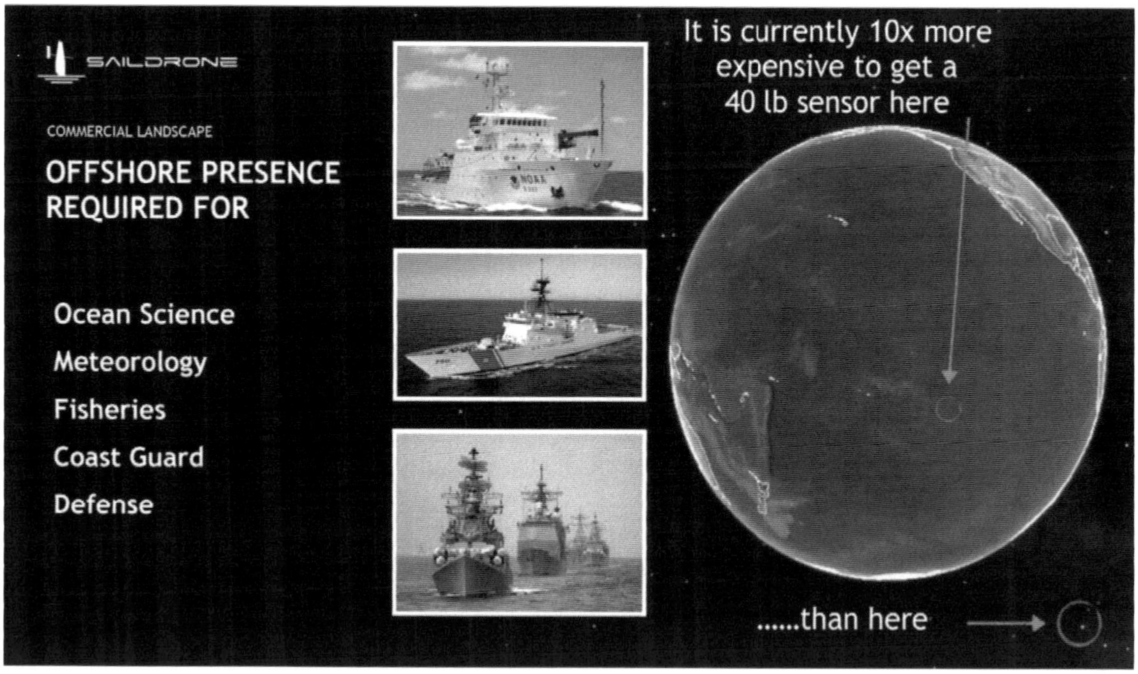
SAILDRONE
COMMERCIAL LANDSCAPE
OFFSHORE PRESENCE REQUIRED FOR
Ocean Science
Meteorology
Fisheries
Coast Guard
Defense
It is currently 10x more expensive to get a 40 lb sensor here
......than here

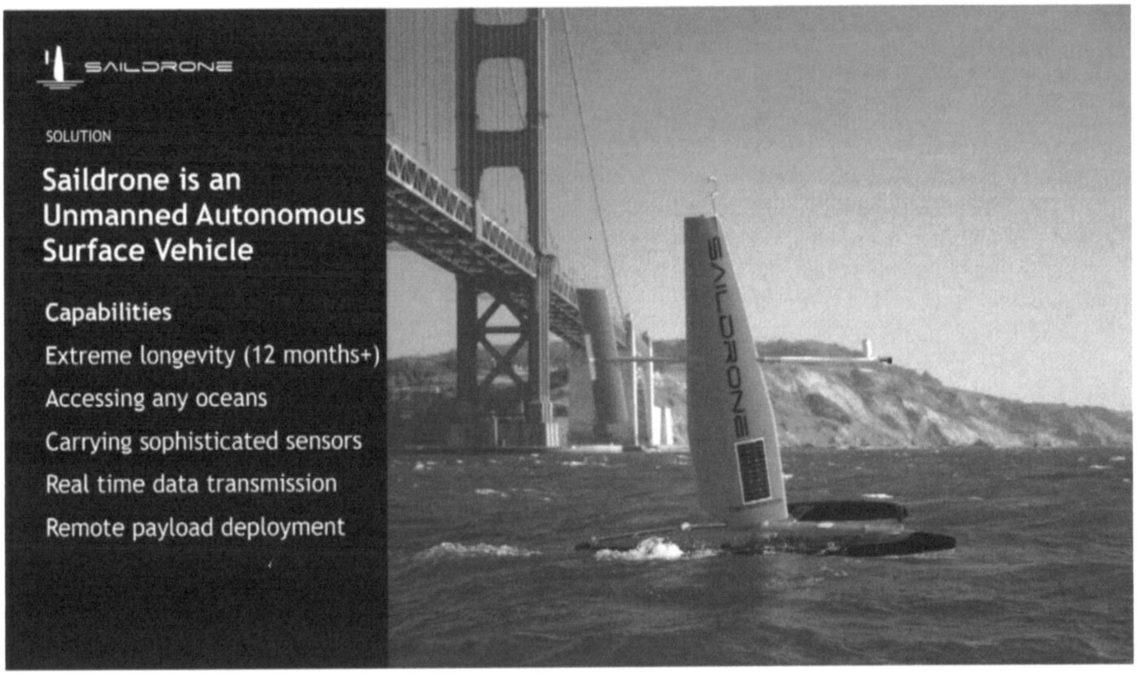
SAILDRONE
SOLUTION
Saildrone is an Unmanned Autonomous Surface Vehicle
Capabilities
Extreme longevity (12 months+)
Accessing any oceans
Carrying sophisticated sensors
Real time data transmission
Remote payload deployment
SAILDRONE

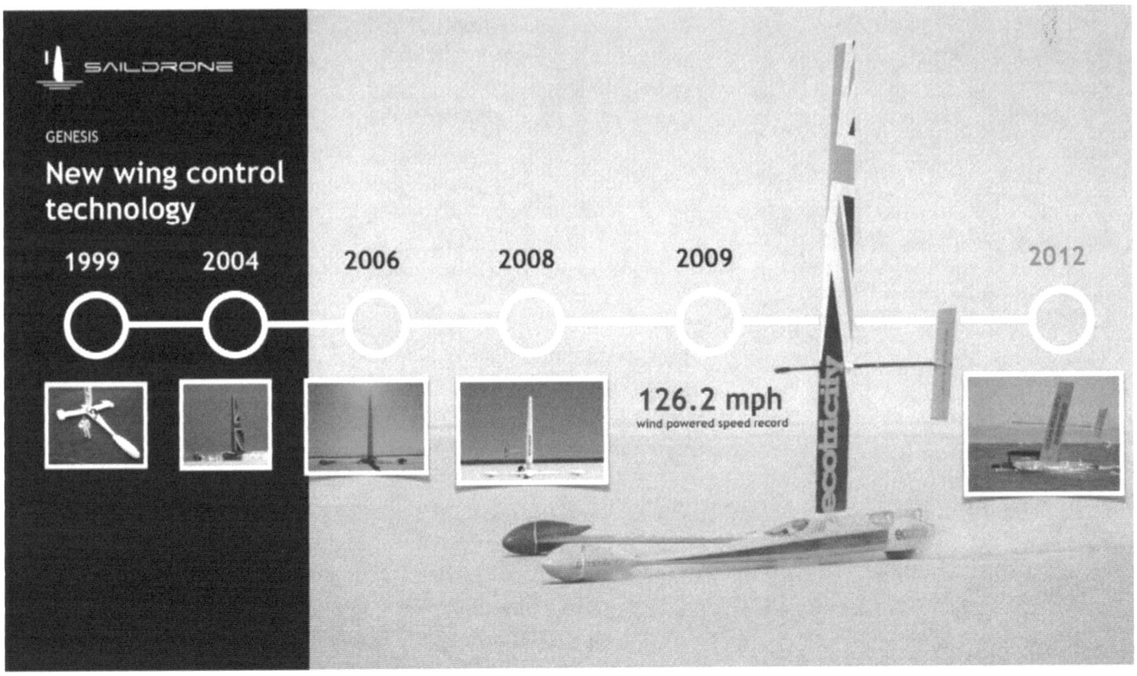
SAILDRONE
GENESIS
New wing control technology
1999
2004
2006
2008
2009
2012
126.2 mph
wind powered speed record
ecotricity

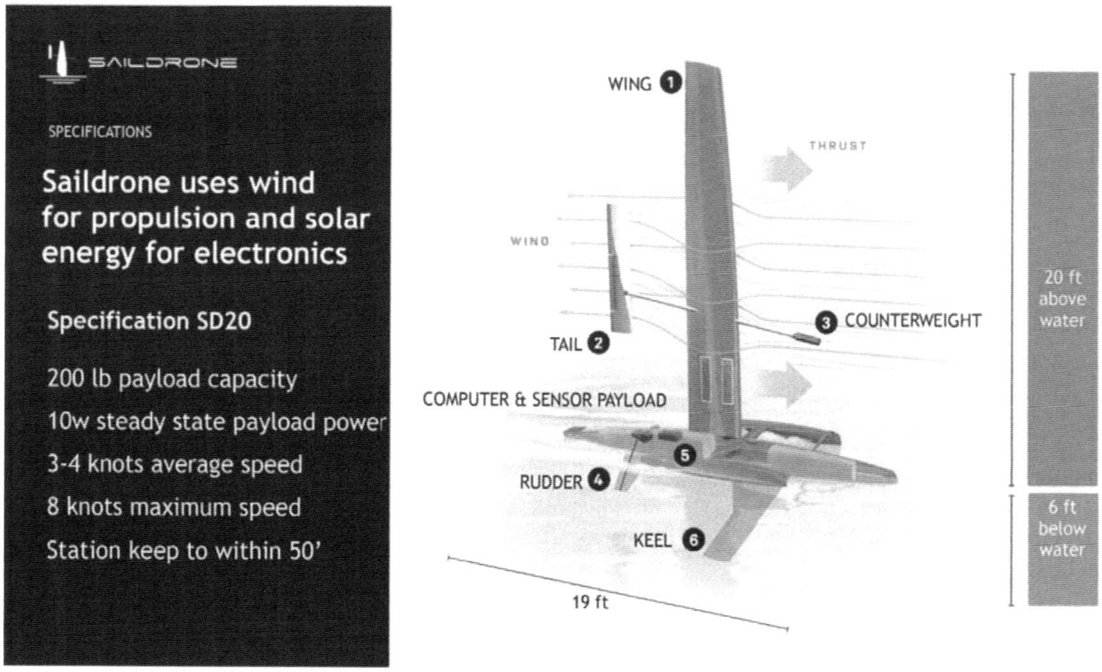
SAILDRONE
SPECIFICATIONS
Saildrone uses wind for propulsion and solar energy for electronics
Specification SD20
200 lb payload capacity
10w steady state payload power
3-4 knots average speed
8 knots maximum speed
Station keep to within 50'
WING 1
THRUST
WIND
TAIL 2
3 COUNTERWEIGHT
COMPUTER & SENSOR PAYLOAD
5
RUDDER 4
KEEL 6
19 ft
20 ft above water
6 ft below water

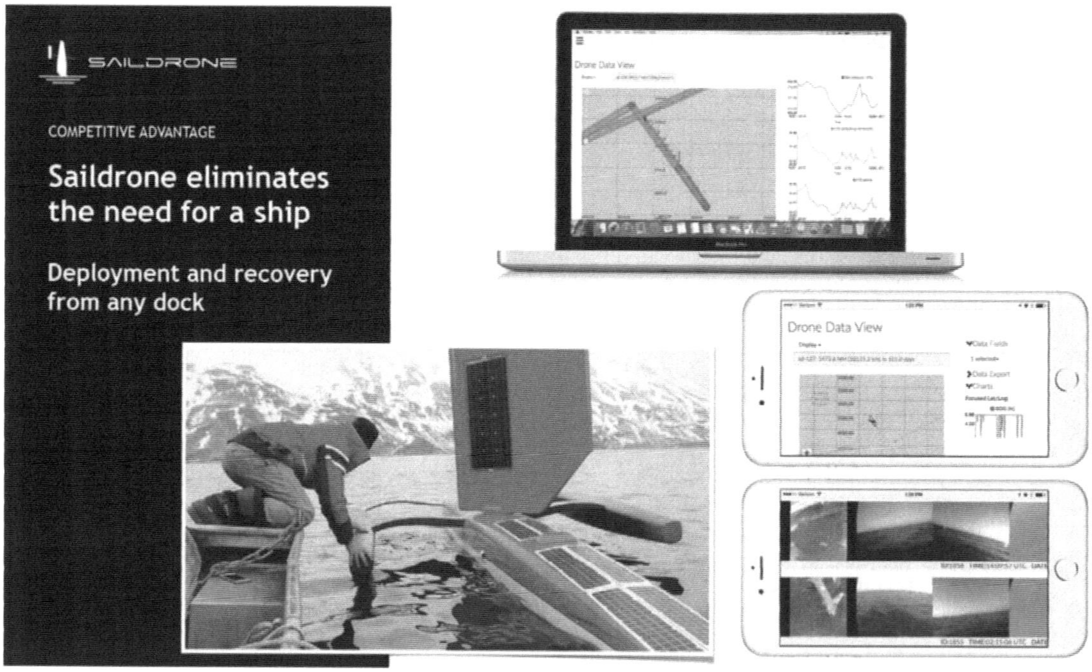
SAILDRONE
COMPETITIVE ADVANTAGE
Saildrone eliminates the need for a ship
Deployment and recovery from any dock
Drone Data View

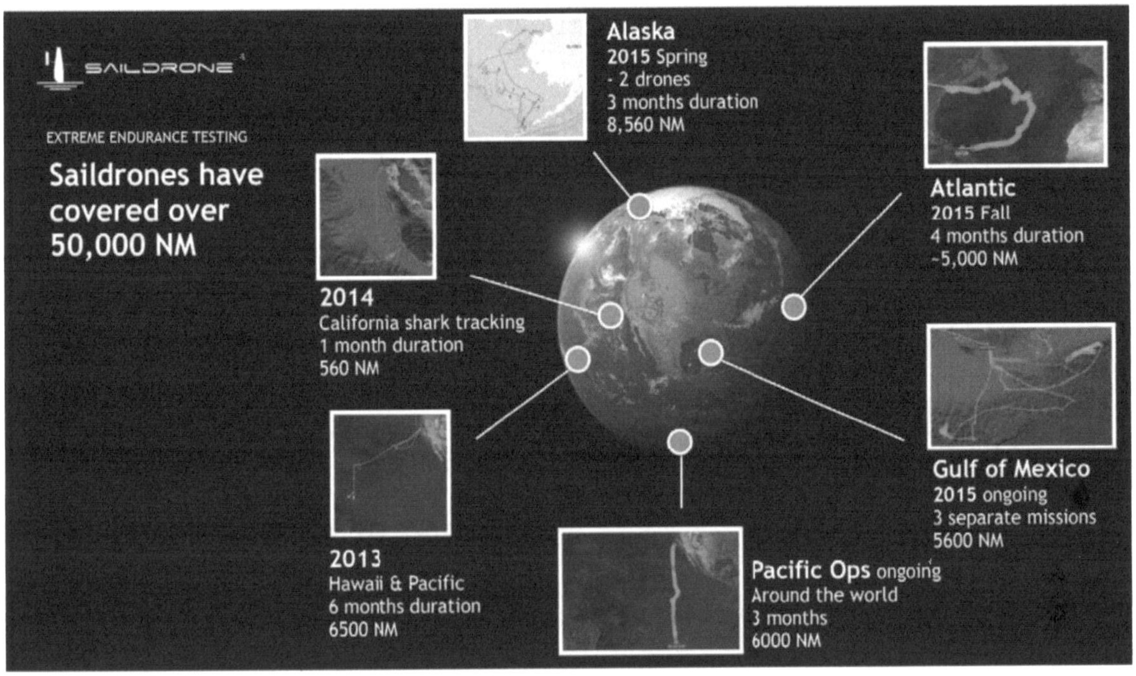
SAILDRONE
EXTREME ENDURANCE TESTING
Saildrones have covered over 50,000 NM
Alaska
2015 Spring
- 2 drones
3 months duration
8,560 NM
Atlantic
2015 Fall
4 months duration
~5,000 NM
2014
California shark tracking
1 month duration
560 NM
Gulf of Mexico
2015 ongoing
3 separate missions
5600 NM
2013
Hawaii & Pacific
6 months duration
6500 NM
Pacific Ops ongoing
Around the world
3 months
6000 NM

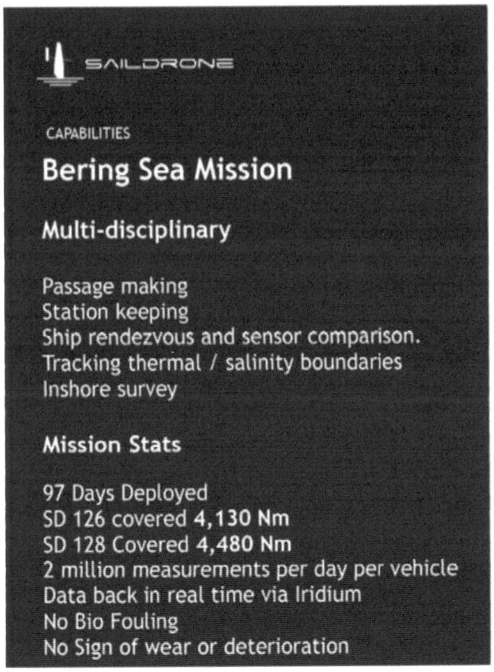
SAILDRONE
CAPABILITIES
Bering Sea Mission
Multi-disciplinary
Passage making
Station keeping
Ship rendezvous and sensor comparison.
Tracking thermal / salinity boundaries
Inshore survey
Mission Stats
97 Days Deployed
SD 126 covered 4,130 Nm
SD 128 Covered 4,480 Nm
2 million measurements per day per vehicle
Data back in real time via Iridium
No Bio Fouling
No Sign of wear or deterioration

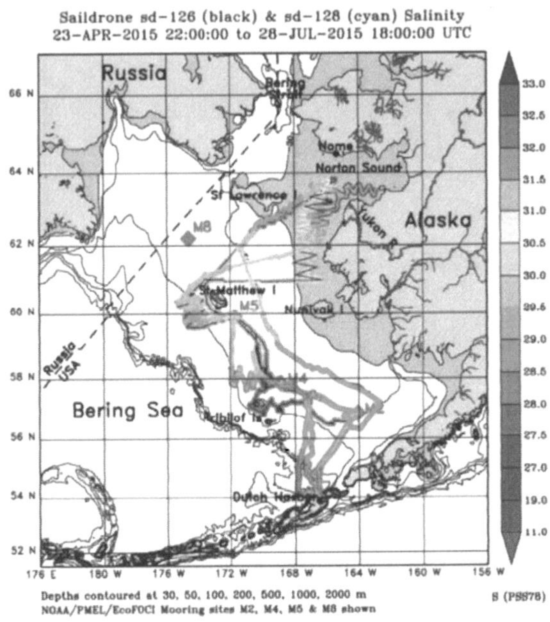
Saildrone sd-126 (black) & sd-128 (cyan) Salinity
23-APR-2015 22:00:00 to 28-JUL-2015 18:00:00 UTC
Russia
Alaska
Bering Sea
Nome
Norton Sound
St Lawrence
St Matthew I
Nunivak I
Pribilof Is
Dutch Harbor
M8
M5
Russia
USA
66 N
64 N
62 N
60 N
58 N
56 N
54 N
52 N
176 E
180 W
176 W
172 W
168 W
164 W
160 W
156 W
33.0
32.5
32.0
31.5
31.0
30.5
30.0
29.5
29.0
28.5
28.0
27.5
27.0
19.0
11.0
Depths contoured at 30, 50, 100, 200, 500, 1000, 2000 m
NOAA/PMEL/EcoFOCI Mooring sites M2, M4, M5 & M8 shown
S (PSS78)

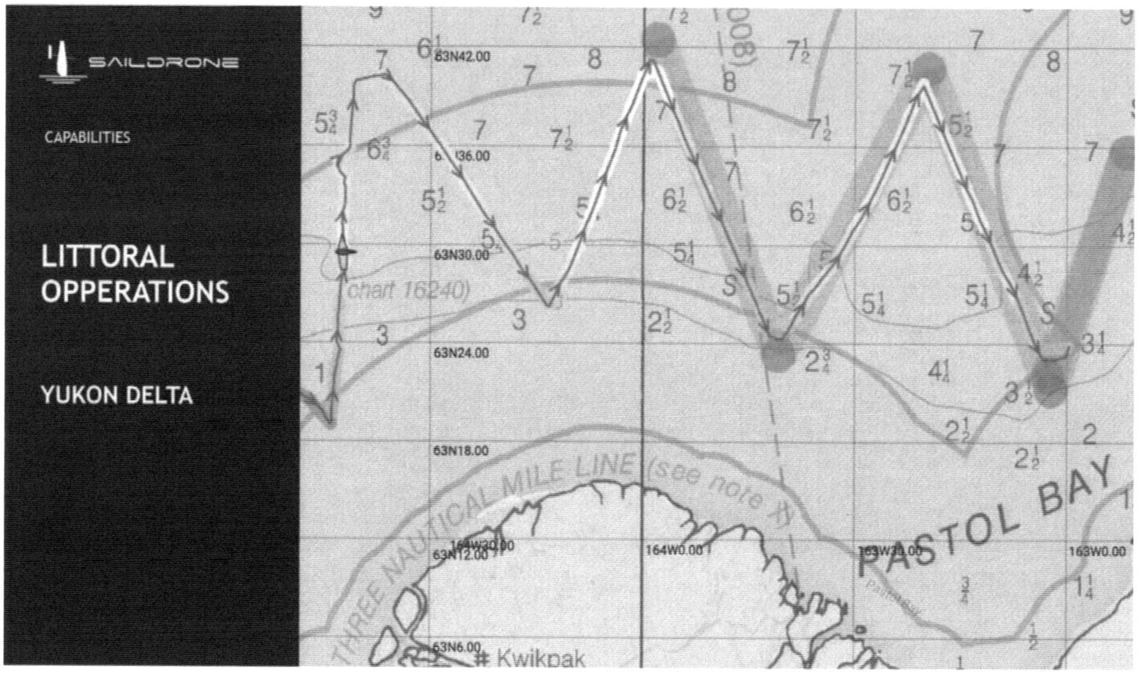
SAILDRONE
CAPABILITIES
LITTORAL OPPERATIONS
YUKON DELTA
63N42.00
63N36.00
63N30.00
63N24.00
63N18.00
63N12.00
63N6.00
164W30.00
164W0.00
163W30.00
163W0.00
chart 16240
THREE NAUTICAL MILE LINE (see note X)
PASTOL BAY
Kwikpak

SAILDRONE
CAPABILITIES
97 DAYS AT SEA
NO FOULING
NO WEAR
NO MAINTENANCE

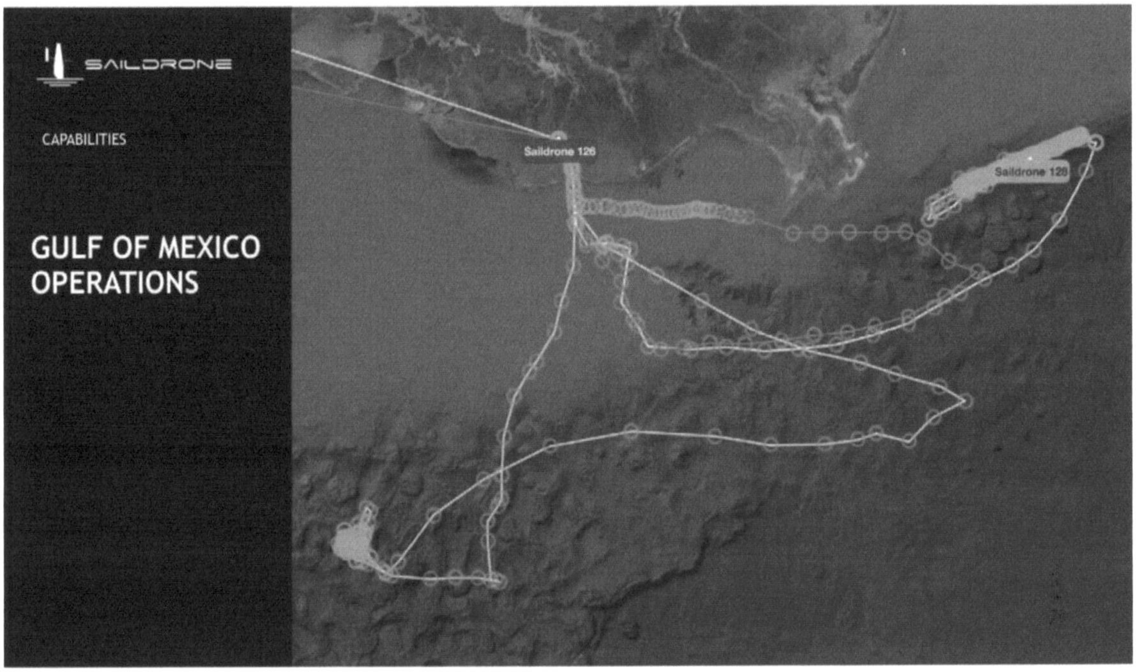
SAILDRONE
CAPABILITIES
GULF OF MEXICO OPERATIONS
Saildrone 126
Saildrone 128

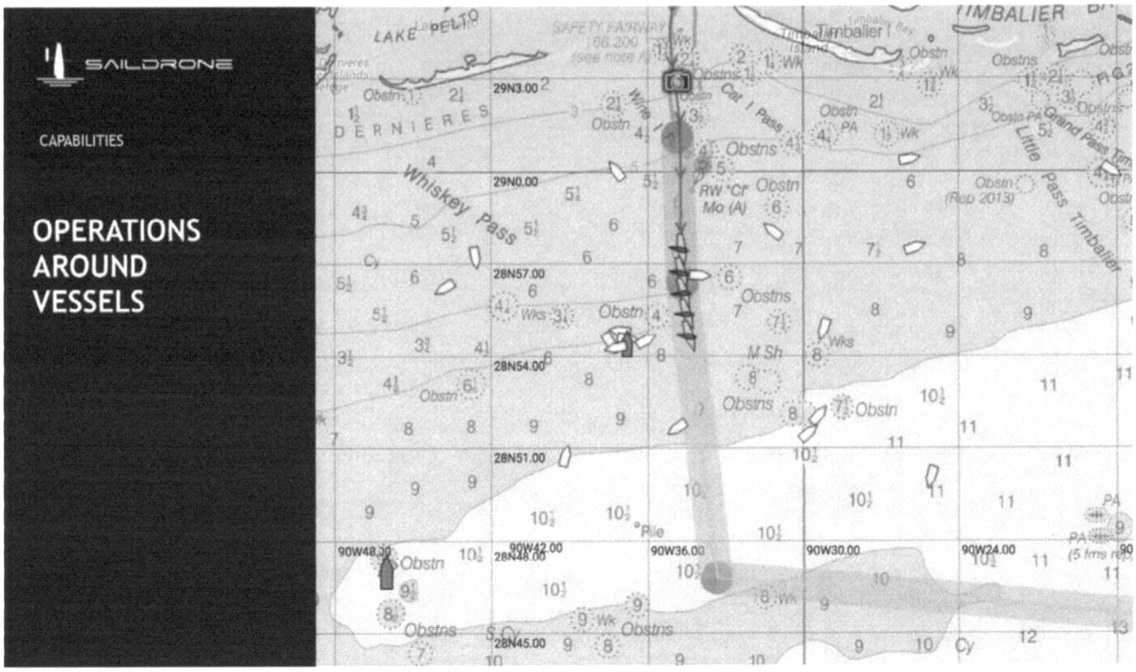
SAILDRONE
CAPABILITIES
OPERATIONS AROUND VESSELS
LAKE PELTO
SAFETY FAIRWAY
Timbalier I
TIMBALIER
DERNIERES
Whiskey Pass
Cat I Pass
Little Pass Timbalier
Grand Pass
29N3.00
29N0.00
28N57.00
28N54.00
28N51.00
28N48.00
28N45.00
90W48.00
90W42.00
90W36.00
90W30.00
90W24.00

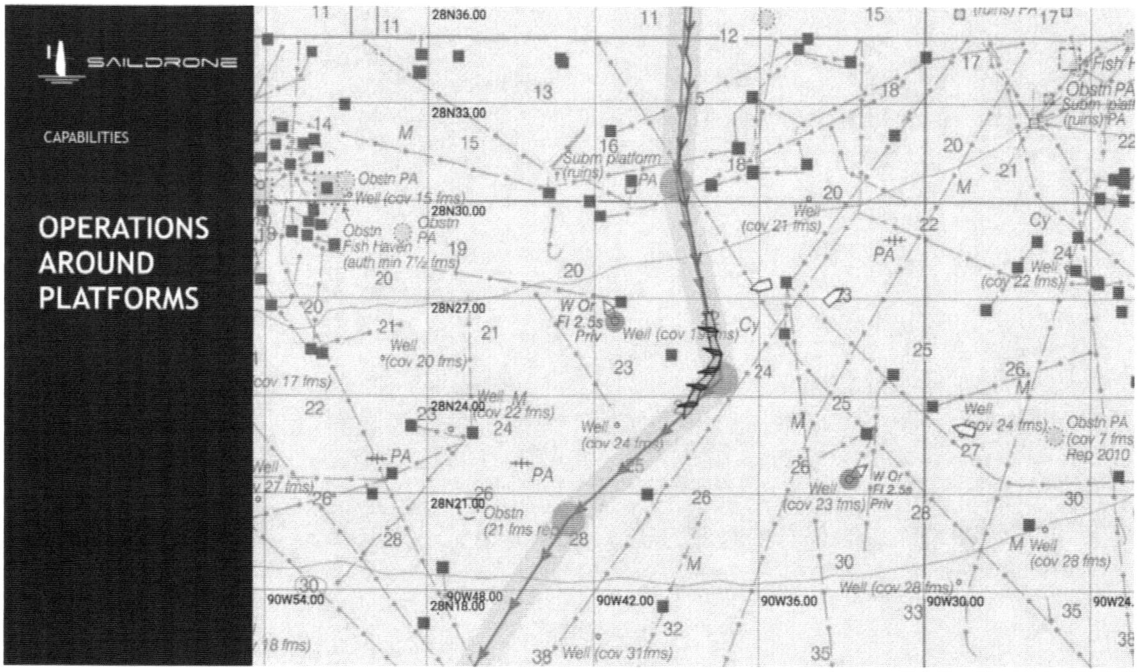
CAPABILITIES
OPERATIONS AROUND PLATFORMS

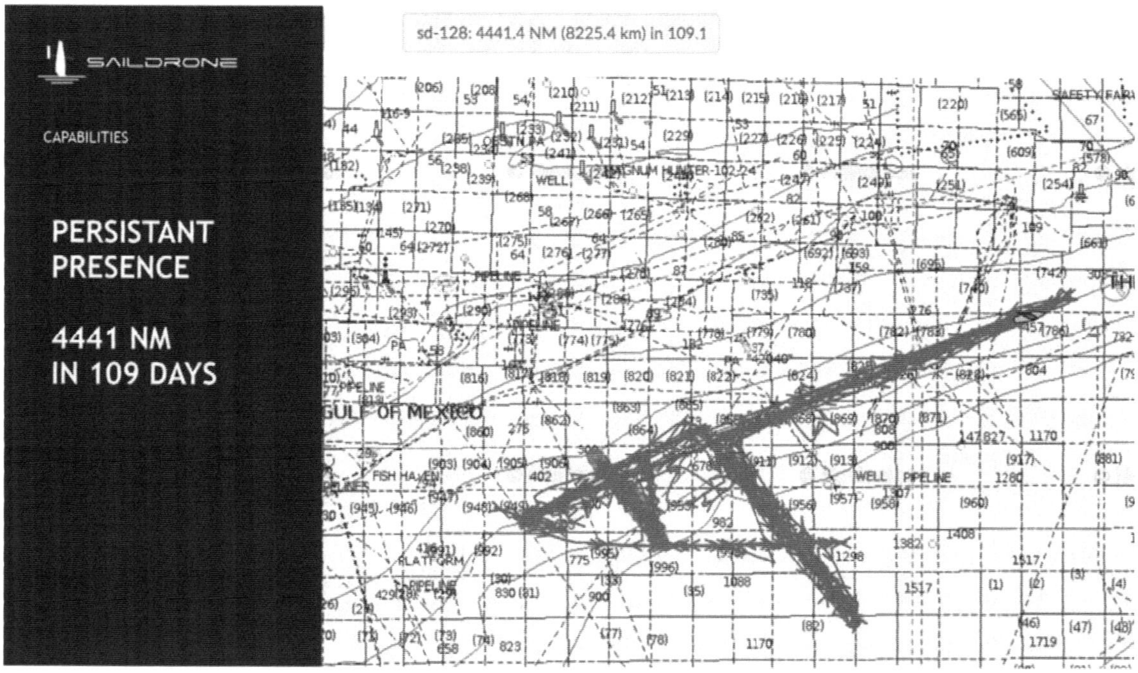
CAPABILITIES
PERSISTANT PRESENCE
4441 NM IN 109 DAYS
sd-128: 4441.4 NM (8225.4 km) in 109.1
GULF OF MEXICO

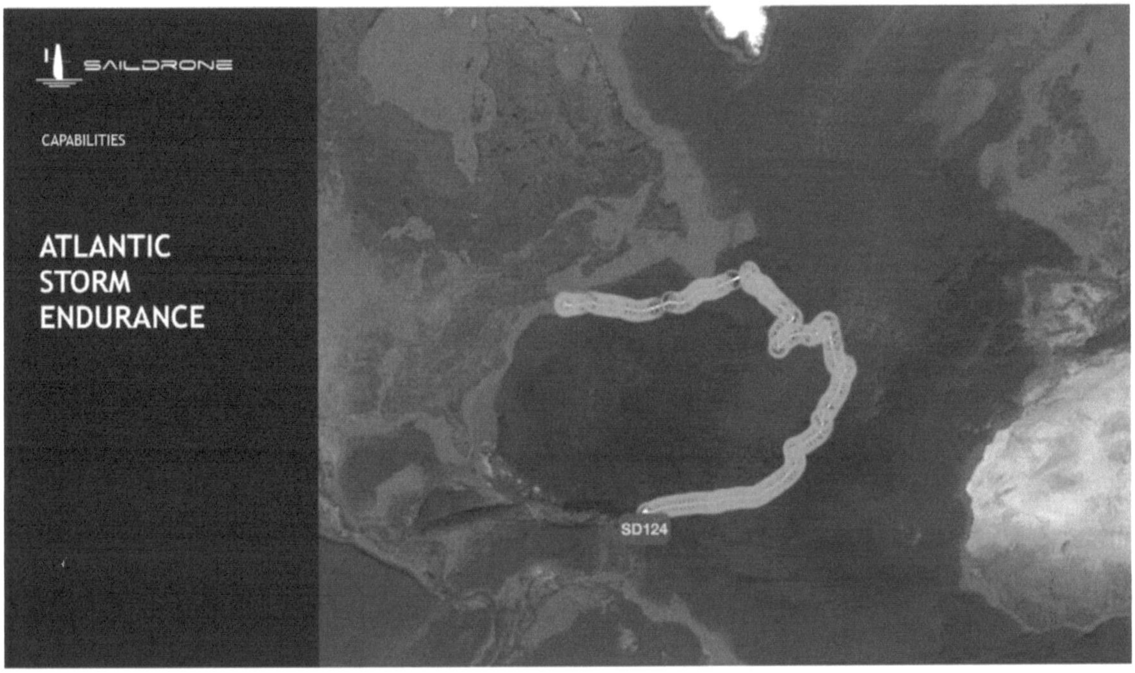
CAPABILITIES
ATLANTIC
STORM
ENDURANCE
SD124

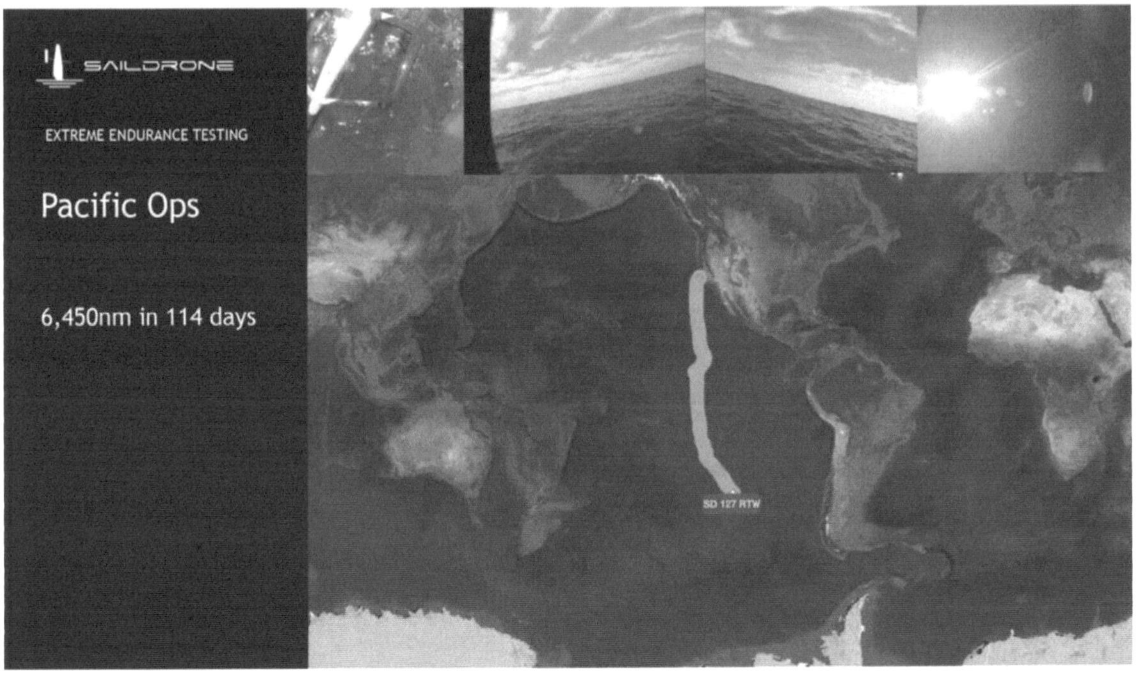
EXTREME ENDURANCE TESTING
Pacific Ops
6,450nm in 114 days
SD 127 RTW

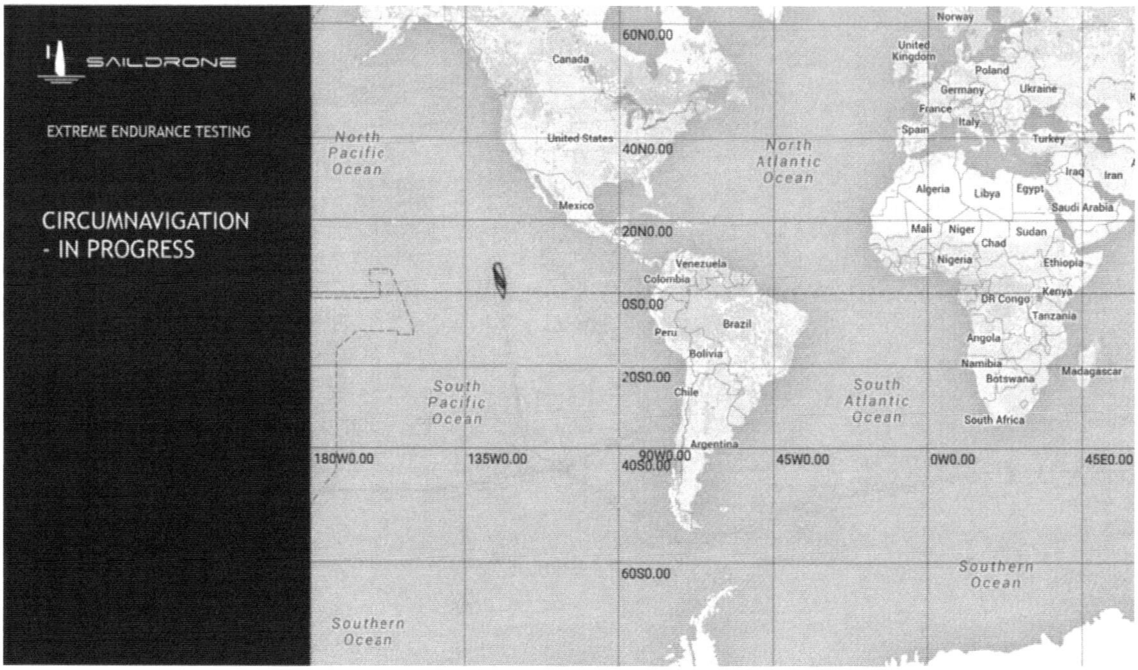
SAILDRONE
EXTREME ENDURANCE TESTING
CIRCUMNAVIGATION
- IN PROGRESS
North Pacific Ocean
North Atlantic Ocean
South Pacific Ocean
South Atlantic Ocean
Southern Ocean
Southern Ocean
Canada
United States
Mexico
Venezuela
Colombia
Brazil
Peru
Bolivia
Chile
Argentina
Norway
United Kingdom
Poland
Germany
Ukraine
France
Italy
Spain
Turkey
Iraq
Iran
Algeria
Libya
Egypt
Saudi Arabia
Mali
Niger
Chad
Sudan
Nigeria
Ethiopia
DR Congo
Kenya
Tanzania
Angola
Namibia
Botswana
Madagascar
South Africa
60N0.00
40N0.00
20N0.00
0S0.00
20S0.00
40S0.00
60S0.00
180W0.00
135W0.00
90W0.00
45W0.00
0W0.00
45E0.00

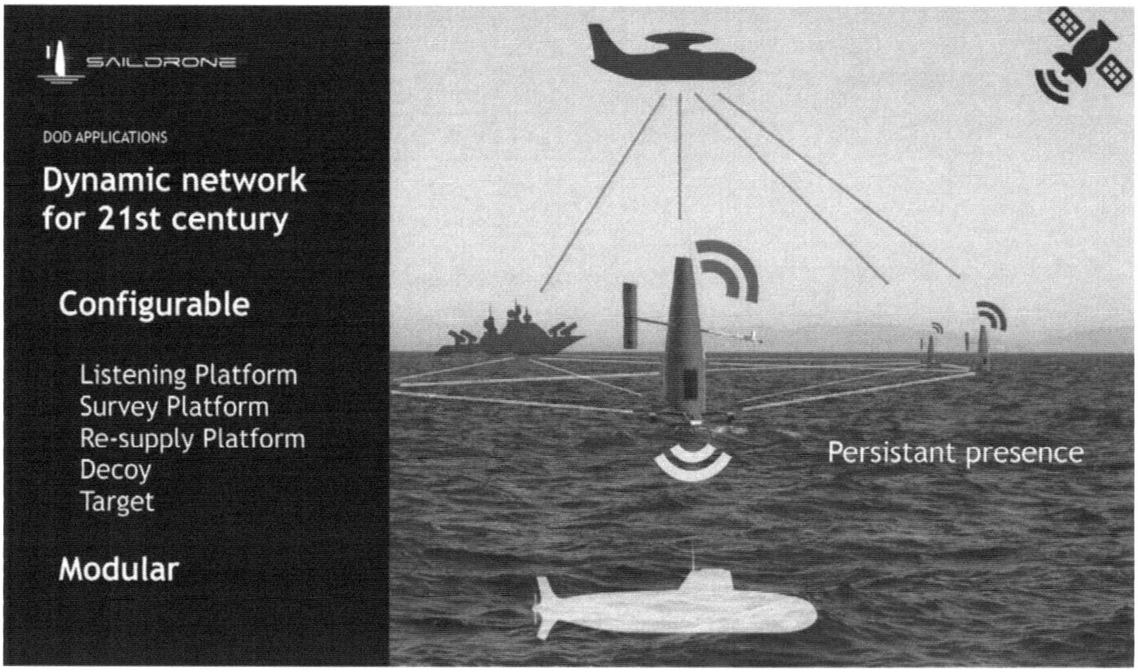
SAILDRONE
DOD APPLICATIONS
Dynamic network for 21st century
Configurable
Listening Platform
Survey Platform
Re-supply Platform
Decoy
Target
Modular
Persistant presence

SAILDRONE
Deploy
Transit
Loiter
Task
Service
Re-deploy
CONTACT
richard@saildrone.com

Teledyne Marine Unmanned Vehicles: Case Studies in Oil and Gas Applications

Arnar Steingrimsson & Erik White

IMAGING • INSTRUMENTS • INTERCONNECT • SEISMIC • VEHICLES

TELEDYNE MARINE
Everywhereyoulook™

Approved for public release: FAL-16-003

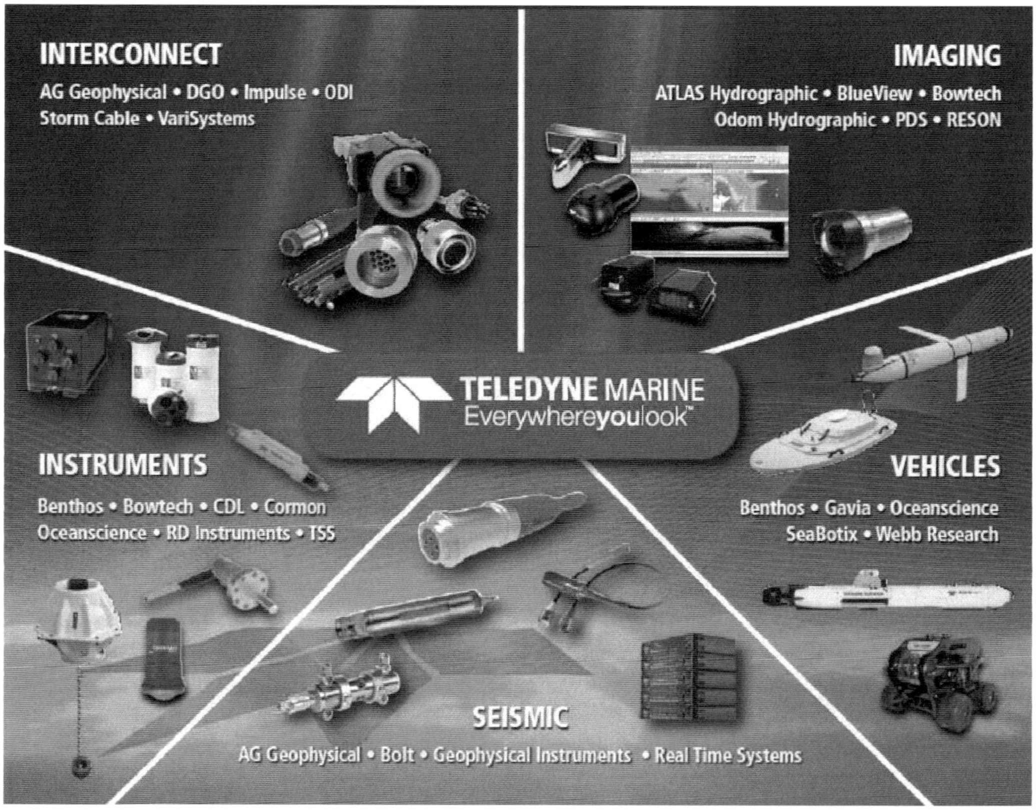

Why Use Unmanned Systems?

- Lower cost than traditional means
- Exceptional data quality (and quantity in some cases)
- Risk Reduction: HSE and project cost
- Force Multipliers :"Doing more with less"

IMAGING • INSTRUMENTS • INTERCONNECT • SEISMIC • VEHICLES

TELEDYNE MARINE Everywhereyoulook™

Approved for public release: FAL-16-003

Vehicles
Slocum G2 Gliders

World Leader in Long-Endurance Glider AUVs

- Over 570 delivered
- Over 40 demonstrated payloads
- 300 - 1,000m depth rating
- Endurance over one year
- Hybrid option provides thruster for demanding environments
- Currently in use with Oil and Gas Service Providers

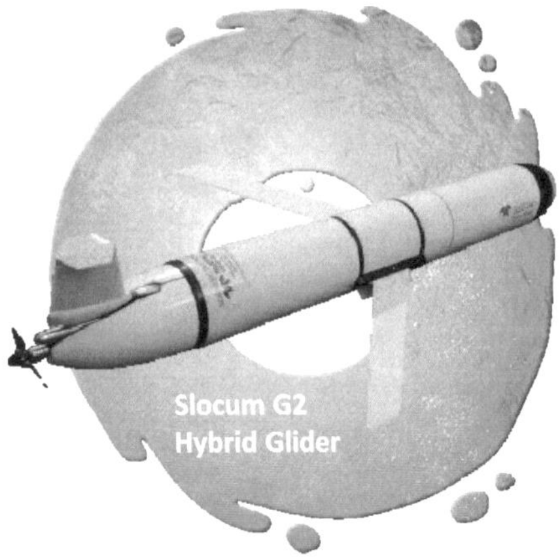

Fleet leading long endurance AUV

IMAGING • INSTRUMENTS • INTERCONNECT • SEISMIC • VEHICLES

TELEDYNE MARINE Everywhereyoulook™

Approved for public release: FAL-16-003

Low Logistics & Autonomous

- Man-portable
- Remotely Managed , Vessel-free Operation
- Iridium-based Control & Reporting
- LARS-free Deployment from Vessels of Opportunity
- LARS option available
- Persistent surveillance at a fraction of the cost of surface vessels.
- High Endurance up to 365 days

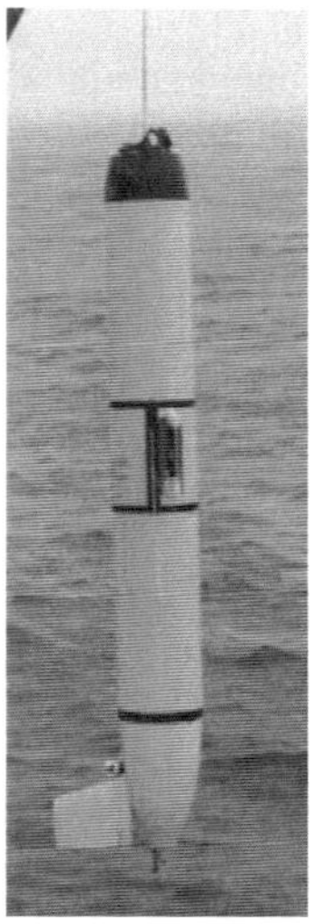

Courtesy Clayton Jones, Teledyne Webb Research

Courtesy Christopher Ordóñez

Courtesy Ben Allsup, Teledyne Webb Research

Approved for public release: FAL-16-003

Sensor Suites: Over 43 Acoustic, Optic, and Chemical Sensors Integrated

Rockland Scientific Microrider Turbulence Sensor

Wet Labs Fluorometer with CDOM

RDI Explorer ADCP

VEMCO Fish Tracker

Three puck bay with Satlantic Irradiance and Wet Labs Eco Puck, Sea Bird Pumped CTD

What do you need to detect? The solution you require may already be available. If not, we are experts at new sensor integrations.

Satlantic SUNA Nitrate

Teledyne Webb Research images

IMAGING • INSTRUMENTS • INTERCONNECT • SEISMIC • VEHICLES

TELEDYNE MARINE Everywhereyoulook™

Approved for public release: FAL-16-003

Typical Slocum Applications for Oil & Gas

- Ocean Observation and Environmental Monitoring
- Hydrocarbon Detection and Monitoring
- Data Harvesting from Subsea Nodes
- Marine Mammal Mitigation

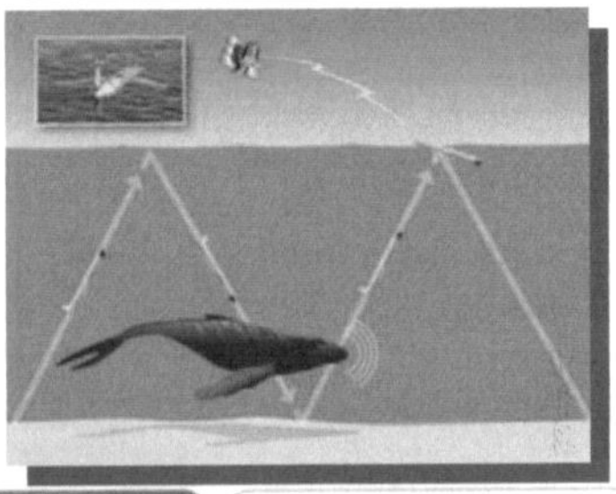

Case Study
Naval Oceanographic Office Glider Use

According to the NAVO brochure entitled "Glider Operations at Naval Oceanographic Office" the Naval Oceanographic Offices utilizes gliders due to:

1. <u>Cost</u>. Each glider profile costs less than $100, compared to over $10,000 for a ship-based profile. The glider itself costs roughly that of three days of survey ship operations

2. <u>More data, faster</u>. Gliders can provide hundreds of profiles per month, compared to only dozens of ship-based profiles per month. Persistent data sampling results in more data over similar timeframes compared to using other platforms. LBS gliders collect thousands of profiles around the globe every year. The same number of profiles collected via ship would take many years and cost many millions of dollars.

3. <u>Force multiplier</u>. After being deployed from a ship, a glider can sample for months with no local ship presence. This frees the ship to perform other functions at the same time.

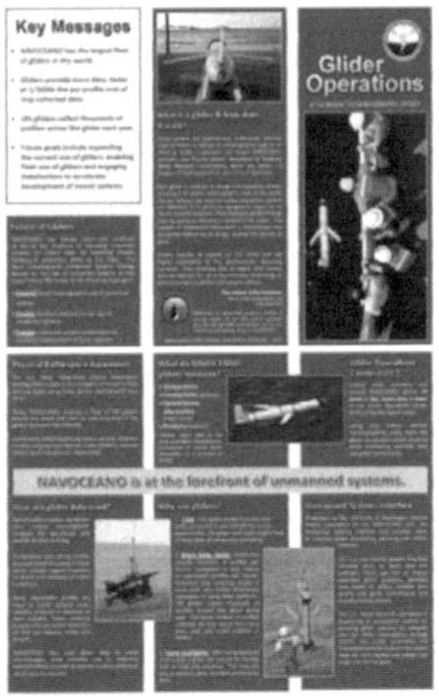

DOF Subsea : GoM Loop Current Monitoring

- Measure Near-Surface Currents
  - ADCP and GPS
- Characterize Water Column Heat and Salinity
  - CTD
- Station-Keeping in Fast Currents
  - Glider Thruster

DOF Subsea

KNOW YOUR OCEAN
SKANDI EXPLORERS

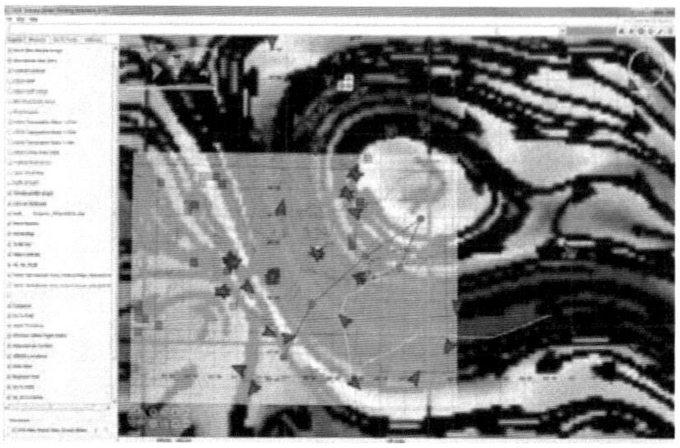

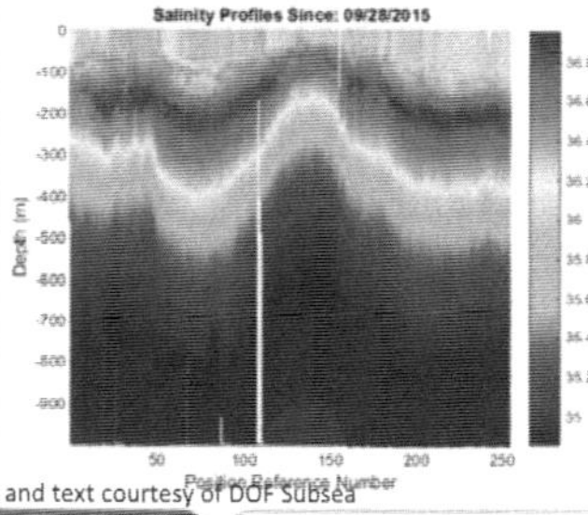

Images and text courtesy of DOF Subsea

IMAGING • INSTRUMENTS • INTERCONNECT • SEISMIC • VEHICLES

TELEDYNE MARINE
Everywhereyoulook™

Approved for public release: FAL-16-003

DOF Subsea: GoM Loop Current Monitoring

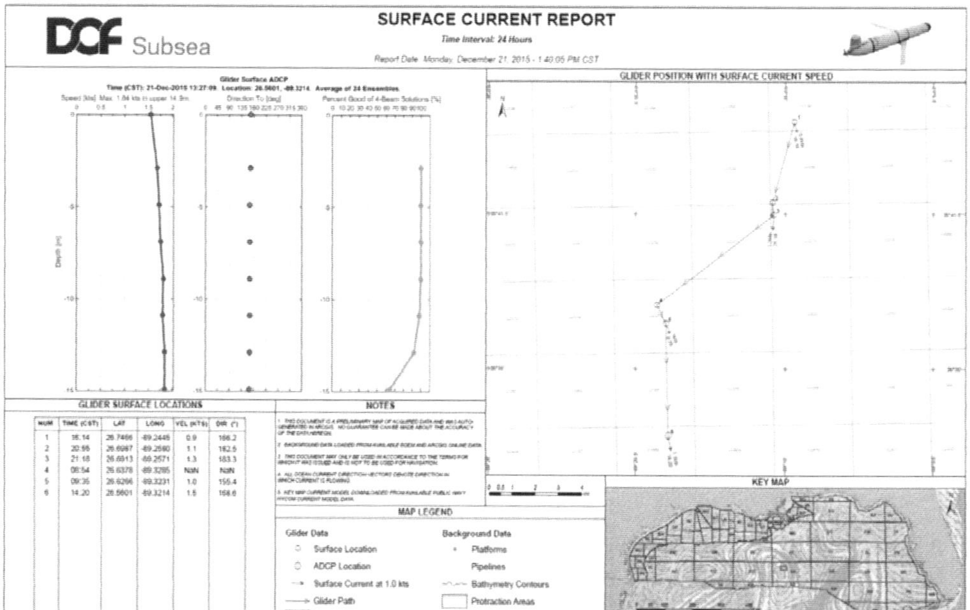

Images and text courtesy of DOF Subsea

IMAGING • INSTRUMENTS • INTERCONNECT • SEISMIC • VEHICLES

TELEDYNE MARINE
Everywhereyoulook™

Approved for public release: FAL-16-003

Blue Ocean Monitoring: Glider Operations

BlueOcean MONITORING

Glider data return

Data captured is received in near-real time for processing and interpretation.

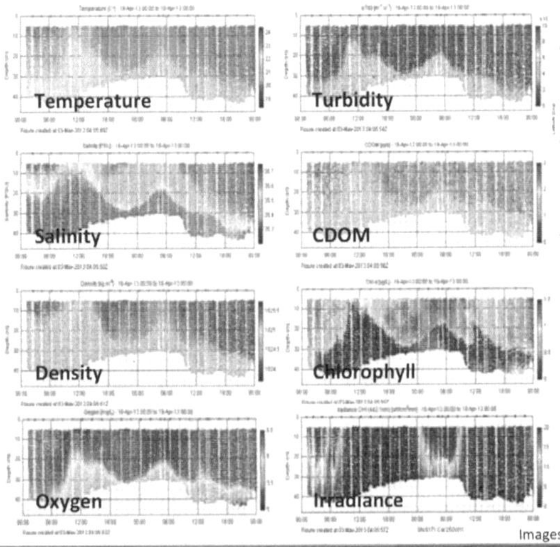

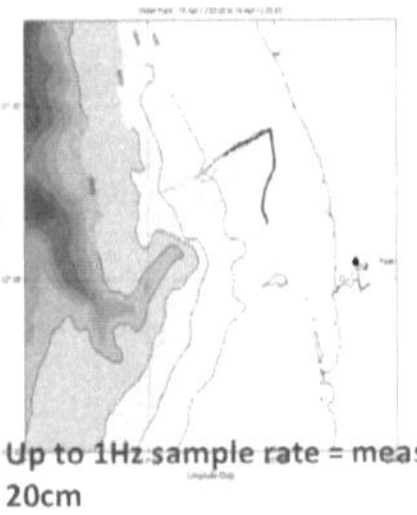

Up to 1Hz sample rate = measurement every 20cm

Sensors:

•Seabird Pumped CTD
•WETLabs FLBBCD
•AADI Oxygen Optode
•Satlantic Irradiance

Images and text courtesy of Blue Ocean Monitoring

Blue Ocean Monitoring: Example Glider Operations

BlueOcean MONITORING

Produced formation water (PFW)

Gliders provide valuable PFW data across the water column, crucial for meeting environmental regulations.

- Environmental regulations on PFW are being enhanced. PFW will be treated the same as other emissions and discharges from offshore petroleum facilities with a requirement for monitoring and reporting
- Glider's play in an important role in PFW survey's by:
  - Supporting field sampling program
  - Ensuring samples are taken within the plume
  - Tracking the plume location
  - Collect depth averaged & surface current measurements in near real-time

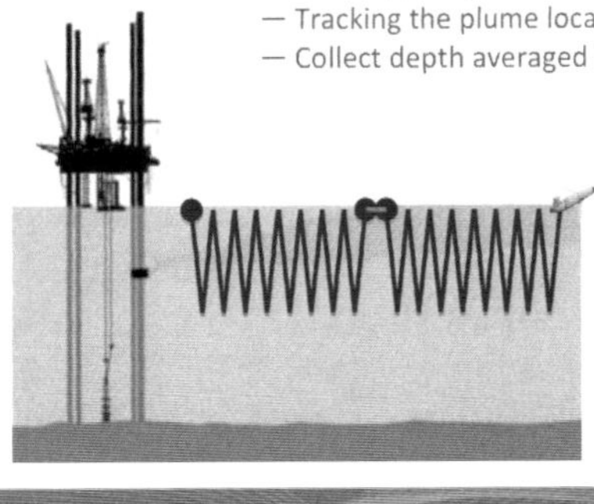

Operation modes:

▪Transect survey

▪Virtual mooring

▪Near surface horizontal survey

Images and text courtesy of Blue Ocean Monitoring

Vehicles
Gavia AUVs

World Leader in Low-Logistics AUVs

- Fully modular design
- Wide variety of payload options
- 1,000m depth rating
- Widely used in O&G applications with numerous service providers

Low-Logistics High Performance AUVs

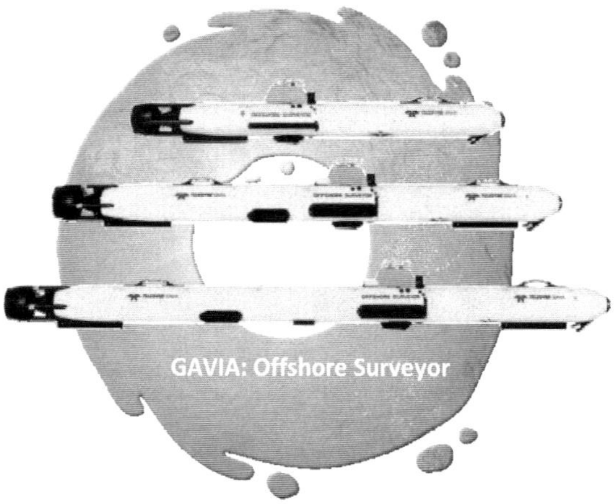

FAL-1502

IMAGING • INSTRUMENTS • INTERCONNECT • SEISMIC • VEHICLES

TELEDYNE MARINE Everywhereyoulook

Approved for public release: FAL-16-003

Gavia Survey Suite

- Side scan:
  - Marine Sonic dual frequency 900 / 1800 kHz
  - Edgetech 2205: 600 / 1600 kHz
  - Klein 3500: 450/900 kHz
- Swath Bathymetry:
  - 500kHz GeoSwath Plus interferometric sonar
- Sub-bottom profiler:
  - Teledyne Benthos CHIRP-III Technology
- MBES: Blueview MicroBathymetry
- Camera:
  - High-performance low-light variable frame rate digital camera: various resolutions, 12-bit depth

14

IMAGING • INSTRUMENTS • INTERCONNECT • SEISMIC • VEHICLES

TELEDYNE MARINE Everywhereyoulook

Approved for public release: FAL-16-003

Enhanced Productivity: Blueview MB 2250

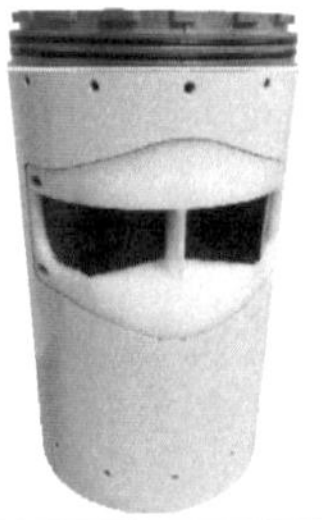

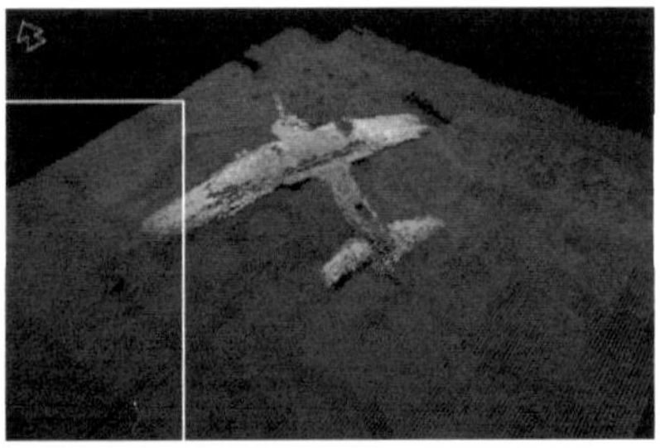

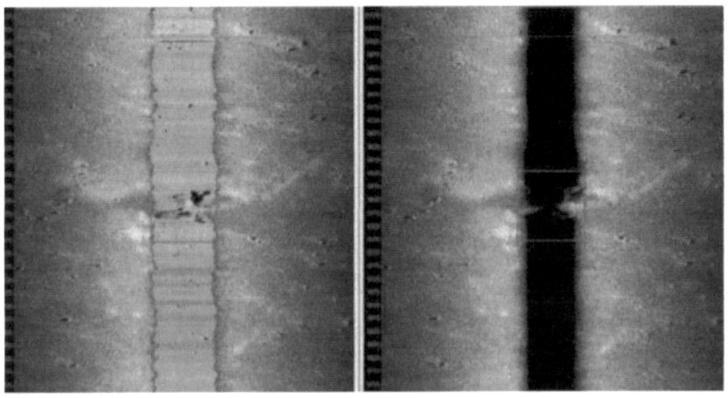

- High resolution bathymetry (2250 kHz)
- Low power and compact
- Side Scan Sonar nadir gap fill improves operational efficiency
- Downward looking, ideal for tying into pipe tracking and co registering with camera

Ideal for a variety of applications including:

- Pipeline inspections and Site survey
- MCM
- Search and Salvage
- Bottom imaging

IMAGING • INSTRUMENTS • INTERCONNECT • SEISMIC • VEHICLES

TELEDYNE MARINE Everywhereyoulook

Approved for public release: FAL-16-003

Attractions of Low Logistics AUV Operations

- Fast deployment anywhere in the world
- Rapid data acquisition after arrival on-site
- Small survey and operating team required on-site
- Low survey cost ; operations from vessels of opportunity rather than dedicated vessels.
- Reduced ship board footprint compared to larger AUVs
- Data quality can be better than vessel-mount or towed systems
- Ability to take sonars to survey sites where other systems have restricted access (rigs, shallows, deep sites)

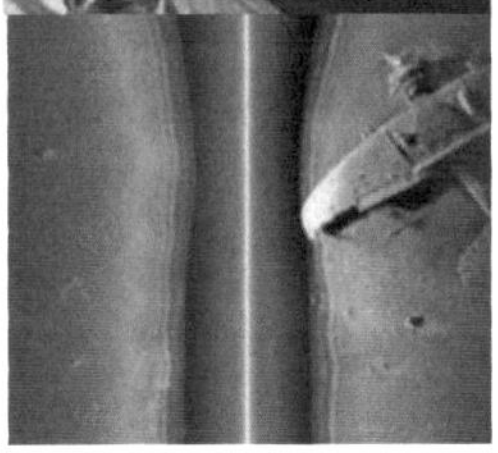

16

IMAGING • INSTRUMENTS • INTERCONNECT • SEISMIC • VEHICLES

TELEDYNE MARINE Everywhereyoulook

Approved for public release: FAL-16-003

Gavia Real World Examples

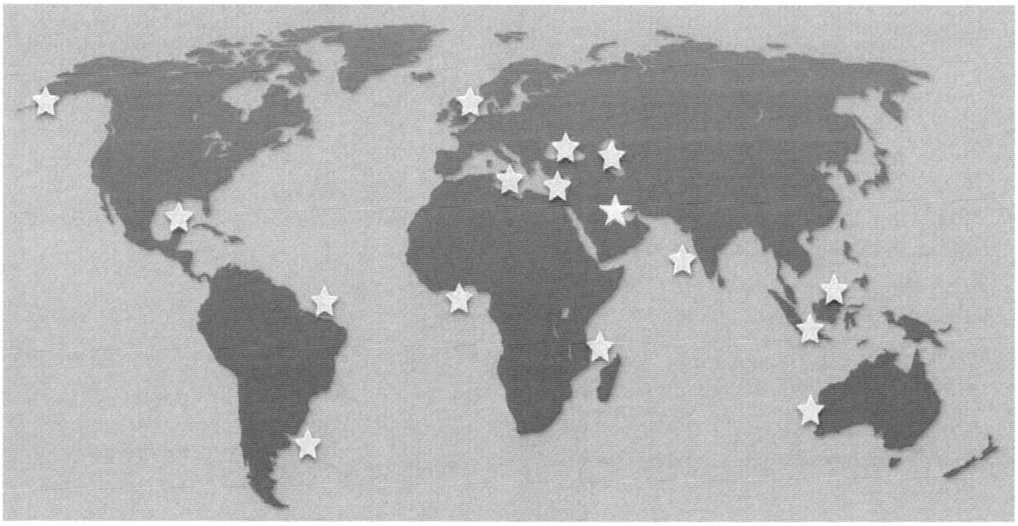

IMAGING • INSTRUMENTS • INTERCONNECT • SEISMIC • VEHICLES

TELEDYNE MARINE Everywhereyoulook™

Approved for public release: FAL-16-003

Operational Excellence Case Study:

UTEC NCS Survey AUV Operations for Quadrant Energy Australia Limited

UTEC NCS SURVEY

- From June 12, 2014 to July 27, 2014, UTEC Survey performed pipeline inspection and platform site surveys for Quadrant Energy Australia Limited (Quadrant) off the northwest coast of Western Australia
- <u>Survey Areas</u>
- The survey area consisted of forty-three pipeline routes with a total length of 571km and twenty platform surveys. Two passes were made on each pipeline, which ensured complete coverage and allowed for visualization of both sides of the pipe.
- A 500m x 500m site survey was completed around each platform.

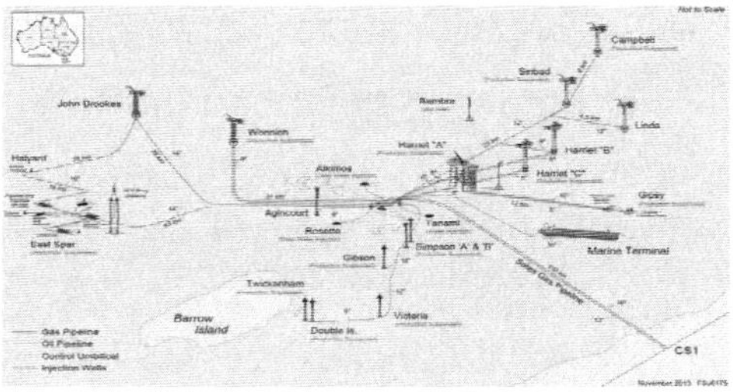

Images and text courtesy of UTEC NCS Survey

IMAGING • INSTRUMENTS • INTERCONNECT • SEISMIC • VEHICLES

TELEDYNE MARINE Everywhereyoulook™

Approved for public release: FAL-16-003

Case Study: UTEC NCS Survey AUV Operations for Quadrant Energy Australia Limited

Project Staffing

- 24-Hour operations with on-board data processing required nine UTEC Personnel; 1 AUV Party Chief, 3 AUV Operators, 1 AUV Engineer, 1 AUV Data Processor, 1 AUV Field Geophysicist, and 2 Online Surveyors

Vehicles

- 2 Gavia AUV systems
- MSTL SSS
- GeoSwath
- 1 Kearfott T24 and 1 iXBlue ROVINS INS

Launch and Recovery

- From (*M/V Yardie Creek*) and RHIB
- RHIB used in very shallow areas for support

Images and text courtesy of UTEC NCS Survey

IMAGING • INSTRUMENTS • INTERCONNECT • SEISMIC • VEHICLES

TELEDYNE MARINE Everywhereyoulook

Approved for public release: FAL-16-003

Case Study: UTEC NCS Survey AUV Operations for Quadrant Energy Australia Limited

Productivity

- Over the course of the twenty-six day project, UTEC averaged approximately 45km of AUV Line km per day.
- This average includes downtime where data could not be collected due to weather, transits between survey locations, calibrations, and vessel and equipment maintenance.
- On July 11, 2014 UTEC set a personal best record for data collection, by collecting 80.5km of survey data within a 24-hour period.

Images and text courtesy of UTEC NCS Survey

IMAGING • INSTRUMENTS • INTERCONNECT • SEISMIC • VEHICLES

TELEDYNE MARINE Everywhereyoulook

Approved for public release: FAL-16-003

Case Study: UTEC NCS Survey AUV Operations for Quadrant Energy Australia Limited

- Near shore operations, which allowed the AUV to work in as shallow as 2m water depth, totaling 571km of pipeline with 1 pass each side (1,142km of AUV track-line).
- Twenty platforms and subsea assets were also surveyed
- Survey duration was 27 days during the winter; including mobilization and demobilization,
- Experienced less than 2% operational AUV downtime and only experienced 18% downtime due to the winter weather associated with Launch and Recovery.
- On a pipeline per kilometer basis, AUV surveys may be less than 50% the cost of a ROV survey.

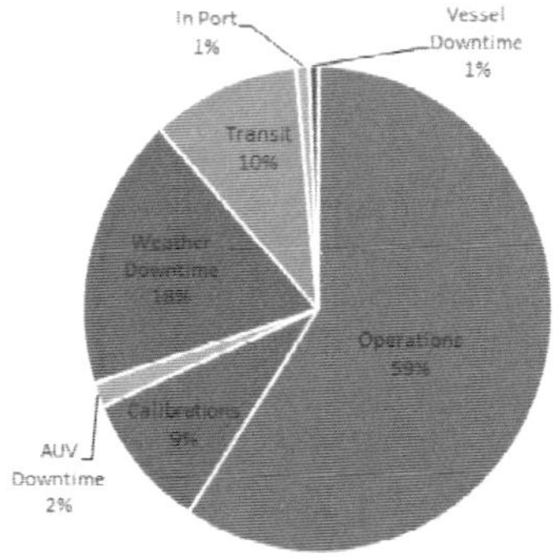

UTEC NCS SURVEY

Images and text courtesy of UTEC NCS Survey

IMAGING • INSTRUMENTS • INTERCONNECT • SEISMIC • VEHICLES

TELEDYNE MARINE Everywhereyoulook™

Approved for public release: FAL-16-003

JAVA OFFSHORE : PLATFORM COMPLEX SURVEY

JAVA OFFSHORE

Images and text courtesy of Java Offshore

IMAGING • INSTRUMENTS • INTERCONNECT • SEISMIC • VEHICLES

TELEDYNE MARINE Everywhereyoulook™

Approved for public release: FAL-16-003

JAVA OFFSHORE GAVIA SURVEY OBJECTIVES

Assess seabed conditions and gas seepage around platform complex

JAVA OFFSHORE

SURVEY EXECUTION & MISSION PLANS

- Total mission plans : 17
- Maximum time consumed during a mission : 5 ½ hours
- Minimum time consumed during mission : 3 hours
- Maximum AUV recovery time : 35 minutes
- Minimum AUV recovery time : 10 minutes
- Average data download time : 30 minutes

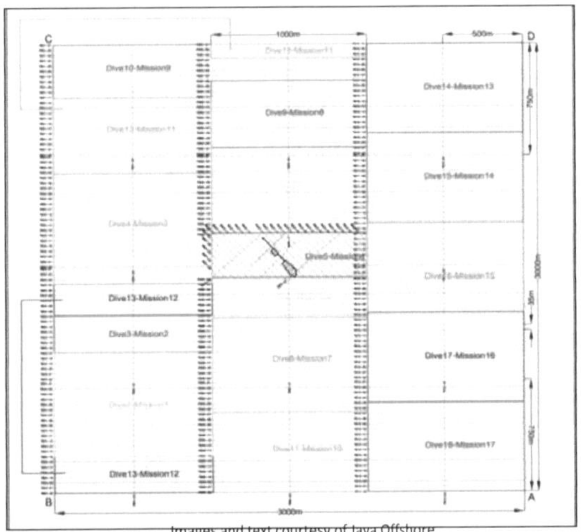

Images and text courtesy of Java Offshore

IMAGING • INSTRUMENTS • INTERCONNECT • SEISMIC • VEHICLES

TELEDYNE MARINE Everywhereyoulook™

Approved for public release: FAL-16-003

MBES DATA COVERAGE COMPARISON

JAVA OFFSHORE

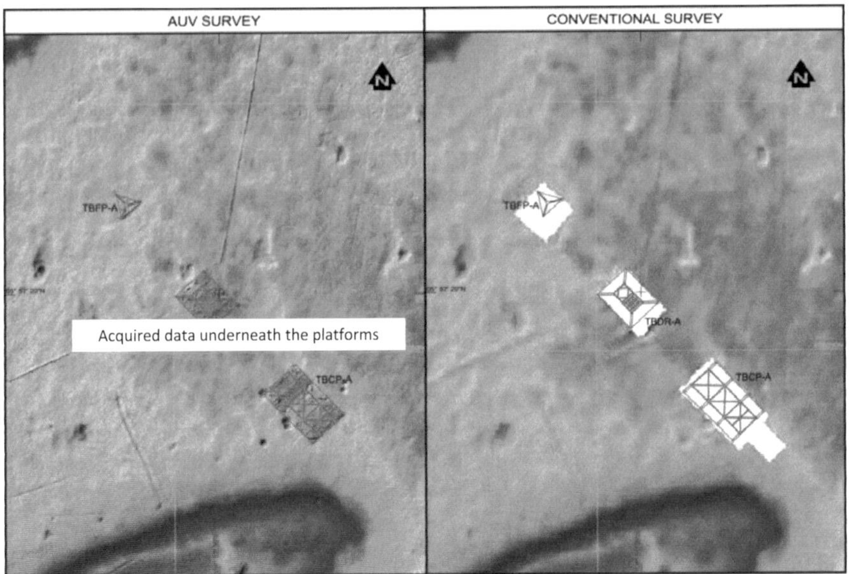

NOTE: COORDINATES ARE IN KERTAU 1968 DATUM, EVEREST 1830 (MOD) SPHEROID, UTM 48N PROJECTION

Images and text courtesy of Java Offshore

IMAGING • INSTRUMENTS • INTERCONNECT • SEISMIC • VEHICLES

TELEDYNE MARINE Everywhereyoulook™

Approved for public release: FAL-16-003

MBES DATA DETAILS COMPARISON

JAVA OFFSHORE

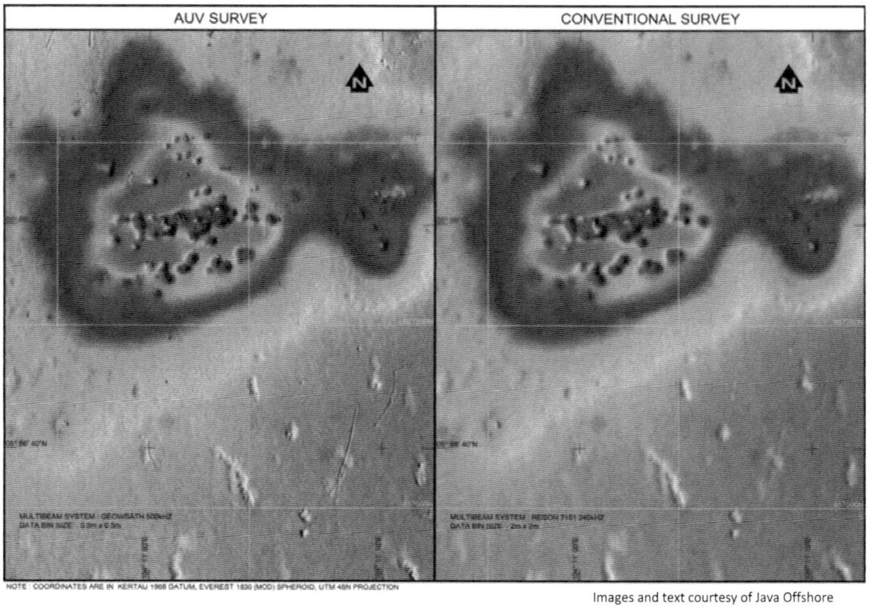

Images and text courtesy of Java Offshore

IMAGING • INSTRUMENTS • INTERCONNECT • SEISMIC • VEHICLES

TELEDYNE MARINE Everywhereyoulook

Approved for public release: FAL-16-003

MBES DATA RESULTS COMPARISON

JAVA OFFSHORE

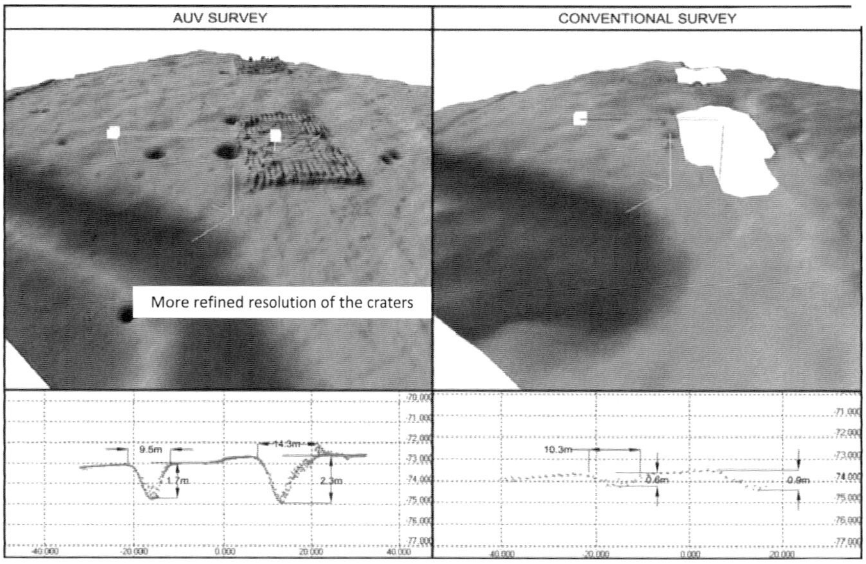

Images and text courtesy of Java Offshore

IMAGING • INSTRUMENTS • INTERCONNECT • SEISMIC • VEHICLES

TELEDYNE MARINE Everywhereyoulook

Approved for public release: FAL-16-003

SSS DATA COMPARISON

JAVA OFFSHORE

Conventional Survey - EdgeTech FS4200 SSS frequency 100 kHz

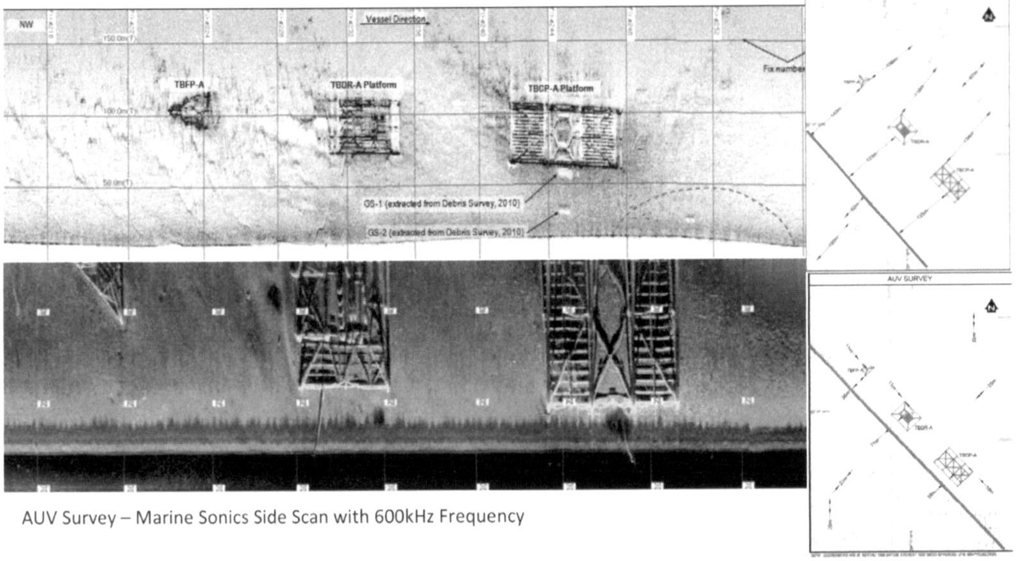

AUV Survey – Marine Sonics Side Scan with 600kHz Frequency

Images and text courtesy of Java Offshore

IMAGING • INSTRUMENTS • INTERCONNECT • SEISMIC • VEHICLES

TELEDYNE MARINE Everywhereyoulook

Approved for public release: FAL-16-003

SSS DATA COMPARISON (Zoom)

JAVA OFFSHORE

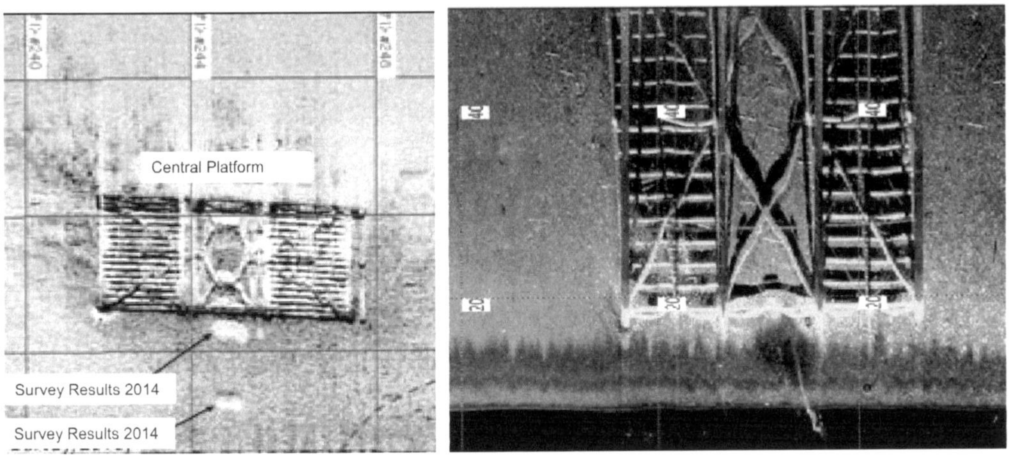

Gas seepage is observed from AUV survey but not from conventional survey

Images and text courtesy of Java Offshore

IMAGING • INSTRUMENTS • INTERCONNECT • SEISMIC • VEHICLES

TELEDYNE MARINE Everywhereyoulook

Approved for public release: FAL-16-003

CENTRAL PLATFORM DETAILS

JAVA OFFSHORE

Images and text courtesy of Java Offshore

IMAGING • INSTRUMENTS • INTERCONNECT • SEISMIC • VEHICLES

TELEDYNE MARINE Everywhereyoulook

Approved for public release: FAL-16-003

*** Formatting Issue - Best Available Paper/Graphic ***

Common Discrepancies Uncovered on ADCI Diving Contractor Audits
&

Ways to Set Your Company up for Success on ADCI Diving Contractor Audits and Reduce Costs

U.I. 2016

New Orleans, LA

UI 2016

FEBRUARY 23-25, NEW ORLEANS

The following presentation is based upon documented observations from ADCI audits conducted by
Mr. Jon Hazelbaker, ADCI Designated Auditor (Hammerhead Marine Services, LLC)
and
Mr. Phil Newsum, ADCI Executive Director.

*** Formatting Issue - Best Available Paper/Graphic ***

In most instances, all discrepancies uncovered during an ADCI Diving Contractor Audit could have been avoided with greater familiarization with the most current edition of *International Consensus Standards for Commercial Diving and Underwater Operations.*

Safe Practices and Operations Manual

In many cases, Safe practices and Operations Manuals are not compliant with ADCI Standards. They are either not reflective of the latest version of the Consensus Standards or they are just a bulky "cut and paste" of the USN Dive Manual or regulatory guidelines.

*** Formatting Issue - Best Available Paper/Graphic ***

Safe Practices and Operations Manual

A copy of the Safe Practices and Operations Manuals should always be at the dive site. All Safe Practices and Operations Manuals should include the five basic components as outlined in Section 5 of the Consensus Standards.

-Safety Procedure Checklist
-Team member assignments and responsibilities
-Equipment procedures and checklist
-Emergency procedures for fire, equipment failure, adverse environmental conditions and medical illness/injury
-Specific individual procedures for tools, equipment and associated systems

Procedures and Documentation

Designated Diving Supervisor –
ADCI Diving Contractor Audit Report Chapter I

-No company letter designating the diving supervisor (at the job site or in personnel file).
(Appointment Letter of Supervisors required.)

*** Formatting Issue - Best Available Paper/Graphic ***

Procedures and Documentation

Commercial Diver Log Book
ADCI Diving Contractor Audit Report Chapter II

"It is recommended that all divers maintain a personal dive log book (ADCI Commercial Diver Log Book or equivalent) to detail hyperbaric exposures." (Section 3.1.5 and Section 5.5)

Equipment Procedures and Documentation

Personal Protective Equipment (PPE) – Section 5 ADCI CS

-Personnel are not donning proper PPE at the dive site or in the company shop (The appropriate ANSI (or standard used within a particular nation) approved personnel protective equipment shall be worn when required.)

*** Formatting Issue - Best Available Paper/Graphic ***

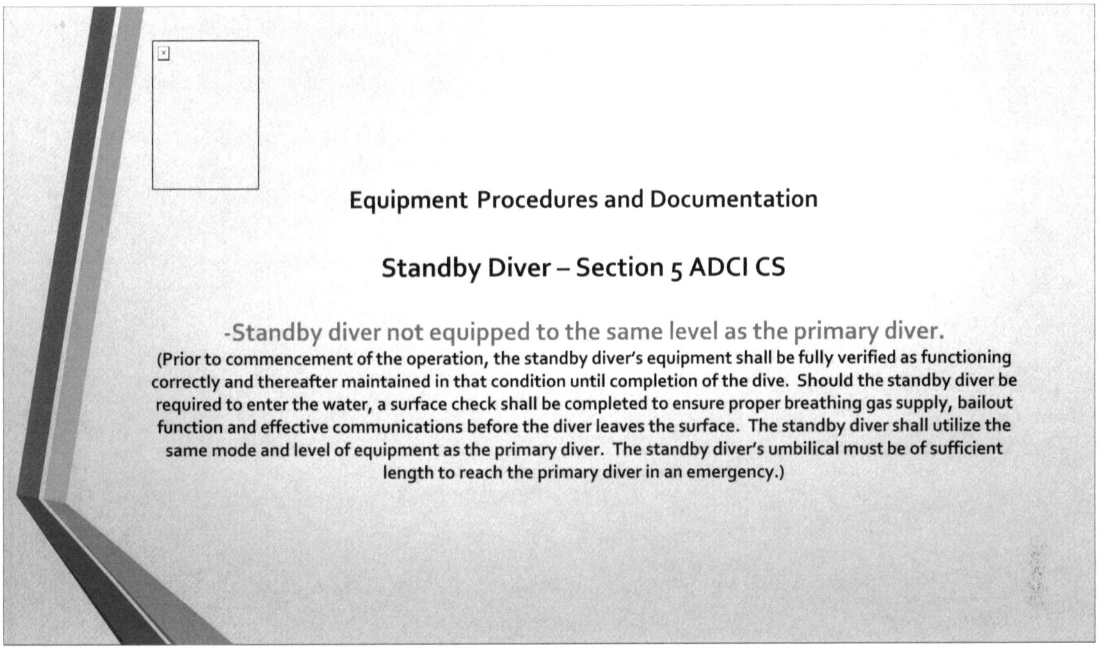

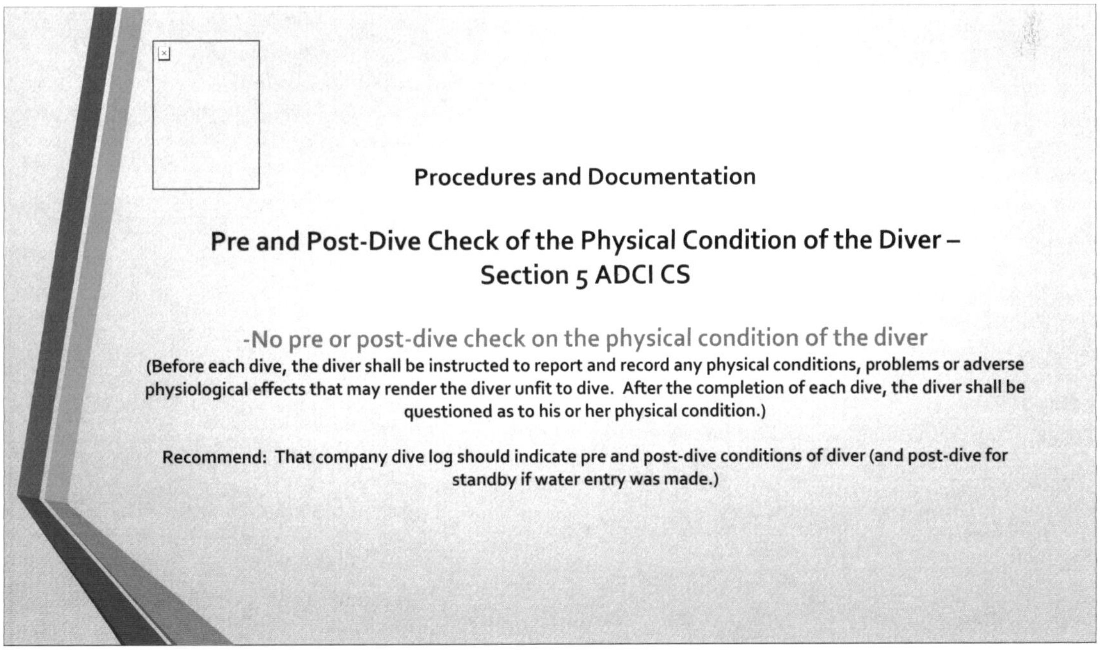

Procedures and Documentation

**Job Hazard Analysis (JHA) –
Section 5 ADCI CS**

-No real-time JHA conducted
-JHA forms do not assign responsibility for hazard mitigation
(Identify the persons preparing, reviewing and approving the JHA. Assign a specific person the responsibility of implementing the safety procedures or protection required.)

Equipment Procedures and Documentation

First Aid Kit – Section 5 ADCI CS

-First Aid Kit contents does not meet ADCI inventory requirements.
-First Aid Kit expiration dates not tracked and items replaced as required
-No First Aid Handbook accompanying First Aid Kit
-No bag type manual resuscitator (or AED)
(First Aid Kits should be hyperbaric compatible for operations where a chamber is onsite.)

Compliant

Non-compliant

*** Formatting Issue - Best Available Paper/Graphic ***

Equipment Procedures and Documentation

Maintenance Records – Section 6 ADCI CS

Suitable equipment logs shall be established and maintained in a correct and current condition.

Life-support equipment shall have a unique identity assigned by the manufacturer or contractor, be easily visible, permanently affixed and traceable to the equipment/maintenance log.

Entries made in the equipment log shall describe the nature of the work performed, including the dates of modification, repair, or test; the name of the individual performing the work or test; and the particular piece of equipment involved.

A preventive maintenance program is required for all life-support equipment.

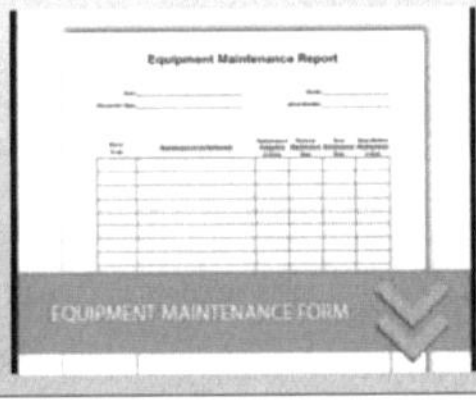

Equipment Procedures and Documentation

Air Purity Requirements – Section 6 ADCI CS

-No record to indicate the performance of semi-annual (6 mos.) air purity checks for all compressors, transfer pumps or booster pumps used for breathing service.

(All compressors, transfer pumps or booster pumps used for breathing air service will be subjected to an air quality test every six months. Documentation of the latest test[s] shall be kept on file and available upon request. This includes the requirement to procure air quality certifications from 3rd party vendor supplied HP air.

*** Formatting Issue - Best Available Paper/Graphic ***

Equipment Procedures and Documentation

Helmets and Masks – Section 6 ADCI CS

-No record to indicate the annual the internal and external inspection of helmets and masks, if required by manufacturer. (All helmets and masks and their associated diver-carried regulators shall be maintained and inspected in strict compliance with the manufacturer's recommendations. Suitable logs shall be maintained to reflect compliance.)

-Annual inspections not performed by qualified and certified technician

-Diver-owned helmets/masks maintenance and inspections not being tracked by the company

Equipment Procedures and Documentation

Depth Gauges – Section 6 ADCI CS

-No record to indicate the semi-annual (6 mos.) calibration. (Be marked with a label, tag, or sticker indicating date of last calibration and date due, which will not interfere with full-scale visibility.)

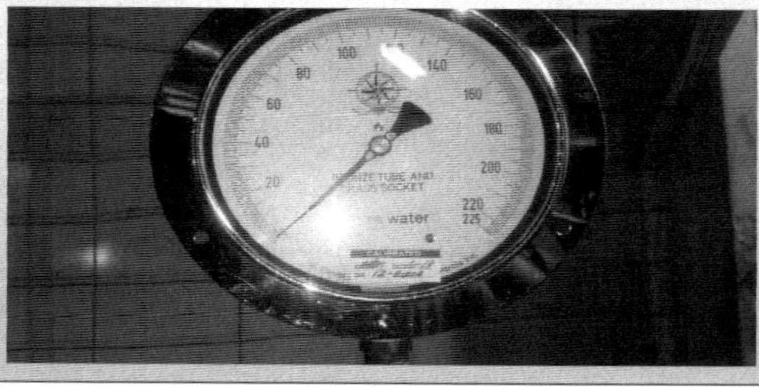

*** Formatting Issue - Best Available Paper/Graphic ***

Equipment Procedures and Documentation

Breathing Gas Hoses(s) LP – Section 6 ADCI CS

-Pressure and pull testing not performed or performed incorrectly. (Each hose assembly will be subjected to an annual pressure test to one-and-one-half times the design working pressure of the system. The pressure test should be maintained [when corrected for temperature] for 10 minutes.)

-Failure to record the above test and maintain the data for future review.

-<u>Recommendation</u> that hose fittings not be covered with electrical, duct tape, or any adhesive when not in use.

Equipment Procedures and Documentation

Umbilical(s) –Section 6 ADCI CS

-Not marked for length. (Be marked from the diver/bell end in 10-foot intervals up to 100 feet and marked in 50- foot intervals thereafter.)

-Pressure and pull testing not performed or performed incorrectly. The company's "Design Working Pressure of the System" is not correctly determined. (Be subjected to an annual pressure test to one-and-one-half times the "design working pressure of the system" with a 200-pound axial load applied on fittings while initial test pressure is applied. The test pressure should be maintained with out loss of pressure [when corrected for temperature] for 10 minutes.)

-Leak test of pneumofathometer hose not performed. (Pneumo hose shall be annually pressure-tested for leakage.)

-Failure to record the above test and maintain the data for future review. (Entries made in the equipment log shall describe the nature of the work performed, including the dates of modification, repair or test; the name of the individual performing the work or test; and the particular piece of equipment involved.)

-Recommendation that umbilical fittings not be covered with electrical or duct tape when not in use.

*** Formatting Issue - Best Available Paper/Graphic ***

Equipment Procedures and Documentation

Gas Storage Cylinders – Section 6 ADCI CS

-No record to indicate the annual external examination for damage and corrosion. (Be visually examined externally at least annually for damage and corrosion.)

-Cylinders hydrostatic testing not up to date. (Be hydrostatically tested every fifth year to the requirements of the code of the manufacturer by an authorized test facility and stamped with the date of the test.)

-Cylinders not assigned a unique identifier and tracked in the equipment log. (Have a unique identity with results of all tests being recorded in the equipment log).

-Cylinders not clearly marked as to the contents and pressure. (High pressure gas cylinders or tubes shall be labeled as to contents. A record shall be kept in a designated place of the contents and pressure of each cylinder, quad, or bank. These records should be updated daily when the system is in use.)

-Cylinders not properly secured and/or stored. (If rack-mounted into banks of cylinders or tubes, have valves and regulators protected from damage caused by impact or from falling objects. Be stored in a well-ventilated area, protected from overheating and secured from falling.)

Equipment Procedures and Documentation

Gas Storage Cylinders (Bailouts) – Section 6 ADCI CS

-No record to indicate the annual external examination for damage and corrosion. (Be visually examined externally at least annually for damage and corrosion.)

-Bailouts hydrostatic testing not up to date. (Be hydrostatically tested every fifth year to the requirements of the code of the manufacturer by an authorized test facility and stamped with the date of the test.)

-Bailouts and regulators not assigned a unique identifier and tracked in the equipment log. (Have a unique identity with results of all tests being recorded in the equipment log).

- Bailout regulators not fitted with relief valve in LP port. (Diver-worn EGS systems shall have a depth-compensating regulator on the cylinder capable of delivering the proper pressure and flow to the diver's helmet or mask in accordance with the helmet or mask manufacturer's recommendations. High pressure bottles used for scuba and EGS (bailout) shall be equipped with an overpressure relief device.)

- Diver-owned bailouts and regulators maintenance and inspection not being tracked by the company.

Equipment Procedures and Documentation

Surface Diving Decompression Chambers – Section 6 ADCI CS

-No record to indicate the annual testing of the relief valve pressure setting. **(Have the relief valve pressure setting tested annually and the test recorded in equipment log.)**

-No record to indicate the annual pressure test of the chamber. **(Pressure test the chamber and associated piping annually to MAWP, as stamped on the chamber name plate, and recorded in the equipment log.)**

-No back-up communication provided for the chamber. **(Have a primary and secondary two-way voice communication system between the occupants and the operator.)**

-No means for the extinguishing of fires in the chamber. **(Be equipped with a readily available means for extinguishing fire.)**

-BIBS not available in both the inner and outer locks (or insufficient number of BIBS). **(Have an installed breathing system with a minimum of one mask per occupant per lock, plus one spare mask assembly per lock.)**

Ways to Set Your Company up for Success on ADCI Contractor Audits and Reduce Costs

Give the Auditor Your Full Cooperation and Responsiveness

-Submit all requested information for review to the auditor in advance of the actual audit.

***Safe Practices and Operations Manual**
***List of diving personnel and their certifications**
***List of equipment and documentation (inventory, testing, and inspection)**
***Sample dive plans**
***Sample Emergency management plans**
***Sample Job Hazard Analysis (JHA)**

-Return required/requested revisions promptly.

*** Formatting Issue - Best Available Paper/Graphic ***

Ways to Set Your Company up for Success
on
ADCI Contractor Audits and Reduce Costs

Keep Lines of Communication Open Between You and the Auditor

-Direct any questions to your auditor or the ADCI Office.

-If you need assistance to prepare for the audit, your auditor or the ADCI Office can provide you with a list of vendors, consultants, or other resources.

-Outline all potential conflicts with scheduling of the audit in advance.

- Address all potential areas of deficiency prior to the actual performance of the audit.

Ways to Set Your Company up for Success
on
ADCI Contractor Audits and Reduce Costs

Budgeting Time

-Ensure you have allotted time for the auditor with as little interruption as possible.

-Have key personnel available as needed (office manager, shop foreman, dive supervisor[s]) for those portions of the audit which directly involve them. This would entail attendance at debrief meetings at the office and after the dive site portion of the audit.

*** Formatting Issue - Best Available Paper/Graphic ***

Ways to Set Your Company up for Success
on
ADCI Contractor Audits and Reduce Costs

Additional Keys to Success

-After the audit, address all identified areas of deficiency promptly and outline timelines for completion.

-Try to set up the dive site portion of the audit within close proximity to the company office/facility.

-Make sure that you are aware of all costs and potential additional costs associated with the audits.

Ways to Set Your Company up for Success
on
ADCI Contractor Audits and Reduce Costs

Finally....

Your cooperation in providing requested documents up front and on-site, along with addressing non-compliant items promptly will have a direct effect on the final cost of your audit!

*** Formatting Issue - Best Available Paper/Graphic ***

Thank You

Questions?

Deepwater Gas Fields: a Review of Current Field Development Options, Enabling Technologies and Main Challenges

Francesco Beltrami, Granherne Inc. (a KBR Company)

Abstract: Large, and sometimes very large, gas discoveries in deep and ultra deep waters have occurred over the past years in different geographic areas across the world. Some of these discoveries are being developed at present using 'Subsea to Beach', floating LNG or other options that require combinations of subsea tie-backs and offshore host(s). The choice on how to develop one or more deepwater gas reservoirs depends both on technical factors, including the readiness level and the applicability ranges of some key enabling technologies, and on non-technical factors, such as geopolitics considerations, market conditions and risks.

This paper presents the most frequently encountered field development options and discusses their advantages and disadvantages. Examples of real projects are also presented, making use of public domain information when available. Scientific literature has been searched to collect information on past and ongoing projects. Press releases and other sources have been investigated to double-check and ensure consistency of the gathered information.

The key technologies considered 'true enablers' for this kind of project are presented and briefly discussed. More focus is given to the readiness level of these technologies and to their current applicability range. For 'subsea to beach' this range is perceived to be within 200km from the shore and up to 1,500m water depth. Recent projects are understood to push the water depth limit up to 2,500m, while it might be more difficult to extend the step-out distance from shore beyond the 200km limit. New technologies are also discussed as possible future enablers.

Non-technical factors are then studied due to the fact that some of the challenges that can be encountered in developing large offshore gas reservoirs are not technical in nature. Examples of these risks are delays in the project schedule and overall cost overruns. These non-technical challenges need to be taken into account while selecting the proper development option.

1.0 Introduction

In recent times the offshore industry has looked with increasing interest to the practical and affordable ways to develop large gas reservoirs located in a deepwater environment, in 300m to 1,500m water depth (WD), or in an ultra-deepwater environment exceeding 1,500m WD. This definition of deepwater is not consistent with API 17A (deepwater ranges between 600m and 1,830m), but is used in the industry and will be followed in this paper. Several projects are currently under execution across the world, in mature areas with existing processing and exporting capabilities and in new provinces completely lacking production infrastructure.

In particular, 'Subsea to Beach', coupled to onshore LNG and Floating LNG (FLNG) has been frequently considered for offshore locations with limited or non-existing production infrastructure. These are not the only options available for the development of deepwater or ultra-deepwater gas fields: in principle, a Wet-Tree Host connected to a subsea production system (SPS) or a

Dry-Tree Host may be utilized in deepwater and ultra-deepwater gas field development. Floating hosts definitely have a longer history than FLNG, but this is likely to change when the planned shallow water and deepwater FLNG vessels start their operations.

Present and future applications of 'Subsea to Beach', FLNG and Hosts are reviewed in the first part of this paper as a contribution to the more general discussion on the optimal applicability ranges of these different development options. The advantages and disadvantages of each option are identified and briefly discussed. More information on the projects under examination can be found in (Beltrami, 2015), (Beltrami et al., 2015) and (Barrero et al., 2016). In particular, (Barrero et al., 2016) reviewed sixty-nine (69) notable offshore projects, thus providing updated information on the same projects in a more detailed way.

The second part of this paper focuses on the technologies required by 'Subsea to Beach' developments. This option is attractive because a host is not initially required and, in turn, the project is more affordable, but may not be applicable when the offshore field is very distant from the shore. The perceived technical limit to the maximum allowable step-out distance as identified during the study is discussed with respect to the readiness status of different technologies and components of the required subsea production system.

The last part of the paper examines the main challenges that have to be faced and overcome to successfully develop deepwater and ultra-deepwater gas fields. These challenges are related to, and represented by, technical and non-technical factors that have to be taken into account while selecting the 'best' field development option. This section of the paper is based on more extensive work done over the past months; see (Beltrami et al., 2015).

The status of FLNG-related technologies and of marine technologies for floating hosts goes beyond the scope of this paper and is intentionally not discussed here. FLNG cannot be considered as a field-proven concept yet, but the first FLNG is expected to commence operation in 2016. A discussion on marine technologies for deepwater and ultra-deepwater floaters would necessarily require more space and is not included in this paper.

2.0 Field Development Options

2.1 Overview

The main development options for deepwater and ultra-deepwater gas fields are:

- 'Subsea to Beach' coupled to onshore LNG plant or to gas terminal
- FLNG vessel
- Subsea tie-back to floating host (Wet-Tree Host)
- Dry-Tree Host.

2.2 Subsea to Beach

The typical 'Subsea to Beach' project is shown in Figure 1, see the subsea wells in the bottom left corner. The subsea tie-backs to a shallow water host (i.e. fixed platform) and to a deepwater floating host are also shown in the same figure.

The main difference between these development options is that 'Subsea to Beach' requires multiphase export to the onshore processing plant while in the subsea tie-backs the gas is processed offshore: treated, single-phase gas is then exported to the onshore plant. 'Subsea to Beach' projects are unique because of their:

- Lack of offshore host (i.e. fixed platform in shallow water or floater in deeper waters), production risers and export risers
- Multiphase flow, wet gas with liquids - not processed offshore
- Natural flow, no offshore or subsea compression, though may require compression at a later stage of field life, especially if the SPS feeds an onshore LNG plant – see Figure 3 and Figure 4
- No pipeline wet insulation; need to constantly inject mono-ethylene-glycol (MEG) to prevent hydrates formation
- Long step-out distances from the subsea wells to the onshore terminal
- Large volumes of gas to be handled: great number of subsea wells in some cases and very large productivity of individual subsea wells in other cases (big-bore, high rate wells)
- Need for engineered shore approach.

Ultra-deepwater 'Subsea to Beach' projects (1,500m WD and more), see Figure 2, also require:

- Transition from deep offshore to continental shelf affecting pipelines and trunklines design
- Subsea power and communication distribution over long distances.

Notable existing 'Subsea to Beach' projects coupled to onshore LNG production are Snohvit in the Barents Sea and the West Delta Deep Marine (WDDM) concession in Egypt. Notable future projects are for example Gorgon and Jansz-lo under construction in Western Australia. Similarly, 'Subsea to Beach' and onshore LNG are likely to be adopted in East Africa, a new, promising gas province. A more comprehensive list of twenty-four (24) current and future 'Subsea to Beach' projects is presented in (Barrero et al., 2016).

The known applicability limits of 'Subsea to Beach' are shown in Figure 6. This chart plots the water depth (well locations) of a given project versus the step-out distance from shore. The current SPS technical feasibility extends up to approximately 200km step-out distance and up to 1,500m water depth. It is understood that the step-out distance in future projects may not exceed these current applicability limits, while the water depth is going to exceed 2,000m, especially in East Africa.

2.3 FLNG Vessel

A typical illustration for ultra-deepwater FLNG is shown in Figure 5. Floating LNG has a relatively short history: some projects have been sanctioned and are currently in the execution phase, others have been temporarily postponed due to the present market conditions. The IGU World LNG Report 2015, Appendix II, provides information on twenty-six (26) LNG and FLNG projects under execution worldwide. (Barrero et al., 2016) recently investigated the status of fourteen (14) FLNG projects: five (5) of these projects appear to be on hold at the present time.

The success of shallow water and deepwater FLNG projects will open a new market enabling the commercial exploitation of abundant stranded gas reserves. The Prelude FLNG development in Western Australia is an example. It is located in 250m WD and 200km from shore. FLNG is also currently being employed for deepwater gas fields; see for example Abadi, Scarborough and Rotan. It is worth noting that FLNG was initially considered for an ultra-deepwater development in Santos Basin in the Brazilian offshore. Although that project appears to be on hold at this time, it indirectly confirms the FLNG potential in ultra-deepwater.

2.4 Wet-Tree Host

Relevant deepwater gas fields have been developed, or are being developed, by means of shallow water fixed hosts or deepwater floaters connected to subsea wells, see Figure 1. Fixed platforms will not be discussed here, as the focus is on deepwater developments. Not all these projects are finalized to LNG production; nevertheless, the examples discussed in the following notes reflect the current status of the offshore industry.

A floating host might be required for the late field-life compression of the produced gas, especially if the subsea wells feed an onshore LNG plant that contractually needs to deliver consistent, fixed amounts of LNG over the time. This option is presented in Figure 3. It should be noted that, in this case, the floater is a Wet-Tree Host. This is due to the fact that this host is a later addition to a 'Subsea to Beach' project developed with subsea wells.

The floating host option is being used for major capital projects, for example in the Ichthys LNG project that ranks among the largest and most significant energy projects in the world. It includes some of the world's biggest and most advanced offshore facilities, massive onshore processing facilities near Darwin in the northern territory of Western Australia and an 889km-pipeline to unite them. The unusual length of this pipeline is due to environmental constraints that dictate a relatively distant LNG plant location. The Ichthys field covers an area of around 800 km^2 in water averaging depths of around 250m WD.

The Shtokman gas and condensate field is located in the central part of the Russian sector of the Barents Sea shelf, about 600km northeast of Murmansk, where sea depth varies between 320m and 340m. The project is understood to be suspended. Field operations would be performed using deepwater production systems and special floating production units (FPU) equipped with quick-disconnect system and able to swiftly move off icebergs. The produced gas would be conveyed via the offshore trunklines to the coast near the Teriberka settlement where the LNG plant will be located. Gas would be produced by twin four-slot templates and then conveyed by flexible production risers from a template to the FPU. Gas and condensate separation, and gas processing operations would be performed aboard the FPU. After that, processed gas would be exported from the FPU by flexible risers.

The Malampaya field is located 80km off Palawan Island in the Republic of Philippines. The platform consists of a deck supported by a concrete gravity sub-structure (CGS). The platform is located in 43m WD while the deepwater subsea wells are at depth of 850m. The produced gas is processed, compressed and exported through a 504km pipeline to the onshore facility. The field has been developed using manifolded subsea wells tied-back to the platform.

The Pluto facility operates as 'not normally manned' and is controlled from the onshore Pluto gas plant near Karratha in Western Australia. Operations include production of hydrocarbons from the reservoirs, receiving hydrocarbons at the Pluto platform, located in 85m WD, and transport via the trunkline to the onshore facility for processing. Gas is piped to shore along a 180km, 36-in pipeline to the new Burrup LNG Park, located between the North West Shelf Venture and Dampier Port. Also in this case, the long pipeline routing is understood to reflect environmental constraints.

Located approximately 240km southeast of Venice, Louisiana, in more than 2,400m WD, Independence Hub is the deepest offshore gas platform ever installed in the world - the current world record for an oil-producing floater being the Cascade & Chinook FPSO, located in the GoM. Independence Hub is a deep-draft, semi-submersible platform that is connected initially to eight (8) distinct natural gas fields. Production from the host is exported via the 216km Independence Trail natural gas pipeline. This project is mentioned here because it is a remarkable example of ultra-deepwater gas field development, although not directly related to onshore LNG production.

2.5 Dry-Tree Host

A Dry-Tree Host is also possible, at least in principle: for example, a Dry-Tree Host was initially considered for Ormen Lange, see (Gustavsson, 2003). However, only Wet-Tree Hosts have been encountered in the present and past investigation(s); see (Barrero et al., 2016). The possible lack of examples of Dry-Tree Hosts for a deepwater or ultra-deepwater gas field may be due to the fact that, so far, only a few of these large fields have been developed.

2.6 Options' Advantages and Disadvantages

2.6.1 Applicability Ranges

'Subsea to Beach' current applicability range is represented by sector I of Figure 6. In the future, the applicability range is expected to expand in order to include sector II, i.e. same step-out distance but deeper waters. Sector II is also populated with long subsea tie-backs, see (Barrero et al., 2016). This means that, despite the challenges to produce gas reservoirs in 1,500m to 2,000m WD at a 200km distance from the shore, technologies and operational lessons learned are available to the industry as a valid starting point.

FLNG and floating hosts, on the contrary, present wider applicability ranges: their applications populate sectors I, II and III on the same chart. The step-out distance of floating hosts ranges between 35km and 889km, and their water depth between 75m and 2,440m. It is worth nothing that the record values are associated to floaters. FLNG and floaters have the potential to populate sector IV (see Independence Hub and the Santos Basin FLNG project).

2.6.2 CAPEX and OPEX

All the field development options discussed in this paper may result in multi-billion dollar projects, especially in case of very large gas accumulations to be produced. The main costs incurred in these major capital projects are typically for well drilling and completion activities and for the host, if any. The SPS CAPEX for 'Subsea to Beach' projects is going to be relevant, while it will be less relevant for FLNG and Wet-Tree Hosts projects, and relatively marginal for Dry-Tree Hosts (subsea export facilities only).

In case late-life compression is required, it will be easier to add compressors to an existing host rather than having to provide a new host or to adopt subsea wet gas compression, a relatively new technology.

The OPEX for 'Subsea to Beach' will reflect the large amount of MEG to be constantly injected into the produced gas, treated in the onshore plant and reutilized. The OPEX for FLNG and floating hosts will be affected by the living and transportation expenses of the offshore personnel.

2.6.3 Recovery Factor

It has been reported that the recovery from fields produced from fixed platforms is greater than the recovery from subsea-developed fields, even if the quality and features of the reservoirs are comparable, see for example (Utseth, 2007) and (Osmundsen, 2013). This is due to the complexity and costs of intervention on subsea wells. A Dry-Tree Host provides direct vertical access to the wells and is probably the best candidate for a larger recovery factor.

2.6.4 Flow Assurance

It can also be expected that flow assurance challenges will be easier to manage with the smaller size SPS required by FLNG vessels or by floaters than by the larger and more complex 'Subsea to Beach' SPS.

2.6.5 Operational Flexibility

'Subsea to Beach' coupled to onshore LNG appears easier to reconfigure and expand during the field life: the SPS is designed to allow such reconfiguration and expansion, while new LNG trains can be added in the onshore plant. A liquefaction train cannot be added to an existing FLNG vessel.

2.6.6 HSE Considerations

'Subsea to Beach' may be preferable to the other options due to the fact that no personnel are working on rotation on an offshore host. At the same time, FLNG and floating hosts may produce the smaller environmental footprint (limited infrastructure on the seabed). In addition, decommissioning may be easier for FLNG and floating hosts rather than for 'Subsea to Beach'.

2.6.7 Technology Qualification

A 'Subsea to Beach' project may require an extensive qualification program for technologies that are not yet field-proven, see the section that follows. Technology readiness issues are less frequently encountered for the subsea production systems associated to FLNG or floating hosts.

3.0 Status of Subsea Technologies

Subsea hardware and controls are typically designed for 3,000m WD even when used in shallower waters. Technology qualification programs - based on well-established international standards - are carried out to ensure that the subsea technologies that are not field-proven for a given project undergo a complete test program to confirm their fitness for purpose.

3.1 Pipeline System

Flowlines, pipelines and trunklines are designed according to international codes and standards to meet the project requirements. Terminations are provided with hubs to connect jumpers, and valves to be used to isolate a line if required during the field life. The design of long pipelines and trunklines can be challenging especially if the wells are far from the shore and there is a transition from the deep offshore seabed to the continental shelf.

Another challenge in pipeline design is the shore approach and crossing, as this requires specialized technologies to minimize the impact on the environment. Nowadays pipelines need to be piggable with smart pigs. Even if the design and the installation of these ultra-deepwater pipelines may be challenging, the current technologies permit deeper waters and longer step-out distances.

A special pipeline is the MEG pipeline used to transfer MEG from the onshore terminal to the offshore location(s). This smaller size pipeline is usually piggy-backed to a larger production pipeline. The MEG leaving its pipeline enters an infield control umbilical, then reaching the dedicated injection point. For this reason, MEG cleanliness may be an issue.

3.2 Subsea Controls

Electro-hydraulic control umbilicals contain multiple hydraulic and chemical tubes to supply control fluid and chemicals respectively; and power and communication (on copper) quads and fiber optics (for large data volume communications). Considering the long step-out distances of these projects, umbilical costs are relevant.

Multiplexed control systems are mature technologies even if the exchange of data and the power distribution over long distances is still challenging. Subsea transformers may be needed to ensure the required electric power is effectively available on the seabed. Large umbilical cross sections will be required for high power distribution over long distances.

Another aspect for control umbilicals is the maximum length that can be produced without splicing: in some cases it is not possible to manufacture the required length in one run, and splicing cannot be avoided. The size, weight and complexity of some deepwater umbilical terminations have increased over time: it might be necessary to re-think these terminations, and also how subsea distribution units are designed.

3.3 Subsea Hardware

Subsea trees, manifolds and foundations are generally field-proven for the type of projects discussed here. Subsea valves and subsea tie-in systems may require an evaluation of their qualifications due to the great number of combinations of size, pressure rating, materials and past project applications. This also applies to the large flowrate chemical injection metering valves required to inject MEG into the produced gas.

3.2 SPS Main Challenges

The main technical challenges for ultra-deepwater 'Subsea to Beach' gas developments appear to be:

- Possible need for offshore (host) or subsea wet gas compression when the subsea wells start depleting. Offshore compression is technically feasible while, at this time, subsea wet gas compression is not yet a completely mature technology. Ormen Lange subsea compression project has been temporarily suspended, see (Rijnbeek, 2014), while Asgard and Gullfaks subsea wet gas subsea compression projects started their operations in summer 2015. The operational results of these two new projects will help determine if subsea gas compression will become viable for future ultra-deepwater 'Subsea to Beach' projects
- Subsea power distribution equipment, including connectors, and power cables (large cross sections), especially in case of subsea gas compression where tens of MWs are required at the seabed
- Reliable high-speed subsea communications over long distances capable of handling increasingly larger data volumes due to intelligent well completions and subsea sensors such as wet gas flowmeter (WGFM), sand detector, etc.
- MEG dosing to ensure proper hydrate formation prevention and MEG cleanliness control after regeneration, including possible subsea filtration systems
- Cocktailing of MEG and other chemicals to limit the number of chemical tubes in the control umbilicals
- Obsolescence of electronic components and backward compatibility of new controls equipment added to the subsea production system. This is an issue for all SPS's but is here more relevant because the design life of 'Subsea to Beach' projects is shifting from the usual 20-25 years to 30 years and above; as an example, it is understood that the design life for Gorgon pipelines and subsea controls is 50 years.

3.3 New Subsea Technologies

To be successful in deeper waters, over longer step-out distances and in more demanding environments, 'Subsea to Beach' projects will likely employ the following technologies as soon as they are commercially available in full:

- Electric actuation for subsea valves on subsea trees (all-electric tree), manifolds and other hardware. Removing hydraulic lines from the control umbilicals will make these umbilicals smaller and therefore less expensive, while the system reliability is expected to increase
- Subsea chemicals storage. This is another way to reduce the size and the cost of control umbilicals: if chemicals are stored on the seabed and from there injected into the produced gas, the chemical tubes of the control umbilical are not required anymore

- Deepwater high-integrity pressure protection system (HIPPS): this technology is obtaining a wider acceptance and is beginning to be used in deepwater projects in the GoM and in other geographic areas. HIPPS enables a less conservative pipeline design resulting in financial savings without compromising to operational safety. The all-electric HIPPS could simultaneously offer the HIPPS benefits and the electrical actuation benefits: this is of particular interest for ultra-deepwater projects
- Subsea heat exchangers to cool hot gas and avoid more expensive materials, such as corrosion resistant alloys (CRA's), for pipelines and subsea hardware, especially if coupled with corrosion monitoring systems (field-proven technology)
- AUV's permanently resident on the seabed and remotely controlled for inspection, intervention and repair activities on the SPS components
- Riserless well intervention in an ultra-deepwater environment: riserless intervention techniques utilizing slick line, wireline and coiled tubing are field-proven in shallower waters. In the near future, new vessels will be commercially available to perform riserless intervention on ultra-deepwater wells too.

4.0 Field Development Challenges

4.1 Technical Factors and Decision Factors

Technical factors to be considered while selecting the development option for an offshore gas field are those that cannot be modified, for example the metocean conditions of a given offshore location. A recent study, see (Beltrami et al., 2015), has focused on the limited number of 'key' factors that need to be taken into account. The main technical factors identified were:

- Recoverable reserves
- Step-out distance from shore (a true key factor)
- Water depth at wells location.

Decision factors are the results of an intentional design choice, for example between specifying single or dual flowline(s) or defining the number and distance of different drill centers. In the above mentioned paper, these were the main decision factors under examination:

- Number of wells per drill center
- Well flowrate
- Overall production flowrate (key factor)
- Drill center count
- Drill center distance (key factor).

These key factors resulted from an analysis of the estimated SPS CAPEX. Considering SPS CAPEX and drilling and completion costs, the step-out distance from shore retained its importance, while greatly increased the importance of the overall production flowrate, well count, and water depth at wells location.

Being technical factors, step-out distance from shore and water depth cannot be modified and therefore appear to be critical in the decision on how to develop a deepwater or ultra-deepwater gas field. Overall production flowrate and well count are decision factors that can be adjusted and optimized to improve the project economics.

4.2 Non-Technical Factors

Non-technical, risk-related factors will act as a go/ no-go regulator in making the final choice between possible development options. Examples of project uncertainties are:

- Local HSE requirements, including decommissioning
- Oil and gas market conditions
- Global geopolitics
- Relationship with governments, authorities, trade unions, other parties including local content requirements
- Relationship with partners and other stake-holders
- Offshore lease terms and conditions
- Contractual strategy.

Examples of possible financial risks are:

- Project delays
- Cost escalation during project execution
- Plant(s) not performing as expected
- Need for de-bottlenecking and brownfield work
- Late-life addition of compression requirements for 'Subsea to Beach'.

5.0 Conclusions

'Subsea to Beach' is a development option that has been successfully used in shallow water, deepwater and ultra-deepwater gas field development. It is now moving to even deeper waters, especially in new provinces lacking production infrastructure. The current 200km step-out distance limit might be crossed in the future, but this will require intensive technology development work on some enabling technologies, including subsea wet gas compression.

FLNG is a relatively new, but versatile, development option. It is expected to work in deepwater and ultra-deepwater and also for step-out distances exceeding the subsea technology limit (200km). Future FLNG projects already appear in sector II and sector III of Figure 6. Future FLNG projects are expected to populate these two sectors as well as the more challenging sector IV: it has to be seen if an FLNG vessel will be installed in ultra-deepwater at great distance from the shore.

Several Wet-Tree Hosts have been encountered during this investigation. A floater connected to subsea wells will process and compress the produced gas to ensure its export over long, or even remarkably long, distances. Some of the projects that have been examined are currently under execution. Their successful start-up and operation will hopefully confirm the validity of this development option, whose applicability is limited to very large reservoirs due to the cost of the facilities (subsea, floater and host.)

No examples of Dry-Tree Hosts have been identified in conjunction with the development of deepwater and ultra-deepwater gas fields. If confirmed, this may come as an interesting result. Possible reasons for not selecting this type of host may be the large size of the deepwater gas reservoirs, requiring multiple and distant subsea drill centers that are difficult to reach with deviated platform wells, the number and size of the required wells (several conventional wells or a more limited number of large-bore wells).

Nevertheless, floaters appear to be the best candidates to populate sector IV of Figure 6. The technologies required to design, install and operate floating hosts in ultra-deepwater and very far from the shore are currently available. At present, this sector contains two projects, Independence Hub and Lucius, which are not finalized to LNG production and one possible future FLNG project (Santos Basin). There are technical and non-technical reasons to explain why sector IV is only partially populated, in particular the fact that there are only a few ultra-deepwater gas reservoirs discovered beyond the 200km distance from shore and in waters deeper that 1,500m.

As a concluding remark of this discussion, the importance of the 'right' sizing of production facilities should not be underestimated. Offshore projects are sanctioned only when robust economics support the field development. Identifying the right size of the project may be challenging and risky: larger facilities require great investments to be repaid, but at the same time offer the largest rewards. Smaller facilities are less capital-intensive, but the profitability of the development may be limited by this size.

6.0 Acknowledgements

Thanks to Javier Gonzalez for the illustrations in the Appendix and to the Granherne colleagues who have reviewed this paper providing their highly appreciated suggestions and comments.

7.0 References

Barrero, L., and Beltrami, F., "Ultra-Deepwater Gas Field Development Options: A Study on FLNG and 'Subsea to Beach' Optimal Applicability", Proceedings of the 21st Offshore Symposium, February 2016, Houston, Texas.

Beltrami, F., Olugbode, F., Barrero, L., and Subramanian, K., "FLNG vs 'Subsea to Beach': A Study on the Critical Factors governing Concept Selection in Deepwater Gas Field Developments", SPE ATCE Houston, Texas, 2015. SPE paper 174927-MS.

Beltrami, F., "Subsea Technologies for Ultra-Deepwater 'Subsea to Beach' Gas Field Development: A Critical Review of Current Achievements and Future Challenges", 26th World Gas Conference, Paris, France, 2015.

International Gas Union (IGU), "World LNG Report - World Gas Conference Edition", 2015. Available at: http://www.igu.org/sites/default/files/node-page-field_file/IGU-World%20LNG%20Report-2015%20Edition.pdf

Rijnbeek, R., "Sub Sea Compression at Ormen Lange", International Petroleum Technology Conference, Doha, Qatar, 2014. IPTC paper 17649-MS.

Osmundsen, P., "Choice of Development Concept - Platform or Subsea Solution?", SPE Oil and Gas Facilities, Vol. 2, Issue 5, August 2013. SPE paper 163144-PA.

API 17A, Design and Operation of Subsea Production Systems – General Requirements and Recommendations, 4th Edition, April 2011.

Utseth, R. H., and Strøm, S., "Subsea IOR: Keynote Introduction", Offshore Technology Conference, Houston, Texas, 2007. OTC paper 18743-MS.

Gustavsson, F., Eriksen, R., Brekke, G., and Kraggerud, A. G., "Selecting the Field Development Concept for Ormen Lange", Offshore Technology Conference, Houston, Texas, 2003. OTC paper 15307-MS.

8.0 Abbreviations

API	American Petroleum Institute
ATCE	Annual Technical Conference and Exhibition
AUV	Autonomous Underwater Vehicle
CAPEX	Capital Expenditure
CGS	Concrete Gravity Sub-Structure
CRA	Corrosion Resistance Alloy
FLNG	Floating LNG
FPSO	Floating Production Storage and Offloading
FPU	Floating Production Unit
GoM	Gulf of Mexico
HIPPS	High Integrity Pressure Protection System
HSE	Health, Safety and the Environment
IGU	International Gas Union
IPTC	International Petroleum Technology Conference
KBR	Kellogg, Brown and Root
LNG	Liquefied Natural Gas
MEG	Mono-Ethylene-Glycol
MW	10^6 Watt
OPEX	Operational Expenditure
OTC	Offshore Technology Conference
SPE	Society of Petroleum Engineers
SPS	Subsea Production System
WD	Water Depth
WDDM	West Delta Deep Marine
WGFM	Wet Gas Flow Meter

Appendix – Figures and Charts

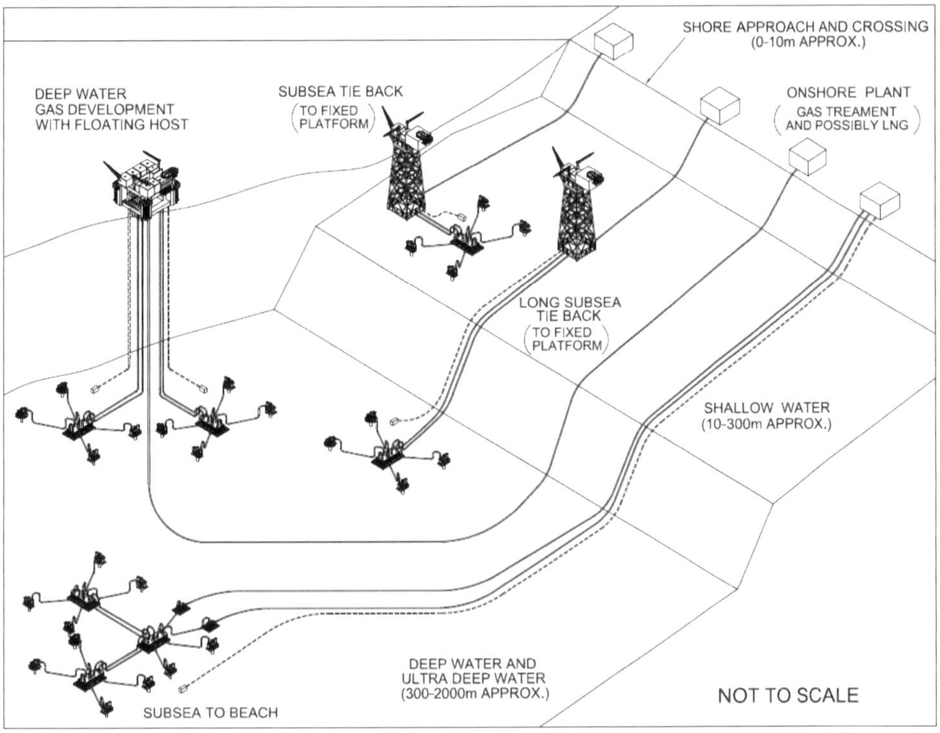

Figure 1 – 'Subsea to Beach' and Subsea Tie-backs

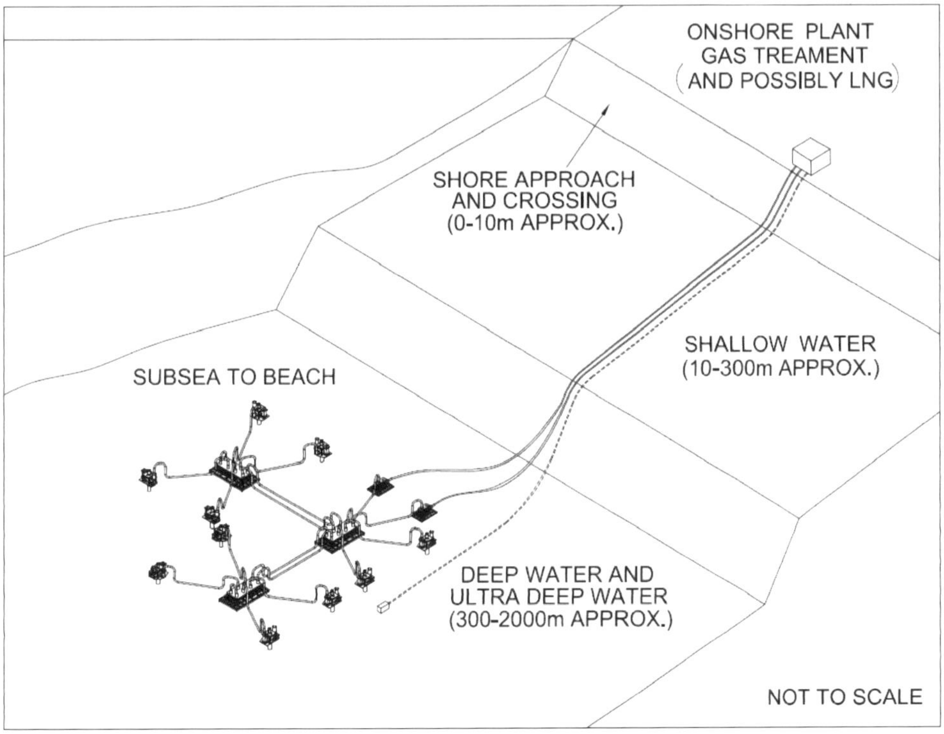

Figure 2 – 'Subsea to Beach' (typical)

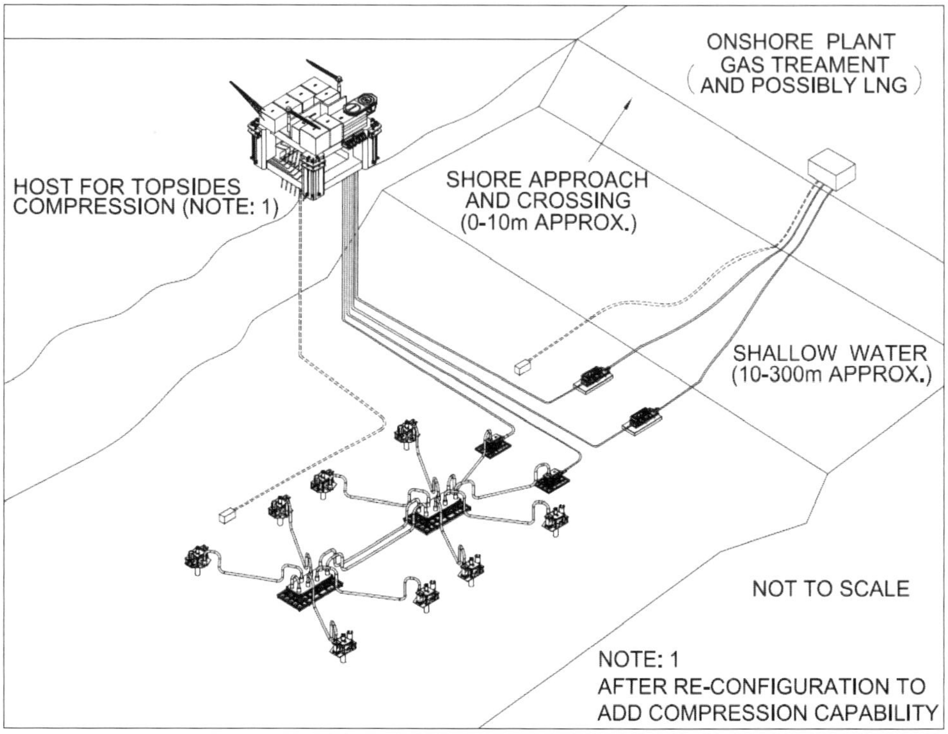

Figure 3 – 'Subsea to Beach' and Late-Life Topsides Compression

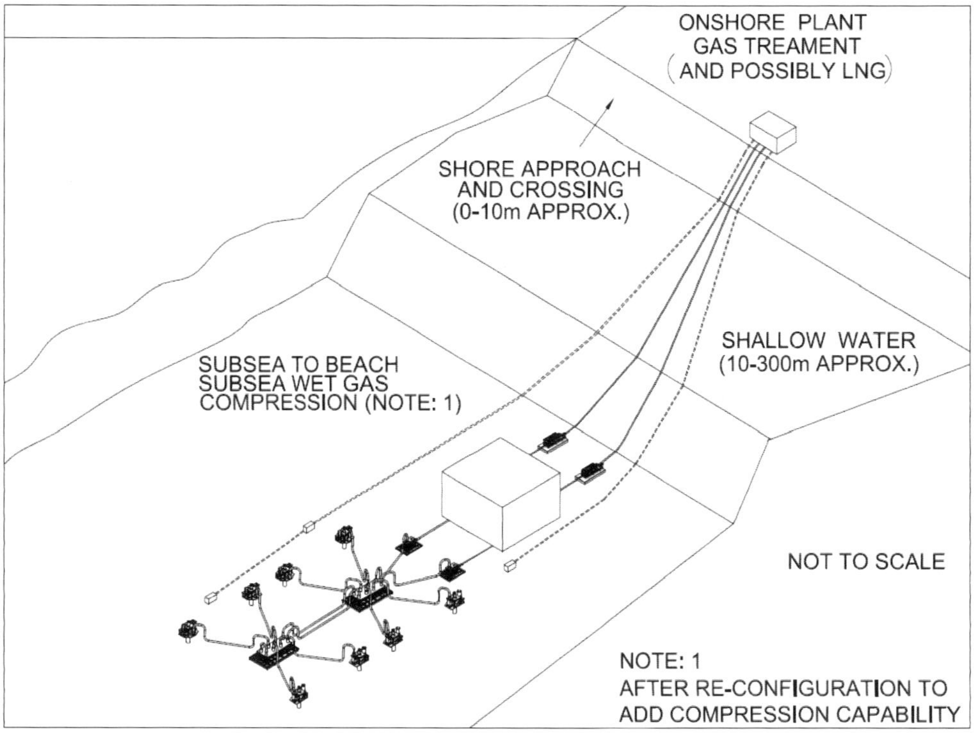

Figure 4 - 'Subsea to Beach' and Late-Life Subsea Wet Gas Compression

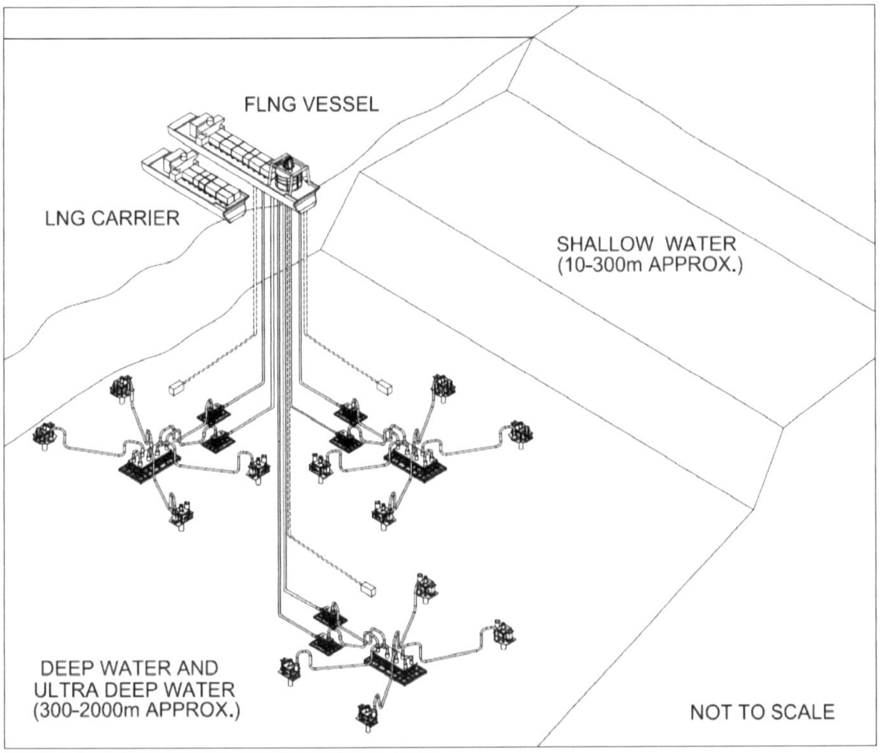

Figure 5 – Floating LNG (typical)

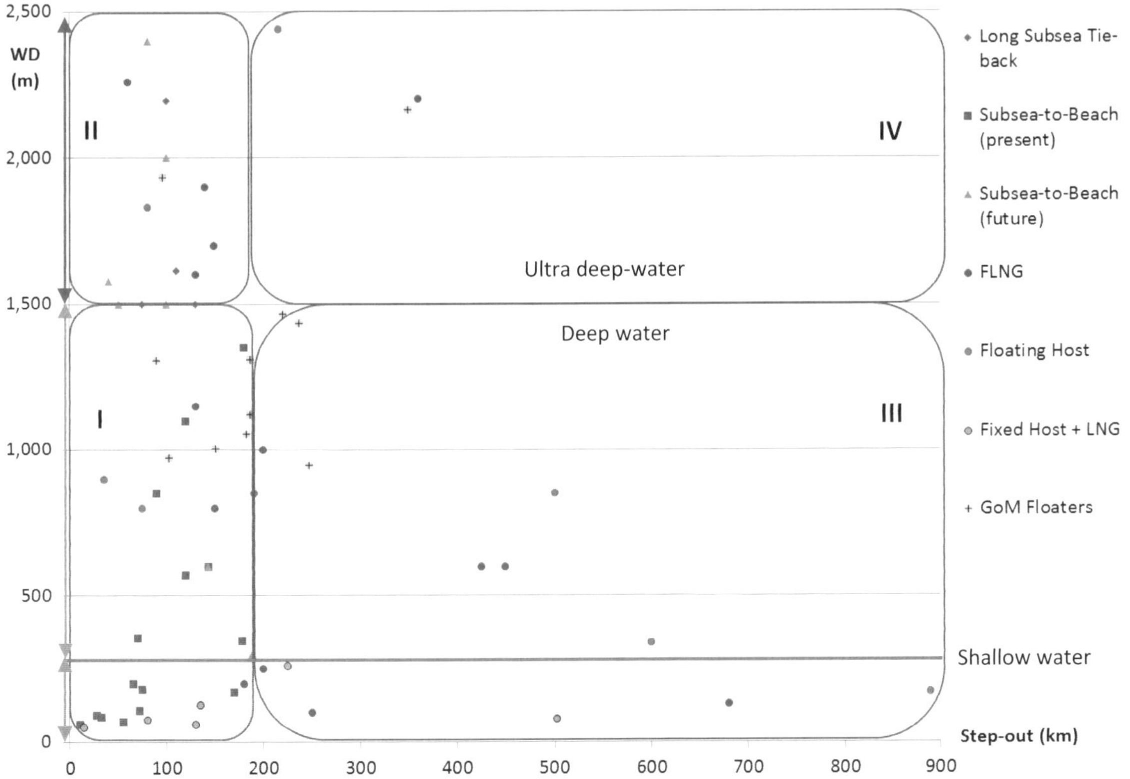

Figure 6 – WD vs. Step-out Distance for 69-off Projects (from Barrero et al., 2016)

Marybeth Gilliam
Greensea Systems, Inc.
Richmond, Vermont USA
mbgilliam@greenseainc.com

Ben Kinnaman
Greensea Systems, Inc.
Richmond, Vermont USA
bkinnaman@greenseainc.com

Doing More With Less

Integrated Solutions Advance Observation Class ROVs

Abstract: *Technology advancements have enabled the industry to field increasingly powerful observation class ROVs. Incremental features and benefits provide an astounding array of possibilities. However, as features and benefits develop, so does the complexity of the system. Operators are challenged to do more with more. More sensors. More data. More tasks. Following the trend of other robotics industries, the subsea industry has focused technology on the development of an "intelligent" machine and has lost focus on the operator and their role in the overall man-machine system. The subsea industry needs a paradigm shift towards the operator and their relationship with the ROV to realize the full potential of the technology available. This paradigm shift will embrace higher levels of coordination and autonomy. It will prioritize systems and solutions over components and create streamlined workflow for the operators. This paper reflects on the current state of the subsea robotics industry as understood through the use of observation class ROVs and provides a roadmap for the future. We present a system for subsea robotics designed for usefulness. A system that places the operator at the center and harnesses "intelligence" to provide superior usability, ease, and productivity. Lastly, we provide details on the technical implementation of this system and several fielded examples.*

1 Doing More with More

Technology has allowed us to create powerful new ROVs. But while advances in sensor technology have made ROVs more capable, the advances have created an increasingly complex environment for the operator. With the number of sensors currently on many vehicles, operators are drinking from a firehose of data.

Illustration 1: Operators are flooded with disparate sensor information.

Operators are doing more with more: more sensors, more capabilities, more tasks. Because each sensor has its own discrete hardware and software, operators are forced to work across multiple individual technologies. Operators manage so many computers that they have little time to manage their mission. Their focus becomes multiple screens, computer mice, and keyboards as opposed to an overall situation. In this environment, it becomes very challenging to manage the mission because at every moment the operator is managing multiple low-level tasks. There is a point of diminishing returns.

Perhaps the biggest challenge though, is the mental management. Besides physically manipulating the disparate vehicle tools, the operator is also mentally integrating all the information. In the current model of observation class ROV operations, the complete and correlated data set for the mission is in one location only - the operator's head. The operator is forced to work harder than ever before but without the tools to help them work smarter.

The industry has focused technological advancement on vehicle components, not the operator's situation or mission. Businesses have rewarded technological advancements of faster, more accurate sensors that can quickly translate into increased sensor sales. The burden of integrating the sensors has largely fallen on the shoulders of the end-user. A person need only take one look at an operator in front of nine monitors to realize "there must be a better way."

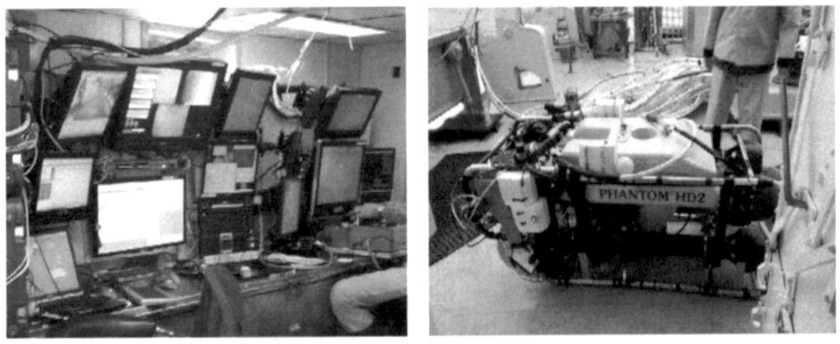

Illustration 2: An operator manages multiple computers to complete a complex scientific mission with their Phantom ROV. A typical scene with many observation ROVs.

The majority of work and research in robotics, including within the subsea industry for ROVs, has been towards developing better, more intelligent sensors. This has lead to more advanced machines but not necessarily more useful machines. To increase productivity, we need to focus on system management. Through technology integration and improved communications with human operators, ROVs can become mission partners and work harder for us.

As technology developers, we must focus on the synergy of the operator-ROV relationship and not just the machine.

2 Doing More with Less

A fundamental elevation of the end-to-end ROV system needs to occur in the marine industry to reduce operator fatigue, reduce costs, reduce risk/error, and fully realize the potential of current and future sensor technologies. The transition requires a fresh look at technology, addressing how we integrate it and how we use it. By putting our focus on the operator and their task, and providing technology that reduces the load on the operator, we can improve their effectiveness offshore. Ultimately, we must enable the operator to do more with less—less computers, less sensors, less work.

The future of ROV technology relies on the improved relationship between man and machine. The next generation of subsea robotics will feature operators supervising machine partners and working together as a team. To make this leap, we need to think about software and hardware in a new way. We can no longer consider sensors solutions. True solutions will reside within a completely integrated system that includes the operator. When man and machine can share high-level knowledge generated from integrated technologies the potential for productivity will be exponentially elevated.

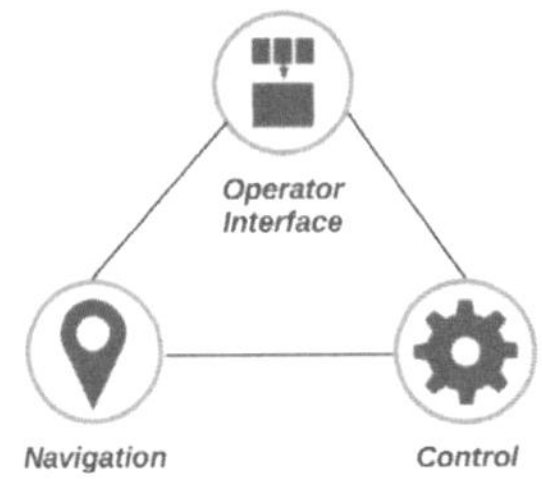

Illustration 3: The three fundamental integrated technologies required for more effective ROV operations.

There are three principle requirements for a cooperative ROV partner.

- The ability to determine its own state (navigation).
- The ability to change its state to a desired or commanded state (control).
- The ability to interface well with their human operators (communications).

While each requirement is important, it is only when the three are fully integrated and working together that they can elevate a system to provide meaningful solutions. Tightly integrated navigation and control creates a more capable vehicle and fully-integrated communications provide operators with a streamlined, efficient workflow.

The important paradigm shift is the integrated implementation of navigation, control, and interface technologies towards the purpose of making the operator more efficient and productive in their mission. Systems-level technology development, based on the tight integration of navigation, control, and communications is revolutionary in our device and widget-focused industry.

2.1 A More Capable Machine Partner

Tightly integrated navigation and control technologies make vehicles more capable by sharing critical data throughout the ROV system. Centralized data increases awareness for both the vehicle and the operator. This awareness, supported by clear, easily understood communications, can dramatically improve productivity. With a fully-integrated vehicle publishing its information to the operator in an integrated, universal interface, the vehicle becomes a significantly more capable partner.

Accurate and robust navigation of ROVs is the first step towards creating a capable partner. While navigation "systems" plentifully exist in our industry, they are more aptly described as strap on sensors rather than systems. While operators may be able to “see” their vehicle’s position using sonar, video, USBL, or an INS, the vehicle itself, including the control system, payloads, and data, lacks this information. The potential is clear. If the operator can see the data, it can be integrated into the vehicle system for broader use throughout the data model and to enable other features. If the vehicle knows its location, it can natively perform the advanced vehicle control functions such as automated positioning, route following, or mission planning, necessary for the vehicle to operate as a contributing member of a team.

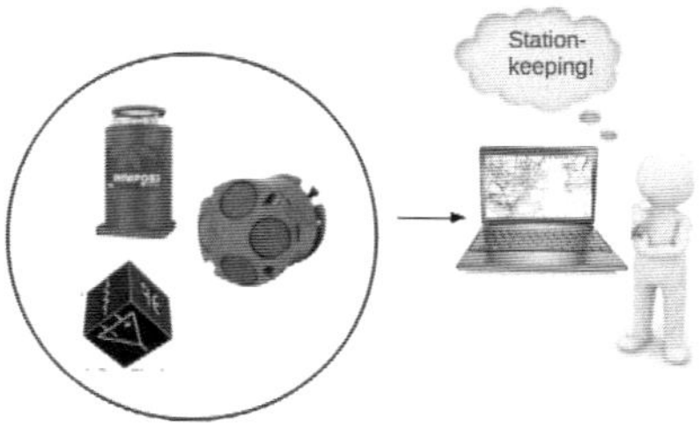

Illustration 4: We use navigation sensors and control components every day offshore to deliver advanced vehicle control. It is fragile, complex, and expensive.

This is the paradigm shift. Sensors exist by the dozens. Systems do not.

In our current model, let us think about what is required to provide station-keeping on an ROV. First the operator will procure a DVL (largely because DVLs have been marketed as navigation sensors). Then, finding the DVL will only really provide sensor frame velocity data accurately in limited conditions, the operator procures an INS and integrates the DVL. Then there are more components including hardware, software, and computers to provide interpretation and visualization of the navigation data. Finally, a closed-loop control function is needed for the vehicle to keep it in one place. Unfortunately, this is a significant integration task involving disparate hardware and software and one many operators have tackled. In our proposed model, the operator would procure a DVL that includes a station-keeping function. Integration into any vehicle would be a hassle-free cable connection.

Illustration 5: An ROV control system.

Let's think about another common application in the subsea industry. Operators rely on sonar to position their vehicle within their work site and near the object they are inspecting or manipulating. Maneuvering their ROV to the work site requires an operator to watch the sonar, interpret the sonar, and then "fly" the vehicle manually or perhaps with some discrete autopilots enabled. Once onsite, the operator holds the vehicle on station by monitoring the sonar or video to keep the ROV correctly positioned relative to the subsea object of interest. The operator could automate keeping the vehicle at the target by using the DVL-aided INS and station

Illustration 6: An ROV control system.

keeping control system described in the previous paragraph. Another option would be to use a sonar target tracking software offered by several vendors, of which Greensea is one, and interface that with the autopilots on the vehicle. This method of operation works. It is used all the time and it is complex and fraught with error. In our proposed system-centric model, imagine a sonar that provides vehicle positioning relative to the features in the sonar data. One sensor, or rather a system, that delivers what the operator really requires—positioning with respect to the sonar data.

The subsea industry currently considers navigation separate from control. One could even argue, the subsea industry does not really even consider control. To elevate our ROVs, bristling with independent and loosely coordinated sensors, to cooperative and capable partners, we must think of navigation and control as a single system.

2.2 A New Way of Communicating

Considered as a single system, navigation and automated vehicle control can dramatically improve a vehicle's capabilities, but only if it can communicate effectively with its operator. Effective communication is critical to a team and it is no less true when the team member is robotic. Unless the operator has immediate and easy access to all the necessary information from the vehicle, they cannot manage operations effectively. An intuitive interface providing a synthesized view of a vehicle's data, is an essential component for a high-functioning system. Through communication technologies, we give ROV supervisors, the human operators, the tools required to be effective and efficient and to realize the full potential of their very capable machine coworker.

Other industries have already implemented a paradigm shift in thinking about the importance of our relationship with highly capable machines. Perhaps the most well-understood example is the 2007 introduction of the Apple iPhone. The iPhone revolutionized how people communicate with their phones by offering a multi-touch interface. The iPhone used a large touchscreen for direct finger input as its main means of interaction, instead of a stylus, keyboard, or keypad, typical of smartphones at the time. And, four years later, they advanced communications again with the introduction of their intelligent assistant, Siri, an integral part of iOS since 2011.

Illustration 7: Perhaps no better example exists for how a paradigm shift in technology focus successfully addressed the man-machine relationship.

The smartphone is a great example of how the power of integration has forged human relationships with machines. Smartphones integrate many disparate tasks, applications, and functions previously found on separate devices. Beyond functional integration, smartphones deeply integrate the data within each task. For example, consider the process of making a dinner reservation on your smartphone. Using the app OpenTable, you can make a reservation at a restaurant. Once the reservation is made, your phone enters it in your calendar and creates a contact for the restaurant in your address book and in your GPS for convenience. While functionally, your smartphone integrates these functions in a single device and makes easy work of doing many things at once, the tight integration of data is what is really powerful.

ROV interfaces can follow this example and transform the traditional graphical user interface of the ROV running on yet another computer topside into a highly integrated "Workspace" for the operator. Most ROVs today, even the most basic observation class vehicles, utilize several individual components to perform work. The components include imaging sonar, navigation, measurement sensors, video, lighting, and tooling. With each component providing their own interface in addition to the ROV's OEM interface, the burden on the operator is not unlike what consumers faced when we used separate electronic devices to manage our music, video, entertainment, calendars, email, addresses, and phone calls. Like the smartphone, a well integrated ROV user-interface provides a single environment through which the operator works and communicates with the ROV. Similar as well to the smartphone, the real power of this integration is the ability for all of the independent devices in the vehicle system to share data.

A well-integrated ROV user-interface provides the entire workflow for the operator across all phases of working with the ROV: maintenance, modifications, diagnostics, and operation. Why should an operator use separate computers, interfaces, and software applications to communicate with a single ROV? Is it not more efficient for an operator to learn and use one interface for an ROV, regardless of the task, payload, or sensor complement? An efficient and effective user-interface enables the operator to perform to their fullest potential by optimizing workflow—the user-interface becomes a "Workspace."

Consider the simple task of inspecting a pipeline with an observation class vehicle. Today, the pilot flies the vehicle and monitors the vehicle's status with the OEM's ROV interface, locates the pipeline with the sonar manufacturer's interface, tracks the vehicle with the navigation system provider's interface (or perhaps a third-party chart plotting package), transits the pipeline and records video using a separate video system, and records inspection notes in a separate dive log package. Through this workflow, the operator is solely responsible for interpreting and transporting data between separate interfaces by making adjustments in the ROV's course based on what he sees in the sonar or video. Should any of these tools change, so does the workflow and interface.

Imagine this same task using the integrated workspace described in our last model. Now the pilot flies the vehicle, operates the sonar, tracks the vehicle, logs data, and manages video within a single software interface that can provide the same exact interface for any sensor or payload regardless of the manufacturer. With this system, the operator locates the pipeline and double-clicks the pipeline feature in the sonar image telling his ROV to track and fly to the feature at a set altitude and speed. With good visibility, the operator can double-click the pipeline in the video stream and have the ROV follow the pipeline using video. Perhaps instead, the operator has created a mission plan which tells the ROV to locate and inspect the pipeline. While the ROV transits the pipeline, the operator logs data and records the job in the same software while supervising the job and ensuring the quality of data. At the end of the job, the operator presents the customer with a single cohesive data package captured from a single interface that coordinates and links all of the data from the job: navigation, video, sonar, and inspection notes. Now imagine we are tasked with repeating that inspection exactly in one year. Which model would be easiest?

3 Field Examples

3.1 Explosive Ordnance Disposal with Observation Class ROVs

Illustration 8: Capability does not have to be big. Spot the navigation and control system."

Teledyne SeaBotix has successfully fielded an extremely capable system built on their vLBV ROV product for subsea Explosive Ordnance Disposal (EOD) called SmartFlight. The only change to the basic vehicle equipped with a sonar and USBL was the addition of a small pressure vessel under the foam pack containing a fully-integrated control and navigation system complete with embedded processor, on-board software, an inertial measurement unit, a magnetic compass, and a pressure sensor. Topside, operators use a graphical workspace that provides vehicle control, mission planning, video capture, sonar display, and positioning relative to sonar features within the standard Teledyne SeaBotix topside console.

With this fully integrated system, operators can survey the seafloor by specifying a search grid and acquire targets for inspection simply by double-clicking on the sonar return. Additionally, all data including video and sonar is logged for archival, post processing, or playback. To reacquire a target, an operator simply has to enter the coordinates or right click on the map display over the

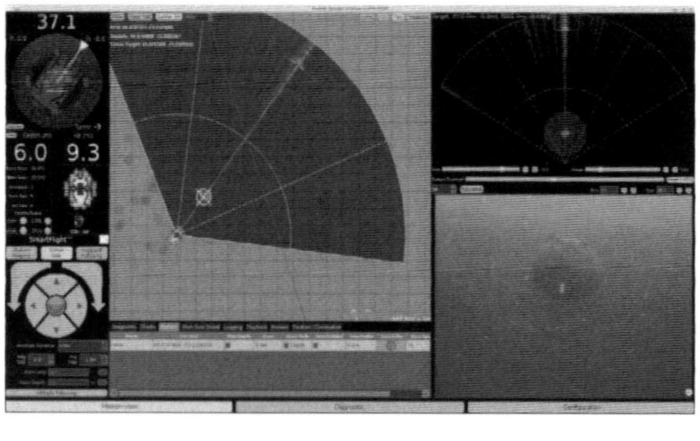

Illustration 9: Teledyne SeaBotix SmartFlight system used for Explosive Ordnance Disposal activities.

location and select "GoTo Position".

3.2 Port and Harbor Security with Observation Class ROVs

Perhaps no more difficult operating environment exists for ROVs than within a port or harbor. Likewise, perhaps no more important operating environment exists for ROVs than a port or harbor. This environment consists of obstructions, complex bottom topography, poor visibility, poor acoustics, currents, and shallow water. All of which are ruinous for subsea navigation.

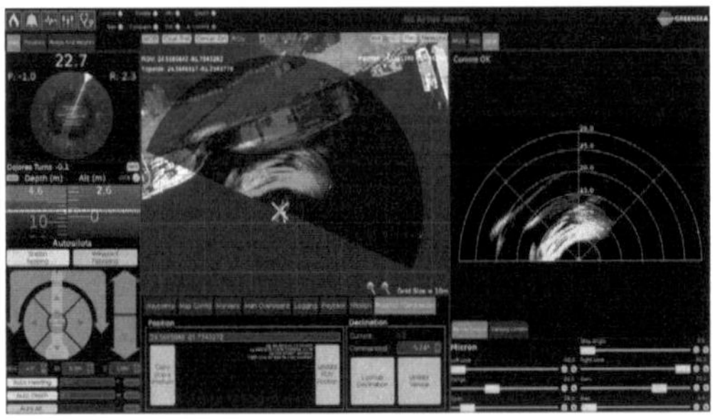

Illustration 10: An integrated navigation and control system coupled with a comprehensive operator workspace allows efficient and productive ROV operations in the most adverse environments.

Greensea has successfully used their Balefire integrated navigation and vehicle control product for port and harbor security applications on several different OEM vehicles. This system provides a turnkey inertial navigation system, advanced control, and sonar-based vehicle localization and mapping using any available multi-beam or scanning sonar. With this system, operators can easily use any observation class vehicle to perform port and harbor security tasks with ease.

3.3 Diver Navigation and Diver Transportation - ROVs with Subsea Operators

Illustration 11: The STIDD DPD uses integrated navigation, control, and operator communications to deliver tactical divers accurately to desired locations.

ROV operators do not have to be above the water and the ROVs they operate do not have to look like ROVs. STIDD Systems has recently developed a diver transportation vehicle enabled by an integrated inertial navigation system, advanced control system, and sonar-based localization. While not a traditional ROV, the Diver Propulsion Device (DPD) equipped with the RNAV2 navigation and control system, provides all the capabilities of a modern ROV system and takes the operator along for the ride. Divers are able to accurately navigate subsea with an on-board inertial navigation system and transit to specified positions by simply clicking the screen or downloaded pre-programmed missions. With the integrated sonar, the DPD uses the sonar data as a complementary navigation system and allows obstacle avoidance and feature target tracking.

4 Conclusion - A New Kind of Partner

Our community has always understood and valued the concept of team. Offshore work is hard at best. Most of us have been humbled by the sea at some point. It is in partnership with our team that we tackle seemingly impossible missions on a regular basis. The next level of robotics embraces ROVs as fully integrated, communicating team members, able to share high-level intelligence and decision-making alongside its operators to dramatically enhance the success of a mission. The industry needs a paradigm shift to realize the full potential of the technology available. This paradigm shift will prioritize systems and solutions over components and create opportunities to accomplish more with less.

5 Acknowledgements

Greensea would like to thank Monterey Bay Aquarium Institute, Naval Postgraduate School, Northwestern Michigan College, Schmidt Ocean Institute, STIDD Systems, VideoRay, Teledyne Marine, and Vulcan for taking the time to understand the power of an integrated system approach and providing feedback that continues to strengthen our products.

6 Authors

Marybeth Gilliam is the CMO and Vice President of Sales at Greensea Systems, Inc. Marybeth worked in advertising, marketing, product development, and market research for over 20 years in both New York and San Francisco before joining Greensea. She has a degree from The University of Vermont.

Ben Kinnaman is the President and CEO of Greensea Systems, Inc. Ben started his career offshore as a diver and ROV technician before becoming an engineer. He has a Physics degree from Davidson College in North Carolina and a Master's Degree in Mechanical Engineering (Control and Robotics) from The Johns Hopkins University in Baltimore, Maryland.

7 Greensea

Greensea develops integrated navigation and vehicle control systems for the marine industry. Since opening in Richmond, Vermont in 2006, the company has successfully installed systems on over 500 vehicles ranging from micro ROVs to high-profile work-class vehicles for Monterey Bay Aquarium Research Institute, National Oceanic and Atmospheric Administration, and the U.S. Navy. Greensea offers commercially available navigation, control, and automation products based on their patent-pending OPENSEA operating platform. OPENSEA currently networks all marine systems (manned, unmanned, subsea, surface) to transform sensors into solutions. To learn more, visit greenseainc.com, find us on Facebook, or give us a call at 802-434-6080.

Intersecting Spherical Hull Form for Manned Submersible Vehicles

Fred C. Jensen, MSME, PE
William M. Simpson, Jr., Ph.D., PE
Capt Jack Ringelberg, MS, PE
Patriot Engineering Co., Chagrin Falls, Ohio USA
UI2016, New Orleans LA, Feb. 24, 2016

Abstract

The most used hull form for the design of underwater vehicles is the single sphere or for larger vehicles a cylinder with spherical caps at each end. The objective of this paper is to show the structural advantages of hull geometry in the form of intersecting spheres for multipurpose submersibles. A theoretical discussion explores how a physical arrangement of intersecting structural spheres, reinforcing rings and compression bulkheads results in a structure that is very competitive with the above mentioned conventional hull forms. Extensive finite element analysis (FEA) of the geometry presented supports the basic structural equations.

The mechanical structural loading on the submerged hull form is the external seawater pressure loading due to diving at depth. Results of the structural FEA and equations are presented in tabular format and also as color plots of stress and deflection. Some of this work was originally started by the late Clifford Ness, President of Manta Research Corporation.

The goal of this effort has been to develop a general-purpose manned underwater vehicle that is adaptable to a variety of deep-water tasks. Some observations from the development of this design conclude that an intersecting spherical hull form results in a more efficient use of usable interior volume especially for smaller submersible vessels. The current standard cylindrical hull form requires a central passageway for crew movement fore and aft of the vessel. Thereby, the entire central volume is unavailable for payload and is not usable space.

The intersecting spherical hull form allows multiple passageways and multiple equipment areas for payload within the interior. Thereby, alteration is easy to customize for tasks such as underwater transport, commercial diving, sub-sea survey work, and ocean research. Another observation is the intersecting spherical hull form lends itself to a saucer like disk hull form that has hydrodynamic advantages in underwater handling and maneuverability similar to a Manta fish. These practical advantages of the proposed hull form are the incentives to proceed with the structural analysis and discussion in the paper.

Discussion

Submersible Design Goals

The primary design goal is to provide a general purpose submersible with a large internal dry payload capacity that only requires a two person crew. This vessel should be ocean capable and if need be can be launched from the well deck of a surface ship yet remain air transportable. Sea vessels must be versatile, dependable, safe and seaworthy for all team members who operate in the marine environment. These factors are integral to this design. This submersible is a simple design. It is versatile in that it has a large internal volume, can operate at depths of 1100 feet (335 m), and make underwater round trip ranges of 90 to 449 nautical miles (167 to 832 km) depending on cruising speed. The concept incorporates arrangements to provide easy access to maintain, replace, and continually upgrade all on-board systems as new technologies become available. Safety is emphasized in structural safety factors and in system design concepts such as battery placement outside the pressure hull. Seaworthiness is enhanced by increased stability both submerged and on the surface.

Submersible Concept Design

An over-all exterior view is shown in figures 1 and 2. The basis of the intersecting spherical pressure hull is the arrangement of six primary spherical shells that intersect in a geometric pattern as shown in figures 3, 4, 5 and over-all in figure 8.

The forward sphere houses the pilot and co-pilot station. The central four spheres provide the payload and accommodation space as shown in Fig. 5. The two aft spheres, one large and one small, provide diver lock-in and lock-out capability with minimal change in buoyancy and weight. This will facilitate salvage operations, construction, underwater archelogy and recreational diving. Propulsion is provided by large twin shrouded propellers that are widely separated to provide exceptional maneuverability. A large prop turns slower and generates less noise underwater while enhancing maneuverability. The reduced vehicle length made possible by the unique pressure hull design results in less wetted area and thus less underwater resistance compared to a cylindrical hull form as has been confirmed in tank testing. The submersible's underwater performance is projected to be five knots average speed, seven knots maximum, with up to eight knots maximum sprint mode. Diving planes forward and inverted-v planes and centerline rudder aft provide for maneuvering and attitude control while making way. Trim tanks provide for attitude control at slow speed or when not making way. Further, because of the flattened oval cross section, which mimics the Manta Ray Fish, the vessel can be "flown" thru turns and is highly resistant to the snap roll that can be a hazard with cylindrical vehicles. The spherical cluster design of the pressure hull offers more internal volume and less overall length compared to a cylindrical pressure hull. The natural divisions provided by the sphere intersections allow optimum arrangement of the internal ribs and bulkheads necessary for resisting pressure at depth. The flattened shape with lesser hull depth allows operation in shallower waters. These are unique bio-mimicry features in a vessel that is only thirty-five feet, ten inches (10.92 m) long can accommodate useful underwater tasks with only two crewmembers.

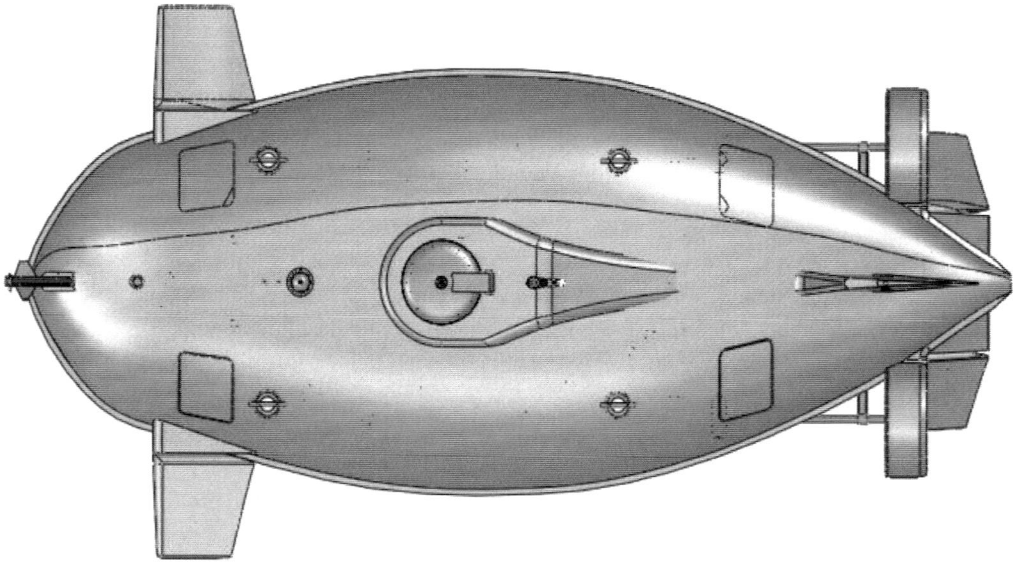

Figure 1, Plan View of Exterior

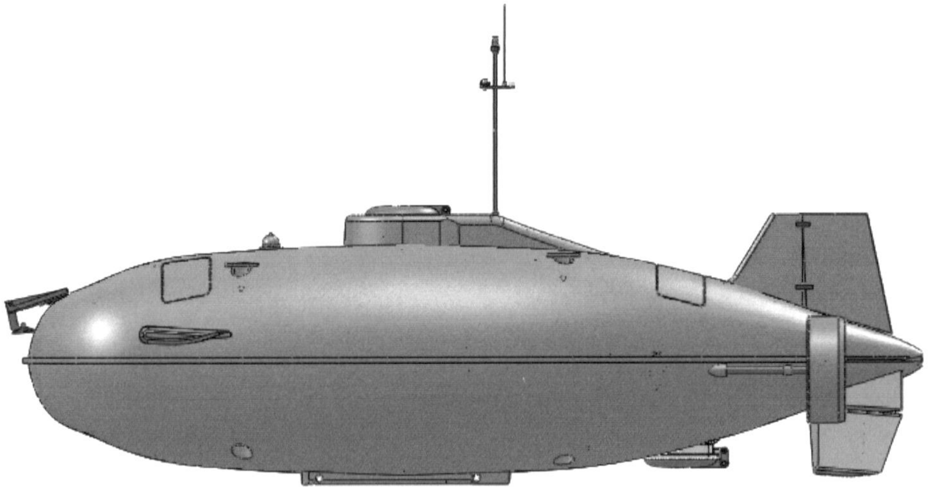

Figure 2, Profile View of Exterior

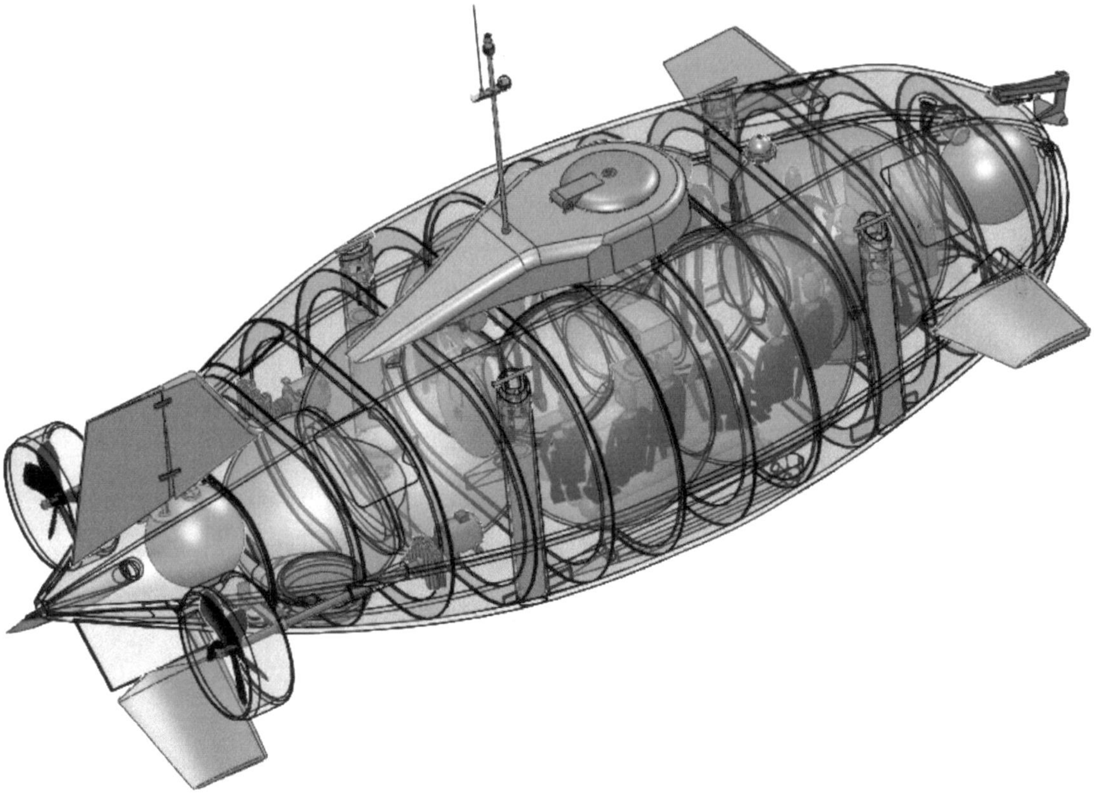

Figure 3, Perspective View of Starboard Side including Payload Areas

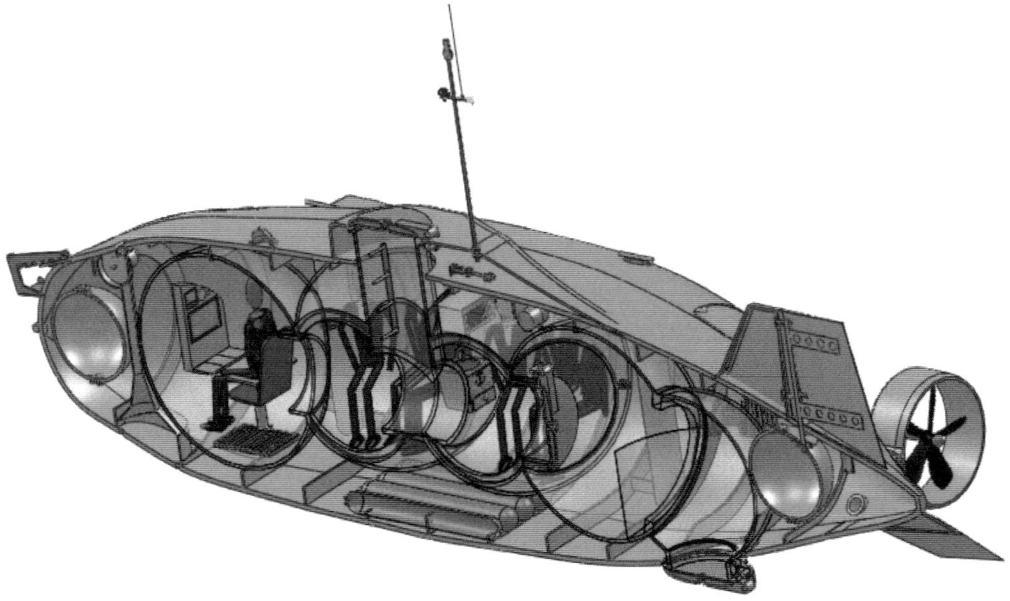

Figures 4, Cut-away Perspective Views of Payload and Accommodation Areas

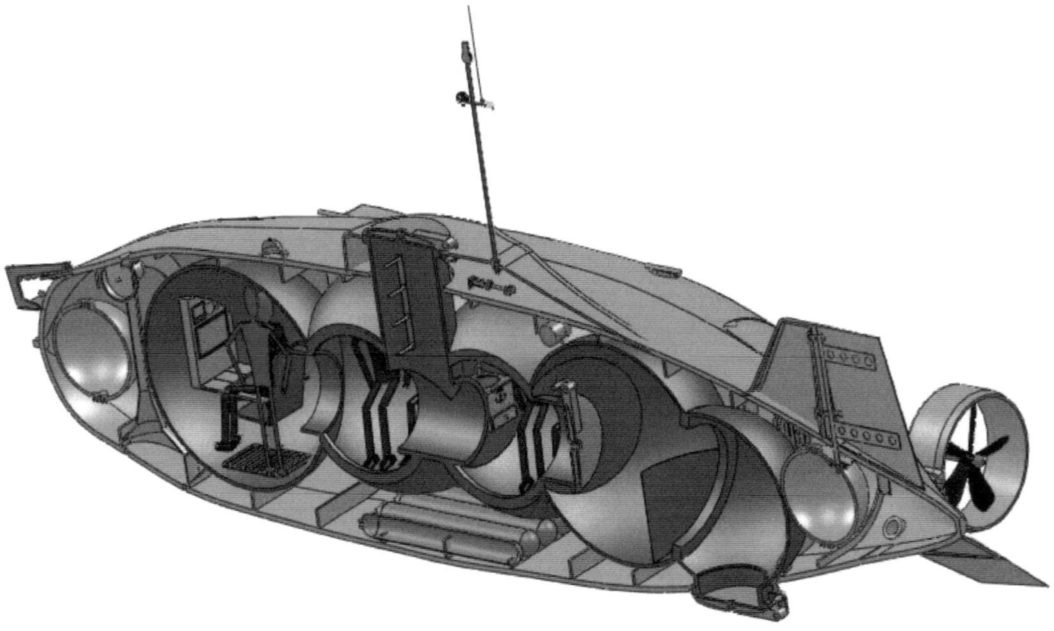

Figures 5, Cut-away Perspective Views of Pressure Hull

Operational Features:

- ABS classed A1 submersible design, design depth 1,100 ft (335 m).

- Over-All Length: 35'-10" (10.92m).

- Surfaced Draft: 7'-0" (2.13m).

- Fits within C-17 Globe master Transport Aircraft: (12'-4" Hi x 18' Wide x 68' Bay (3.73 m x5.49 m x 20.73 m)) with 170,900 lb (77,680 Kg) max cargo weight.

- Loaded dry weight with two crew and 3600 pounds of payload: 107,000 lb (47.77 LT, 48.64 tonnes).

- Pressure Hull: HY-100 Steel (Carbon Fiber – Epoxy Laminate, Optional).

- Outer Hull: Fiberglass Epoxy Laminate.

- Operational depth in sea water: 1,100 feet (335m).

- ID Diver lock-out hatch, 31 in. (78.7 cm) operational at depth.

- Submerged range (assumed hotel load 50 kWh)
  at 7 knots range is 90 nm (13.0 km/hr, range 167 km)
  at 5 knots range is 229 nm (9.3 km/hr, range 419 km)
  at 3 knots range is 449 nm (5.6 km/hr, range 832 km)

- Propulsion System: Two carbon fiber 45 in (114 cm) OD five bladed propellers; Two main drive motors {15 hp (11.2 kW) each @ 2300 RPM base speed and 20 hp (14.9 kW) each in sprint mode};
240 volt DC, Variable speed Inverter drive;
Main drive gearing is a 5 to 1 speed reduction.

- Battery bank, 450 kWh.

- Life Support: 96 hours for 14 persons.

- Relative humidity maintained at < 65% RH (12) high pressure ASME cylinder tanks at 3,000 psi (20.7 MPa) holding 10 ft^3 (0.283 m^3) each for air storage.

- Emergency built in breathing system (BIBS) for crew.

- First Aid Station.

- Marine head {100 gal. (0.379 m^3) holding tank}, small galley with {75 gal. (0.284 m^3)} fresh water tank, sink, and microwave oven.

Pressure Hull Structural Concepts

Spherical Pressure Vessels

Spherical pressure vessels are the most efficient shape for a pressure vessel. In chapter VI of *Submersible Vehicle Systems Design* (Allmendinger 1990) the buoyancy factor for an un-stiffened sphere and an un-stiffened cylinder are compared. The buoyancy factor, B, is defined as the hull weight divided by the displaced water weight. Allmendinger

(1990) shows that conceptually the spherical pressure hull will have 25% less weight compared to an un-stiffened cylindrical pressure hull with the same displaced volume. The analysis supporting this assertion is provided in Appendix A. The advantage of the unstiffened sphere increases when stiffeners are added to the cylinder. Appendix A also shows the deformation incompatibility that results when a cylinder is joined to a spherical end cap. There are other more complex factors to be considered when comparing spherical and cylindrical pressure vessels, but this does show the basic advantage of the sphere over the cylinder.

The stress analysis and structural requirements for a spherical pressure vessel are provided in Appendix B. The American Bureau of Shipping (2013) requirements for spherical pressure hulls are listed and discussed. It is shown that the ABS requirements are based on whether the computed elastic buckling stress is more or less than the yield stress. An example of a spherical shell with r = 42 in. (107 cm) is evaluated to show that yield, not elastic buckling, governs for design depths exceeding 100 m (328 feet). The ABS factor of safety on yield for this example shell is shown to be in the range of 4 to 6 when the shell thickness meets the requirements for a working depth from 150 – 300 m (~500 to 1,000 feet). The ABS fabrication tolerance consisting of an allowable variation in radius over a specified arc length is found to produce only a small change in the computed buckling pressure and thus it is considered not to be governing. From this it is concluded that for realistic operating depths (1) elastic buckling will likely not be governing, and (2) a design stress factor of safety in the 4 to 6 range would likely be adequate based on the judgment and experience of ABS. This also gives an indication that normal fabrication tolerances and small distortions in the spherical shell geometry due to variations in boundary rigidity will not immediately cause a buckling failure. This is considered to be important because, even with the advanced simulation and analysis tools available today, it is likely not possible to imagine and perform a non-linear analysis of all possible combinations of fabrication tolerances and fine mesh detail modeling. New designs involving variations in sphere boundary rigidity will of course require analysis and model validation, but these criteria provide a comfortable initial design space.

Linearly Intersecting Spheres

The advantages of the spherical pressure hull have been argued above, but there are clearly limitations in designing a vehicle with a single spherical pressure hull. The single sphere pressure hull is clearly the choice for deep diving submersibles, but, for more moderate depths, a conflict quickly becomes obvious in attempting to avoid excessive hull depth and draft, and when attempting to make efficient use of interior volume. If vehicle mobility is a requirement, hydrodynamic efficiency with a single sphere pressure hull becomes a major challenge. These issues have been address by linearly joining spherical shells to create a pressure hull with greater length to beam and length to depth ratios. The submersible DEEP QUEST had a pressure hull of two intersecting spheres and the US Navy Deep Submergence Rescue Vehicles (DSRV) had a pressure hull of three linearly intersecting spheres (Garland 1968). A cross section of these pressure hulls is shown in Figs. 6 and 7.

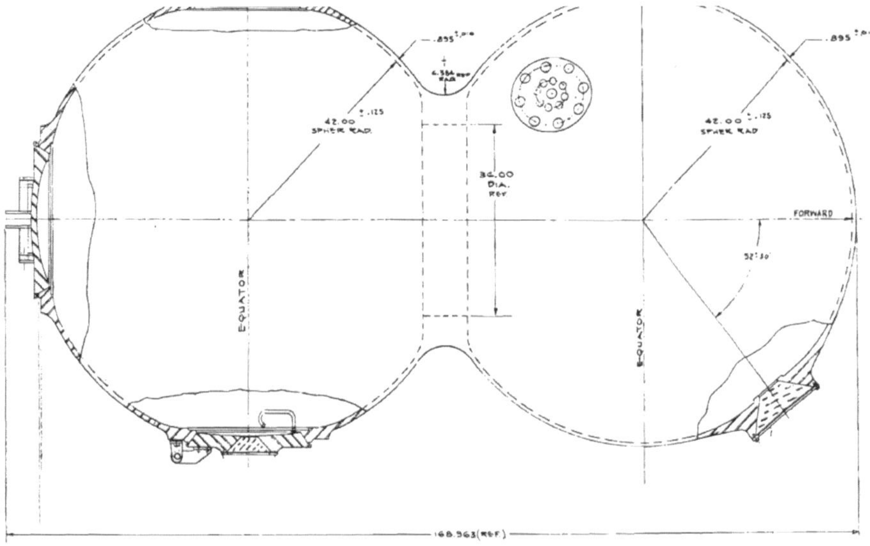

Fig. 6 DEEP QUEST Pressure Hull (Garland 1968

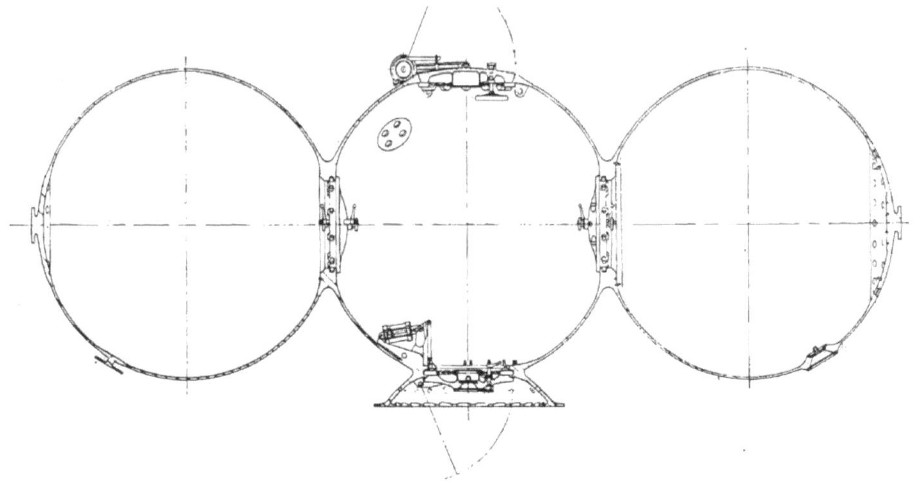

Fig. 7 DSRV Pressure Hull (Garland 1968)

When intersecting spherical shells are used, it is necessary to provide reinforcing rings at the shell intersections to exactly replace the support and stiffness of the shell material cut away to join the spheres. The presence of the required reinforcing rings can be seen in Fig. 9. Allmendinger (1990) provides the computation tools needed to size the reinforcing rings at the shell intersections. An explanation of the computations to size the reinforcing rings is provided in Appendix C.

The reinforcing ring size computations in Appendix C assume the shell and the reinforcing ring are constructed of the same material. Allmendinger (1990) extends the development to show that by using a reinforcing ring with a higher modulus of elasticity significant weight savings are possible. That is beyond the scope of this work except to say the possibility of composite shells and metal reinforcing rings with their significantly higher modulus of elasticity would likely have great potential.

Geometrically Arranged Intersecting Spheres

The advantages of pressure hulls consisting of intersecting spheres are explained above. The natural arrangement for the intersecting spheres is to arrange them in a straight line or linearly. However as the need for volume combined with length to hull depth limitations drive to the use of more than three or four intersecting spheres the length and thus the surface area gets excessive. As underwater resistance is primarily a function of exterior surface area, an excessive length resulting in excessive surface area leads to reduced performance. The means to overcome the limitations of a pressure hull constructed with multiple linearly intersecting spheres has been proposed by Ness (1995) in U.S. Patent No. 5,477,798. Ness (1995) proposed intersecting spheres arranged in a geometric pattern such as a hexagon with a central sphere. Ness (2000) and Ness (2011) presented possible submersible designs taking advantage of this arrangement. Those designs suggested a hexagonal arrangement of six spheres around a central sphere. An alternate arrangement with four central spheres arranged in a square configuration is shown in Fig. 8. As shown in the figure additional spheres can be added at the ends to provide additional buoyant and interior volume to suit the particular requirements of the vehicle being designed.

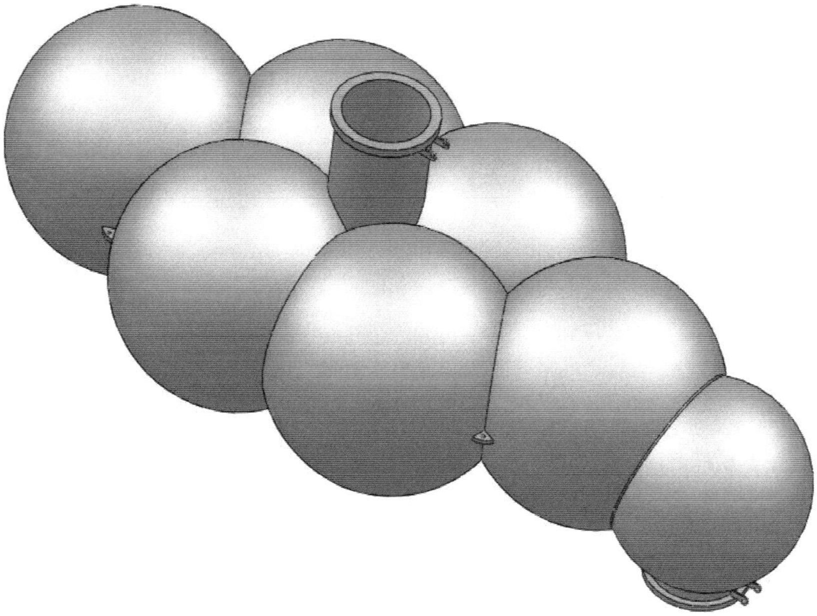

Fig. 8 Geometrically Arranged Intersecting Spheres

In this arrangement the intersections between adjacent spheres would be handled exactly as described above for linearly intersecting spheres. The challenge comes in providing exactly the correct support and rigidity at the end of the reinforcing rings where three come together. The requirement is to ensure that when deforming under external pressure loading the ends of the rings where the three ring come together move vertically exactly the same as the spheres would move if they did not intersect each other. This is easily achieved by adding interior columns or bulkheads at the ends of the rings as shown conceptually in Fig. 9.

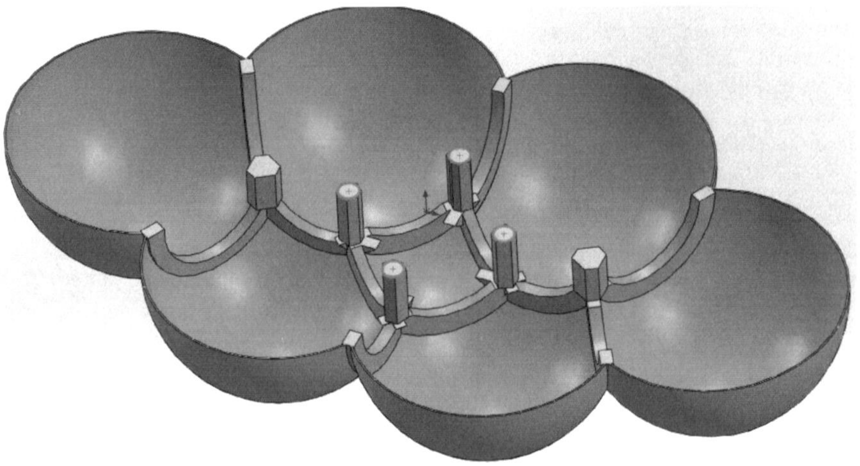

Fig. 9 Reinforcing Ring End Support Columns

The sizing of the columns might seem difficult but in reality is quite simple. The mathematic analysis is provided in Appendix C. In words, a free-body diagram is created sectioning the three intersecting spheres with planes passing thru the tops or vertexes of the three spheres. The vertical force supported by the column at the end of the rings is then the projection of the area of the sectioned spheres on a horizontal plane multiplied by the external pressure. The required deflection in the column is the vertical component of the spherical shell deflection at the location of the column. Knowing the load and the required deflection, the required cross sectional area of the column is known.

The concept of how the intersecting sphere structure can be designed to maintain deformation and thus geometric compatibility has been presented. Theoretically, if the structure is designed as proposed, the buckling pressure of the resulting structure will be the same as the buckling pressure of the component spheres. As with individual spheres, the buckling pressure may or may not be governing, and, as noted above, for practical working depths elastic buckling will most likely not be governing. Because of that, for a given design, it may not be necessary to strictly follow the balanced design of sphere intersection reinforcing rings and ring end supporting columns described. Variations from the balanced design can easily be evaluated and modified to provide stress values with an acceptable safety factor. The discussion above does provide insight into how a structure may need to be modified if analysis shows excess stress in a particular area. Also the resulting structure can be evaluated for elastic buckling to ensure that the elastic buckling pressure is well in excess of pressure that will cause the stress to exceed yield as expected.

Pressure Hull Design Specifications

The pressure hull and reinforcing rings are constructed of HY-100 steel. Pressure hull scantlings are listed in Table 1 below.

Table 1 Pressure Hull Scantlings

Location	Wall Thickness
Forward Spherical Pressure Hulls	3.18 cm (1.25 Inch)
Aft Spherical Pressure Hull	2.54 cm (1.0 Inch)
Forward Bulkhead	3.18 cm (1.25 Inch)
Forward Trunk Passage	3.18 cm (1.25 Inch)
Mid-ship Bulkhead	4.44 cm (1.75 Inch)
Mid-ship Trunk Passage	7.62 cm (3.0 Inch)
Upper Hatch Cylinder Wall	3.18 cm (1.25 Inch)
Hatch Flanges	3.18 cm (1.25 Inch)
Aft Bulkhead	3.18 cm (1.25 Inch)
Aft Trunk Passage	3.18 cm (1.25 Inch)
Aft Sphere Rib	10.16 cm (4.0 Inch)
Mid-ship Center Ribs	12.7 cm (5.0 Inch)
Lower Hatch Cylinder Wall	5.08 cm (2.0 Inch)

Pressure Hull Finite Element Analysis

As is seen in the figures showing the submersible design the use of reinforcing rings and supporting columns described in the PRESSURE HULL STRUCTURAL CONCEPT section above has been modified to better meet the arrangements needs of the vehicle. Recognizing the need to develop a modified pressure hull structural concept with acceptable spherical shell deformation compatibility and acceptable stress levels, structural Finite Element Analysis (FEA) was used to evaluate and modify the pressure hull. That process is on-going in terms of developing structural details to address local stresses and to evaluate the safety margin against shell buckling. The results of the FEA for the current pressure hull evaluation are provided in Appendix D. The analysis was done for external pressure loading of 490 psi (3.38 MPa). Figs. D1, D2, and D3 show the results using shell elements. The general effectiveness of the sphere intersection reinforcing rings and other supporting structure is seen in the uniform stress in the spherical shells. Fig. D4 show stresses in the supporting structure. Fig. D5 is a typical more detailed solid element model that evaluates more fully the high stress areas identified in the shell element models at the central intersections of the reinforcing rings and the central trunk. Fig. D6 shows a higher order solid element detail at the joint between the centerline reinforcing ring and the central trunk. The maximum stresses shown provide a safety factor in excess of 2, but, as the structural details are refined, it highly likely these stresses can be brought down to much closer to the nominal stress in the spherical shells. In that context, there is confidence that, with revised details, the maximum stresses can be brought down to provide a minimum factor of safety of 4 as the ABS requirements incorporate since the highly stressed volumes are actually very small as shown in figure D6..

Carbon Composite pressure Hull Alternative

The intersecting sphere pressure hull is readily adaptable to composite materials in lieu of steel. Modern carbon fiber composites are an obvious choice due to their superior properties and the ease with which they can be fabricated into spherical segments. This ease of producing close tolerance spherical shells would greatly reduce if not eliminate the material cost differential compared to high strength steel which would likely require at least some machining. The composite pressure hull could be laid up on a consumable male mold or spherical composite segments could be joined to reinforcing rings and other supporting structure construction separately. As mentioned above in the discussion about intersecting spheres (Allmendinger 1990) there would potentially be significant advantage in joining composite spherical segments to metallic reinforcing rings that had a higher modulus of elasticity. The application of composites for the pressure hull of a small submersible such as shown here obviously deserves further study.

Conclusion

The conceptual design for a small general purpose submersible has been presented. The intersecting spheres are arranged in a similar flattened shape that bio-mimic the hydrodynamic abilities of the Manta Ray fish. The design presented exploits the inherent advantages and flexibility of a pressure hull constructed with intersecting spheres. It shows the potential for economically constructing an air transportable submersible. The design incorporates exterior placement of the newer high performance battery systems by submersion in ambient seawater thereby reducing risk of fire. As explained it may be possible to significantly increase the advantages of the design by using a composite pressure hull in lieu of a steel pressure hull.

Acknowledgements

The contributions of Barbara Jensen in preparing and reviewing this paper are gratefully acknowledged.

Dedication

This paper is dedicated to the memory of Mr. C. Clifford Ness who after a long illness died on 11-03-2014. For many years, Mr. Ness worked on the design and development of submarines, nuclear and underwater technology. He passed the baton of his learned technology for further evolution; sharing his countless sketches, data, insights, as well critique and encouragement during our ongoing design process. Mr. Ness is gratefully acknowledged; the passion and drive of his guidance and experience will be missed.

Disclaimer

The views and opinions expressed in this paper are strictly those of the authors and do not necessarily reflect those of any organizations with which the authors are associated.

References

Allmendinger, E. Eugene. *Submersible Vehicle Systems Design,* Alexandria, VA: The Society of Naval Architects and Marine Engineers, 1990.

American Bureau of Shipping, *ABS Rules for Building and Classing Underwater Vehicles, Systems and Hyperbaric Facilities 2014*, Houston: American Bureau of Shipping, 2013. http://www.eagle.org/eagleExternalPortalWEB/ShowProperty/BEA%20Repository/Rules&Guides/Current/7_UWVS_2014/UWVSRules (19 May, 2014)

Callister, William D., Jr., and Rethwisch, David G., *Materials Science and Engineering An Introduction*, 8th Ed., Hoboken, NJ: Wiley, 2010.

Garland, Charles, "Design and Fabrication of Deep-Diving Submersible Pressure Hulls" *Transactions, The Society of naval Architects and marine Engineers,* Vol. 76, 1968

Ness, C. Clifford, "High Strength, High Safety Submersible Vessel Resistant to Extreme External Pressures" U.S. Patent No. 5,477,798, 1995.

Ness, C. Clifford, and Simpson, William M., Jr., 2000. "A New Submarine Paradigm" *Naval Engineers Journal,* 112: 4, 2000

Ness, C. Clifford, Ringelberg, John M., and Simpson, William M., Jr., "6,000 Foot Submersible", *Offshore Technology Conference,* Houston, 2011

Philpot, Timothy A., *Mechanics of Materials: An Integrated Learning System,* Hoboken, NJ: Wiley, 2013.

Young, Warren C., *Roark's Formulas for Stress and Strain,* 6th Ed., New York: McGraw-Hill, 1989.

Figure 10, Salvage Version with Forward Multi-Axis Arm,

Extended Feet (Spuds) and Optional Mud Shoes.

Figure 11, Stern View

Appendix A

SPHERICAL AND CYLINDRICAL PRESSURE HULL STRUCTURAL ANALYSIS

In chapter VI of *Submersible Vehicle Systems Design* (Allmendinger 1990), the buoyancy factor for an un-stiffened sphere and an unstiffened cylinder are compared. The buoyancy factor B is defined as the hull weight, divided by the displaced water weight.

For the unstiffened sphere:

$$B = \frac{3}{2}\left(\frac{\rho_s}{\rho_w}\right)\left(\frac{p}{\sigma_{\max}}\right) \qquad \text{(A1)}$$

where

ρ_s = density of shell

ρ_w = density of water

p = hydrostatic pressure

$\sigma_{\max}$ = maximum stress

For the unstiffened cylinder

$$B = 2\left(\frac{\rho_s}{\rho_w}\right)\left(\frac{p}{\sigma_{\max}}\right) \qquad \text{(A2)}$$

This shows that conceptually the spherical pressure hull will have 25% less weight compared to a cylindrical pressure hull with the same displaced volume. The advantage of the un-stiffened sphere increases with stiffeners added to the cylinder. Although not this simple, it does show the basic advantage of the sphere over the cylinder.

For both the spherical and cylindrical pressure hulls both maximum stress and buckling strength or rather factors of safety against both yield stress and buckling failure must be considered. For the externally loaded ideal thin wall sphere with a shell thickness to sphere radius ratio, t/r, of less than 0.1 (Philpot, 2013) the shell stress is:

$$\sigma_s = -\frac{pr}{2t} \qquad \text{(A3)}$$

where

r = shell radius

p = hydrostatic pressure

σ_s = sphere in-plane stress

t = shell thickness

The stress is assumed uniform through the thickness of a thin wall. On the outside of the shell, normal stress that is due to the hydrostatic pressure loading is ignored; the assumption that it will be small compared to the in-plane stress. For a depth of, say 300m (984 feet), the normal stress on the outside of the shell will be equal to the hydrostatic pressure which will be in the order of 3.02 MPa (438 psi). If normal stress is 3.02 MPa and the in-plane stress is 63.61 MPa (9,226 psi) for example (r = 1.07 m, t = 2.54 cm), the resulting absolute shear stress on the outer surface would be:

$$\tau_{absmax} = \frac{1}{2}\left(\sigma_s - \sigma_{normal}\right) \qquad \text{(A4)}$$

$$\tau_{absmax} = \frac{1}{2}\left[-\frac{pr}{2t} + p\right] = -\frac{pr}{4t} + \frac{p}{2} \qquad \text{(A5)}$$

where

τ_{absmax} = absolute maximum shear stress

For example if r = 1.07 m (42.1 in), t = 2.54 cm (1in), and P = $3.02\text{x}10^6$ N/m^2 (438 psi) and the normal stress is included:

$$\tau_{absmax} = -\frac{3.02\times10^6 \, N/m^2 \times 1.07m}{4\times0.0254m} + \frac{3.02\times10^6 \, N/m^2}{2}$$

$$\tau_{absmax} = -30.30MPa \quad (-4{,}394psi)$$

To compute the von Mises or effective stress in the outer surface of the shell, we note the stress in the plane of the shell is equal in all directions. We can take the stress in any two perpendicular directions in the plane of the shell as σ_s, i.e., $\sigma_{sx} = \sigma_{sy} = \sigma_s$. The von Mises or effective stress is then:

$$\sigma_e = \frac{\sqrt{2}}{2}\left[\left(\sigma_{sx} - \sigma_{sy}\right)^2 + \left(\sigma_{sy} - \sigma_n\right)^2 + \left(\sigma_n - \sigma_{sx}\right)^2\right]^{1/2} \qquad \text{(A6)}$$

where

$\sigma_{sx} = \sigma_{sy}$ = in-plane stress

σ_n = stress normal to shell surface

$$\sigma_e = \frac{\sqrt{2}}{2}\left[(0)^2 + \left(-60.59\times10^6\right)^2 + \left(+60.59\times10^6\right)^2\right]^{1/2}$$

$$\sigma_e = 60.59MPa \quad (8{,}787psi)$$

If the normal stress is not included:

$$\tau_{abs\max} = -\frac{3.02\times10^{6}\,N/_{m^{2}}\times1.07m}{4\times0.0254m}$$

$$\tau_{abs\max} = 31.81MPa \quad (-4{,}613psi)$$

$$\sigma_{e} = 63.61MPa \quad (9{,}226psi)$$

For this representative relatively shallow depth, the error in shell stress is in error by 5%, and is conservative. Thus, in these discussions, assume the shell is "thin" and the stress in the direction normal to the shell surface is "0". These assumptions, of course, need reconsideration if the shell is "thick" or the depth is great.

With these assumptions, the shell has plane stress and thus plane strain. For the spherical shell under compression, the in-plane stress, σ_s, is equal in all directions and as stated above, the principal stresses in any two perpendicular in-plane directions are σ_s. The in-plane shear stress is 0. Recognizing the contribution of Poisson's ratio, the in-plane strain is then:

$$\varepsilon_{s} = \frac{1}{E}\left(\frac{pr}{2t} - \upsilon\frac{pr}{2t}\right) = \frac{pr}{2tE}(1-\upsilon) \qquad \text{(A7)}$$

where

$\varepsilon_{s} = \text{in-plane strain}$

$E = \text{modulus of elasticity}$

$\upsilon = \text{Poisson's ratio}$

In contrast to the unstiffened sphere, the unstiffened cylinder has different longitudinal and hoop stresses. As is easily seen from a simple sectioning of an unstiffened thin walled cylinder, the longitudinal and hoop in-plane stresses are:

$$\sigma_{long} = -\frac{pr}{2t} \qquad \text{(A8)}$$

$$\sigma_{hoop} = -\frac{pr}{t} \qquad \text{(A9)}$$

where

$r = cylinder \text{ radius}$

$p = \text{hydrostatic pressure}$

$\sigma_{long} = \text{longitudinal stress}$

$\sigma_{\text{hoop}} = \text{hoop stress}$

$t = \text{cylinder shell thicknes}$

Neglecting the normal stress as discussed above, the absolute maximum shear stress and the maximum effective or von Mises stresses for the cylinder are:

$$\tau_{abs\max} = \frac{1}{2}\left(\sigma_{hoop} - \sigma_{radial}\right) \qquad \text{(A10)}$$

$$\tau_{abs\max} = \frac{1}{2}\left[-\frac{pr}{t} + 0\right] = -\frac{pr}{2t} \qquad \text{(A11)}$$

$$\sigma_{e} = \sigma_{hoop} = -\frac{pr}{t} \qquad \text{(A12)}$$

Following the example above except for an unstiffened cylinder this time if r = 42 in., t = 1in., and P = 450 psi, the stresses are:

$$\tau_{abs\max} = -\frac{450 \,{}^{lb}\!/_{in^2} \times 42in}{2 \times 1in}$$

$$\tau_{abs\max} = -9{,}450\,psi$$

$$\sigma_e = -\frac{450 \,{}^{lb}\!/_{in^2} \times 42in}{1in}$$

$$\sigma_e = -18{,}900\,psi$$

The advantage of spherical shell is thus seen in terms of basic shell stress magnitudes. Another important factor is the complications that arise when a hemispherical end dome is used to close the ends of a cylinder. For compatibility of the shell, the hoop directed strain in the cylinder and the strain in the sphere should be equal. Unfortunately, that is not a natural result. The spherical shell in-plane strain is given in equation (A7). The cylinder hoop strain is:

$$\varepsilon_{hoop} = \frac{1}{E}\left(\sigma_{hoop} - \nu\sigma_{long}\right) \qquad \text{(A13)}$$

$$\varepsilon_{hoop} = \frac{1}{E}\left(-\frac{pr}{t} + \nu\frac{pr}{2t}\right) = \frac{pr}{2tE}(2-\nu) \qquad \text{(A14)}$$

Comparing equations (A7) and (A14), and recognizing that the radius and the thickness must the equal where the cylinder and the hemispherical head meet, the ratio of strains is: $\frac{\varepsilon_{cylinder}}{\varepsilon_{sphere}} = \frac{\frac{pr}{2tE}(2-\nu)}{\frac{pr}{2tE}(1-\nu)} = \frac{(2-\nu)}{(1-\nu)}$ (A15)

For steel or aluminum, $\nu \approx 0.3$, then:

$$\frac{\varepsilon_{cylinder}}{\varepsilon_{sphere}} = \frac{(2-0.3)}{(1-0.3)} = \frac{(1.7)}{(0.7)} = 2.43 \qquad \text{(A16)}$$

This shows the incompatibility of deformations at the hemispherical head to cylinder union and this illustrates one of the disadvantages of cylindrical pressure hull with hemispherical dome end closures. The lack of deformation compatibility affects the buckling strength of the shell; pointing out the need to carefully consider the buckling strength of any shell loaded with external pressure. There are multiple buckling modes for both the sphere and the cylinder pressure hull. This is complicated further by the fact that fabrication tolerances must be considered, thereby avoiding deviations from ideal shell geometry. For practical purposes, algorithms have been developed for allowable design pressures that take into consideration allowable fabrication tolerances that depart from perfect geometry. Allmendinger, 1990, provides a discussion of buckling strength of spherical shells based on the local variation of shell radius due to fabrication tolerances. It also includes a discussion of the varied buckling modes for stiffened cylinders. Allowable loads for both spherical and cylindrical pressure hulls are in ABS Rules for Building and Classing Underwater Vehicles, Systems and Hyperbaric Facilities 2014 (American Bureau of Shipping, 2013). ABS (American Bureau of Shipping, 2013) specifies permitted geometric tolerances and maximum allowable working pressure.

For better understanding of the ABS criteria it is helpful to correlate the ABS criteria with the discussion in Allmendinger (1990) about the buckling strength of spherical shells based on both theoretical buckling stress and buckling stress determined experimentally. Allmendinger (1990) gives the theoretical elastic buckling pressure, $p_{elastic}$, as:

$$p_{elastic} = 2E\left(\frac{t}{r}\right)^2\left[3\left(1-\nu^2\right)\right]^{-1/2} \qquad \text{(A17)}$$

For Poisson's ratio, ν, of 0.3 this reduces to:

$$p_{elastic} = 1.21E\left(\frac{t}{r}\right)^2 \quad \text{(A18)}$$

Allmendinger (1990) reports that tests at David Taylor Naval Research and Development Center indicate a better expression for the elastic buckling pressure for a near perfect sphere with $\nu = 0.3$ is:

$$p_{elastic} = 0.84E\left(\frac{t}{r}\right)^2 \quad \text{(A19)}$$

Allmendinger (1990) further gives the following for the buckling pressure when the stress exceeds yield prior to buckling failure for what is termed inelastic or elastic-plastic buckling:

$$p_{elastic} = 0.84\sqrt{E_{sec}E_{tan}}\left(\frac{t}{r_o}\right)^2 \quad \text{(A20)}$$

where

E_{sec} = secant modulus in compression

E_{tan} = tangent modulus in compression

r_o = outer radius

In this expression, the secant modulus is the initial slope of the stress-strain curve and the tangent modulus is the slope of the stress-strain curve in the plastic region (Callister, 2010).

Allmendinger (1990) further describes a measure of fabrication tolerance as a maximum variation in radius measured over a local arc length as a function of thickness and local radius. The local arc length for which the radius is to be determined is:

$$L_c = 2.42\sqrt{r_l t_a} \quad \text{(A21)}$$

where

L_c = critical arc length

r_l = radius over arc length

t_a = average thickness over arc

$\nu = 0.3$

ABS (American Bureau of Shipping, 2013) specifies an arc length in the form of a graph with values of Lc slightly larger than given by this equation. Allmendinger suggests a reduction in the allowable stress based on the ratio of the local radius versus the design radius. In reality, this is not significant when using the ABS criteria as ABS specifies a maximum variation of the local radius of 0.5%. Therefore, this is a fabrication tolerance to achieve, not a significant stress limiting factor.

Looking now to the ABS criteria (American Bureau of Shipping, 2013), the primary reference is the comparison of the perfect sphere elastic buckling pressure with the pressure that will cause yielding. Using ABS notation:

$$P_{ys} = \frac{2\sigma_y t}{r_o} \quad \text{(A22)}$$

$$P_{es} = \frac{2E}{\sqrt{3(1-\nu^2)}}\left(\frac{t}{r_o}\right) \quad \text{(A23)}$$

where

P_{ys} = pressure that will cause yielding

P_{es} = elastic buckling stress

r_o = outer radius

σ_y = yield stress

ABS applies a safety factor dependent on whether the computed elastic buckling stress is less than or greater than the material yield stress. If the computed buckling stress is less than yield, the allowable working pressure is given by:

$$P_a = 0.2124\eta \frac{P_{es}}{P_{ys}} \qquad \text{(A24)}$$

where

$\eta =$ usage factor

ABS specifies the usage factor, η, to be 0.67. While the origin is not known, it is assumed to be related to the manufacturing process factor discussed by Allmendinger (1990), for HY-80 spherical shell tests. When the usage factor is applied, this results in a safety factor of 7 against the elastic buckling pressure of an ideal sphere. When the computed elastic buckling stress exceeds yield, ABS specifies the allowable working pressure as:

$$P_a = 0.73910\eta P_{ys}\left[1+\left(\frac{P_{ys}}{(0.3P_{es})}\right)^2\right]^{-1/2} \qquad \text{(A25)}$$

In this case, the factor of safety against both buckling and yield will depend on the specifics of the dimensions and the material. As an example, if the material is HY-100 ($\sigma_y = 100\,ksi, 690\,\text{MPa}$) and $r_o = 1.07m\,(42.1in)$, the resulting factor of safety versus the operating depth (t varies with operating depth) is:

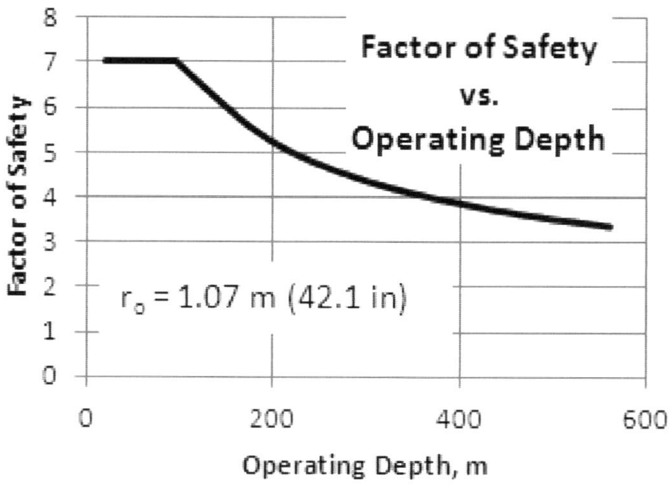

Fig. A1 ABS Factor of Safety vs. Operating Depth

For this example, with $r_o = 1.07m\,(42.1in)$, the computed perfect sphere elastic buckling stress is less than the yield stress for operating depths of less than 100 m ($t_{\text{required}} \approx 0.6$ cm, 0.25 in), with a resulting factor of safety against a perfect sphere elastic buckling of 7. For an operating depth of greater than 100 m, the computed perfect sphere elastic buckling stress exceeds the yield stress, and the factor of safety shown is based on yield stress rather than the elastic buckling stress.

Appendix B

LINEARLY INTERSECTING SPHERE DESIGN AND STRUCTURAL ANALYSIS

When intersecting spherical shells are used as a pressure hulls it is necessary to provide reinforcing rings at the shell intersections to exactly replace the support and stiffness of the shell material cut away to join the spheres. Allmendinger (1990) provides the computation tools needed to size the reinforcing rings at the shell intersections. A conceptual sketch of a cross section thru two intersecting shells and the associated reinforcing ring is shown in Fig. B1.

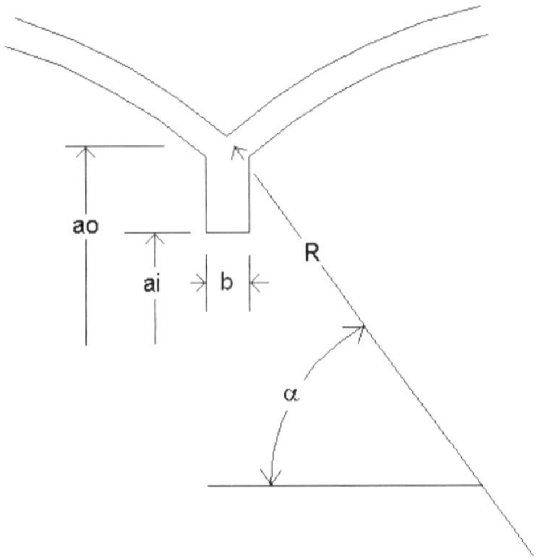

Fig. B1 Symbol Definitions

Using equation A7 above for the in plane strain the change in circumference, δ_c, is:

$$\delta_c = 2\pi r\varepsilon_s = 2\pi r\frac{pr}{2tE}(1-\nu) \quad \text{(B1)}$$

The change in the radius, δ_r, is:

$$\delta_r = \frac{\delta_c}{2\pi} = \frac{pr^2}{2tE}(1-\nu) \quad \text{(B2)}$$

Using the sketch in the Fig. B1 as the orientation reference, the vertical deflection or radial deflection, δ_{radial}, of the shell intersection (perpendicular to the axis of revolution) is:

$$\delta_{radial} = \delta_r \sin\alpha = \frac{pr^2 \sin\alpha}{2tE}(1-\nu) \quad \text{(B3)}$$

where

$r =$ mean radius of spherical shell

$\alpha =$ angle from sphere center to intersection

Assuming the reinforcing rings are the same material as the shells, the deflection of the reinforcing ring due to the load from the intersecting shells is (Allmendinger 1990):

$$\delta_{radial} = \frac{pr^2 \sin\alpha\cos\alpha}{Eb}\left(\frac{1+C^2}{1-C^2} - \nu\right) \quad \text{(B4)}$$

where

$b =$ width of reinforcing ring

$\mathrm{C} = \mathrm{ai}/\mathrm{ao}$

$ai =$ ring inner radius

ao = ring outer radius

In order for the spherical shells to remain spherical while being loaded and deforming, the radial deflection of the ring must equal the radial deflection that the spherical intersection would experience if the spheres were not joined. This is achieved by setting the two deflections equal to each other, and selecting reinforcing ring dimensions that satisfy the equality. The governing equation then becomes:

$$\frac{pr^2 \sin\alpha}{2tE}(1-\nu) = \frac{pr^2 \sin\alpha \cos\alpha}{Eb}\left(\frac{1+C^2}{1-C^2} - \nu\right) \quad \text{(B5)}$$

This reduces to:

$$\frac{b}{2t}(1-\nu) = \cos\alpha\left(\frac{1+C^2}{1-C^2} - \nu\right) \quad \text{(B6)}$$

Following Allmendinger (1990) and using the expression for maximum stress in an externally loaded ring from Young (1989), the stress at the inner radius of the ring is:

$$\sigma_{ring\max} = \frac{2pr\cos\alpha}{b(1-C^2)} \quad \text{(B7)}$$

where

$\sigma_{ring\max}$ = max ring stress, inner radius

Once again, following Allmendinger (1990) and defining the ratio of the ring max stress to the spherical shell stress as K:

$$K \equiv \frac{\sigma_{ring\max}}{\sigma_{shell}} = \frac{2pr\cos\alpha}{b(1-C^2)}\frac{2t}{pr} = \frac{4t\cos\alpha}{b(1-C^2)} \quad \text{(B8)}$$

The expression can then be inverted to express the width of the reinforcing ring, *b*, as a function the ratio of the stresses:

$$b = \frac{4t\cos\alpha}{K(1-C^2)} \quad \text{(B9)}$$

Equation (B9) then allows selection of the reinforcing ring dimensions to provide the needed rigidity at the intersection, while limiting the maximum stress on the inner radius of the ring.

The development above assumed the shell and the reinforcing ring constructed of the same material. Allmendinger (1990) goes on to show that, by using a reinforcing ring with a higher modulus of elasticity, significant weight savings are possible. That is beyond the scope of this work except to say the possibility of composite shells and metal reinforcing rings with their significantly higher modulus of elasticity would likely have great potential.

Appendix C

REINFORCING RING END SUPPORT COLUMNS

The columns supporting the ends of the reinforcing rings where they meet and terminate must be sized so the deflection (contraction) of the columns equal the vertical deflection in the spherical shells at the intersections. As given in equation B3 the vertical deflection of the spherical shell is:

$$\delta_{radial} = \frac{pr^2 \sin\alpha}{2tE}(1-\nu) \quad \text{(C1)}$$

The corresponding deflection of the column is given by:

$$\delta_{Column} = \frac{\sigma_{Column}}{E}h = \frac{\sigma_{Column}}{E}r\sin\alpha \quad \text{(C2)}$$

where

h = height of column from mid-plane to the ring intersections

Setting the vertical component of the shell deflection equal to the column deflection, assuming the reinforcing rings and the columns are of the same material:

$$\frac{\sigma_{Column}}{E}r\sin\alpha = \frac{pr^2 \sin\alpha}{2tE}(1-\nu) \quad \text{(C3)}$$

$$\sigma_{Column} = \frac{pr}{2t}(1-\nu) \qquad \text{(C4)}$$

Given the relationship for the stress in the shell, σ_s :

$$\sigma_s = \frac{pr}{2t} \qquad \text{(C5)}$$

Then:

$$\sigma_{Column} = \sigma_s(1-\nu) \qquad \text{(C6)}$$

With the desired column stress determined, it is only necessary to compute the load on the columns. To do that a free-body diagram is created by sectioning the three intersecting spheres with planes passing thru the tops or vertexes of the three spheres. An example of a resulting segment is shown in Fig. C1. The vertical component of the surface pressure load on this segment is the load on the column.

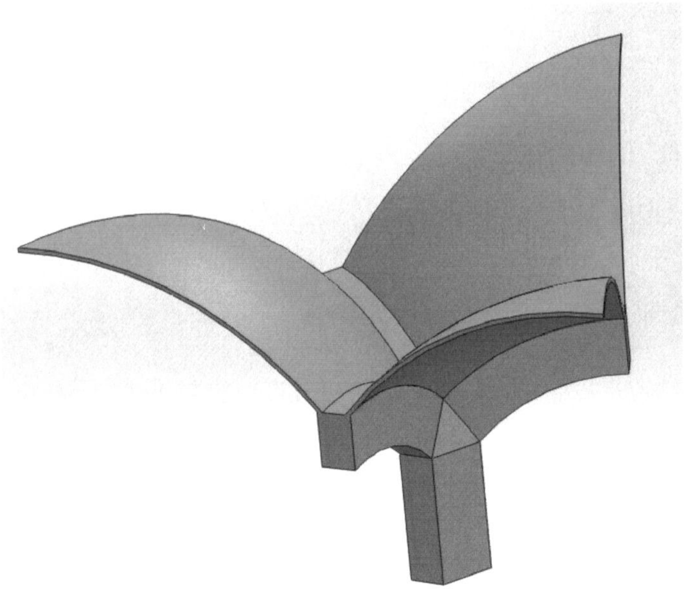

Fig. C1 Column Support Segment

The vertical force supported by the column is the projected area of this segment on a horizontal plane multiplied by the external pressure. The loads on the edges of the segment are ignored when computing the vertical load on the column as all the edges are vertical and there is no shear force on the edges. Because the edges are vertical and there is only normal force on the edges, there is no vertical load component on the edges. Knowing the column load and the stress needed for deformation compatibility the required column cross sectional area is computed.

APPENDIX D
Structural Finite Element Analysis

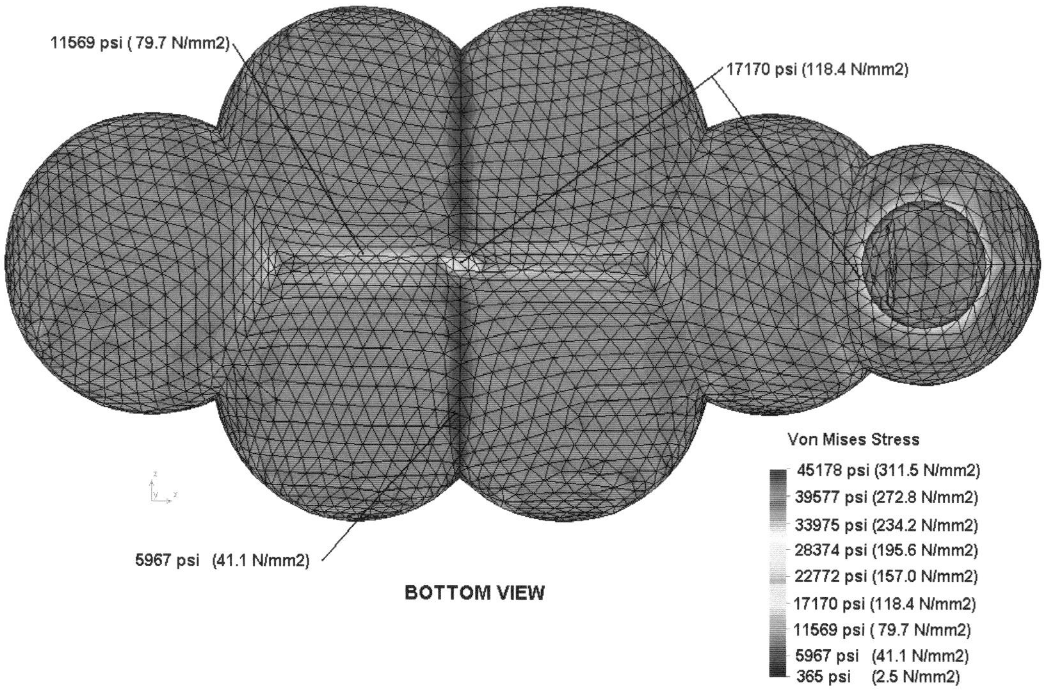

Fig. D1 Pressure Hull Stress – Shell Elements - Bottom View

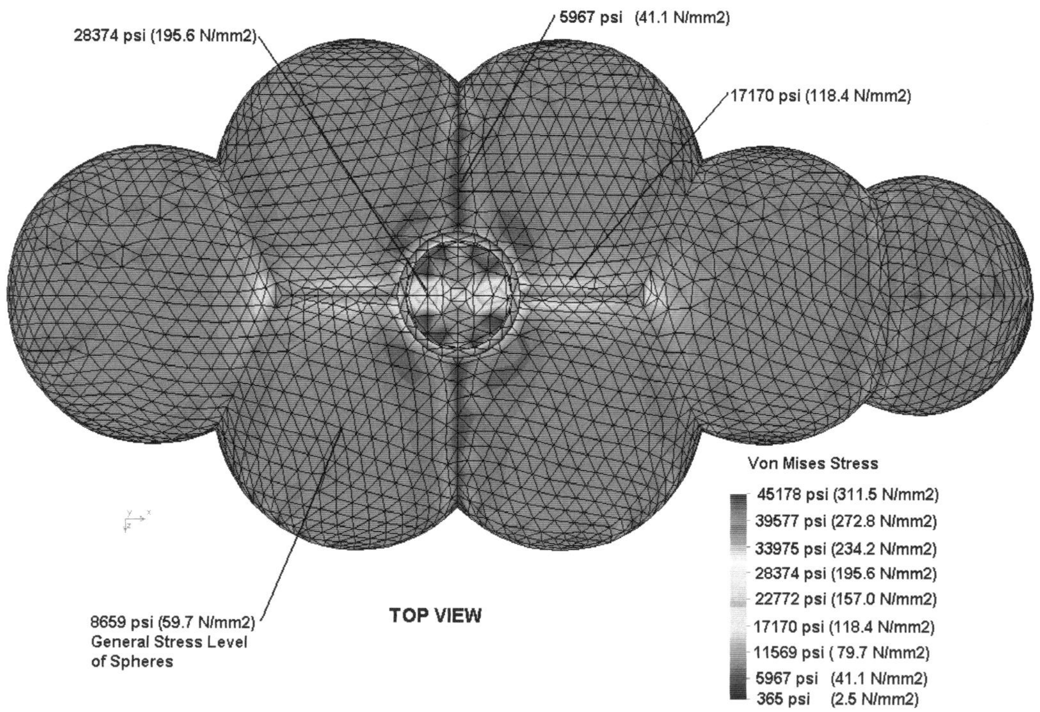

Figure D2, Pressure Hull Stress – Shell Elements - Top View

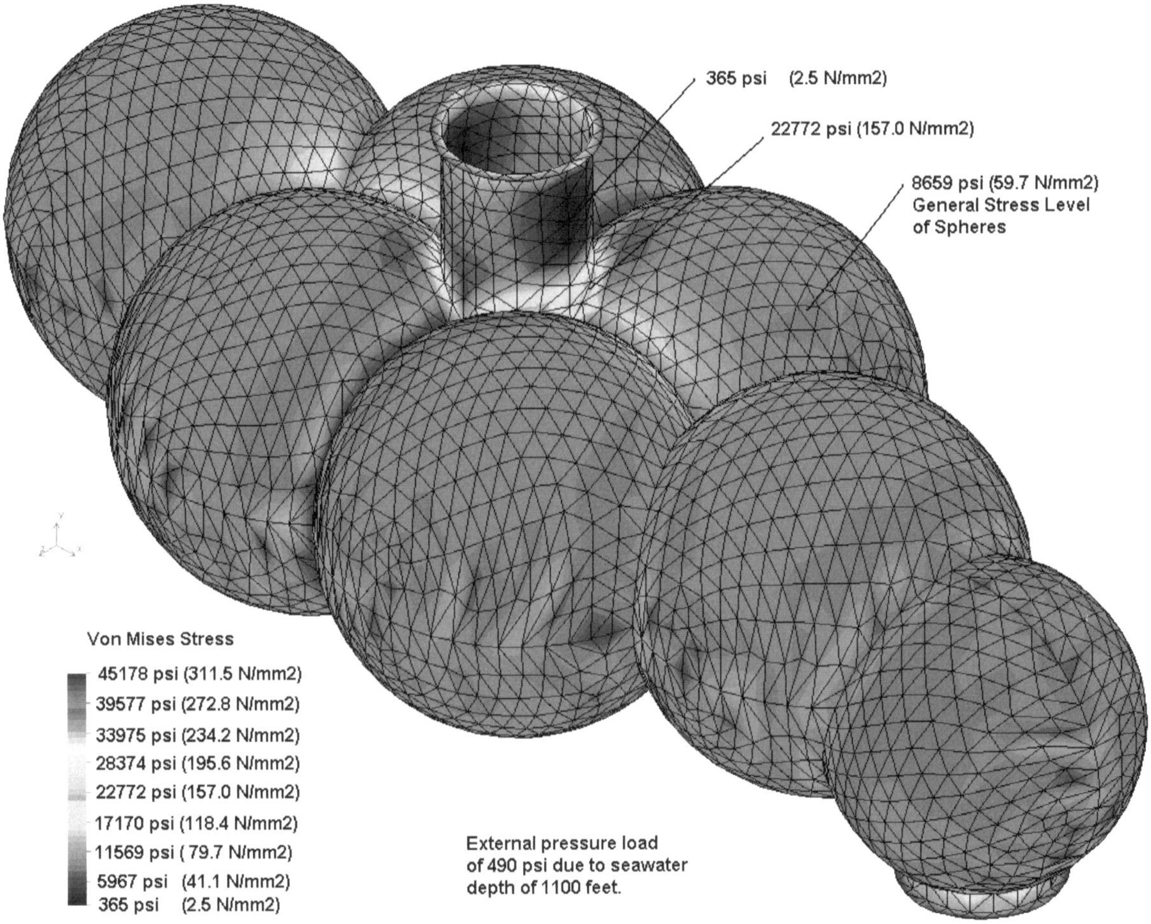

Figure D3, Steel Pressure Hull Stress – Shell Elements - Perspective View

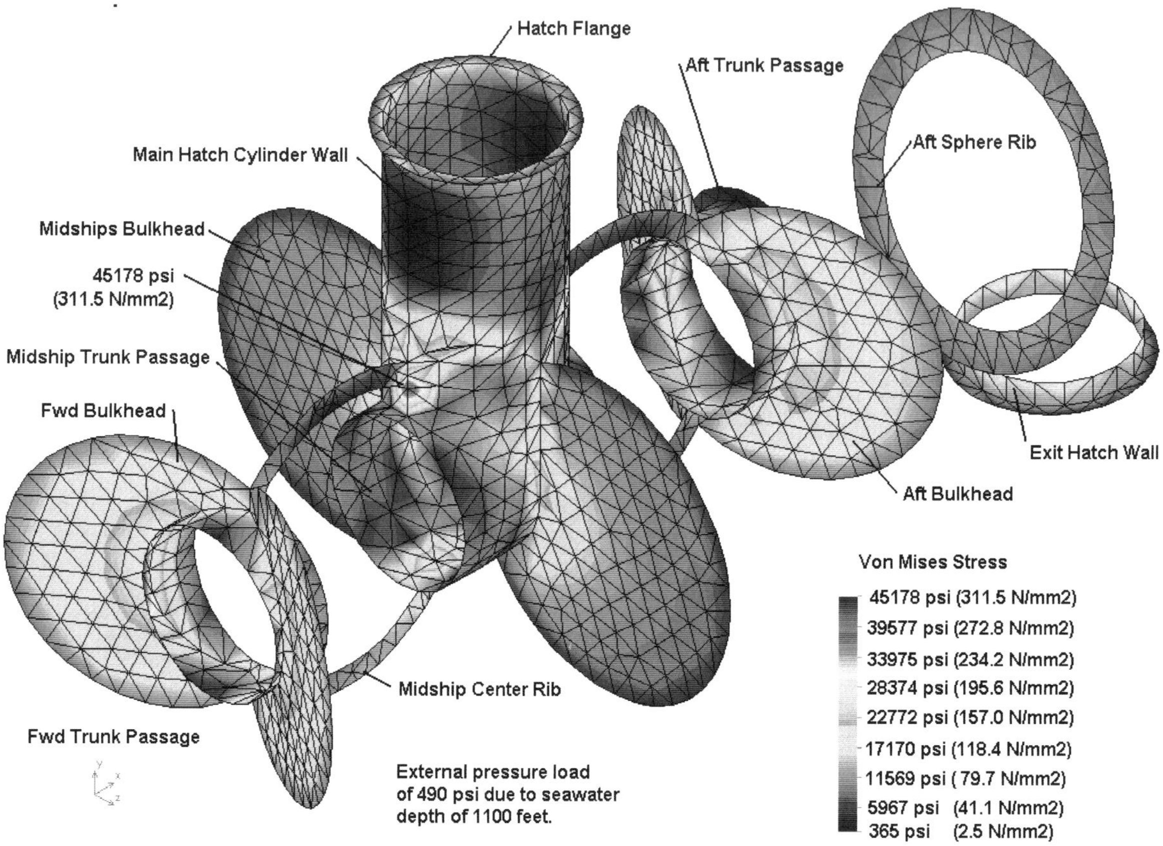

Figure D4, Supporting Structure Stress (Steel Ribs and Bulkheads) – Shell Elements

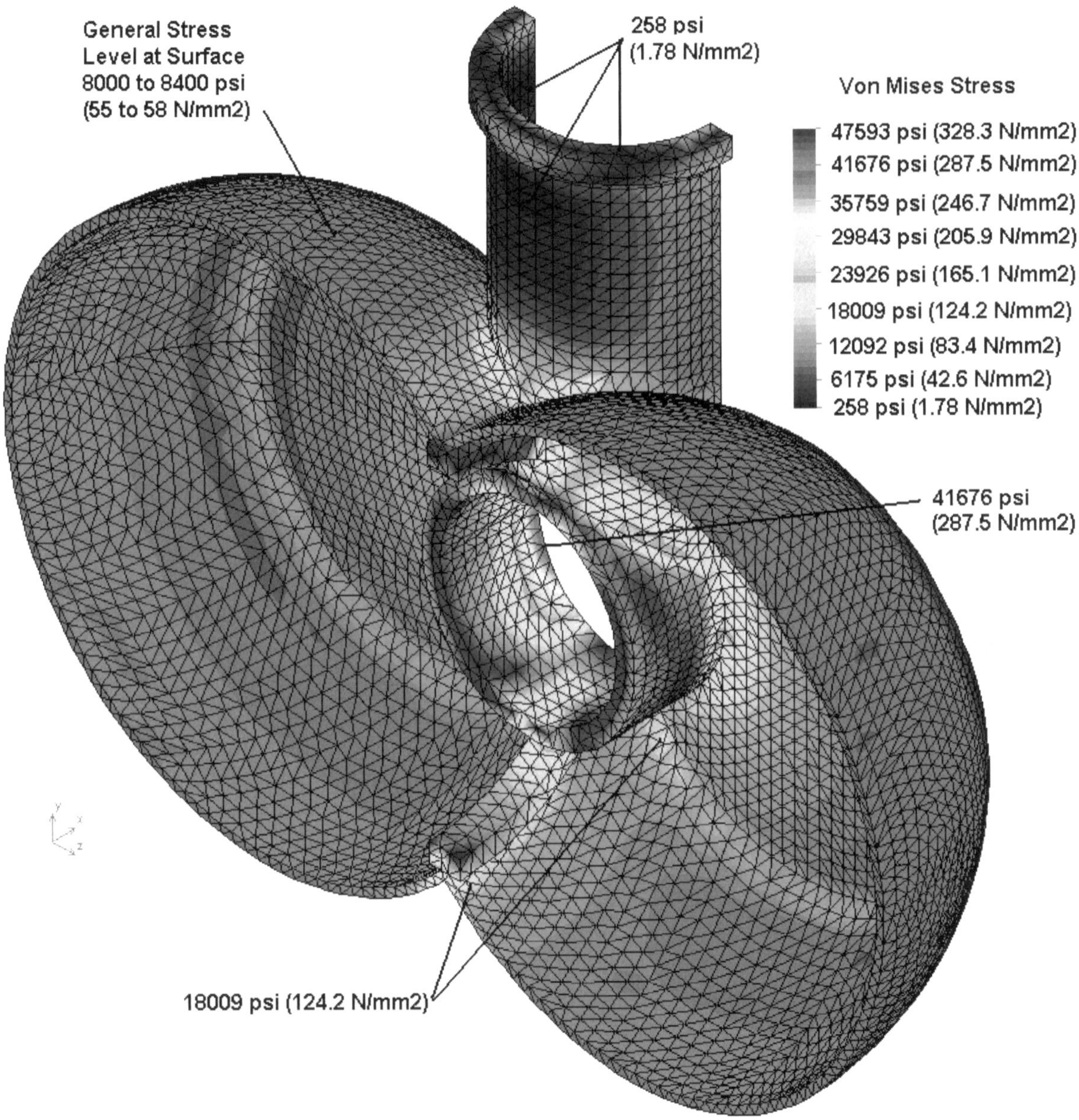

Figure D5, Steel Pressure Hull Stress, Center Bulkhead Detail – First Order Solid Elements

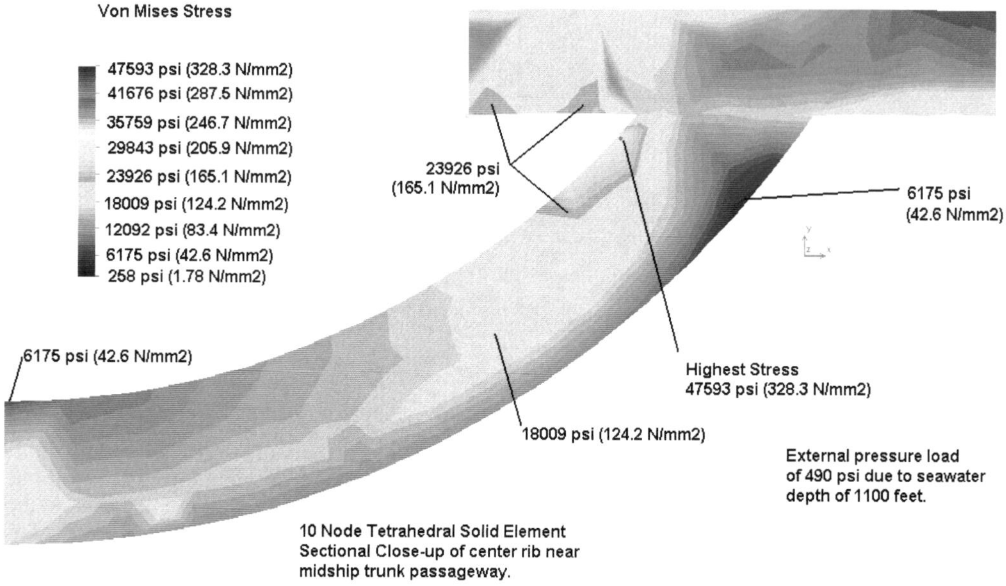

Figure D6, Steel Reinforcing Ring at Centerline Trunk Detail with Higher Order Solid Elements (Note the extremely small High Stress Area)

Author Resumes:

Fred C. Jensen, MSME, PE

Co-founder of Patriot Engineering Company with Barbara Jensen; located in Chagrin Falls, Ohio 44023. He is a graduate of Florida State University with a BS in Engineering Science and a Master's in Mechanical Engineering from Cleveland State University. At Patriot he completed mechanical engineering, design and structural Finite Element Analysis (FEA) of the housing / packaging for the Gamma Ray Spectrometer and Neutron Spectrometer for the Messenger Spacecraft to Mercury. He was the lead mechanical engineer for the (SAX) Satellite Astronomy X-ray telescope that identified numerous gamma-ray bursts from extra-galactic objects. Patriot Engineering has completed over 1200 projects in aerospace, marine, medical, mining and special machinery for industry. He is a lifelong sailor having commercial fished off the East coast, ICYRA championship sailing including the Kennedy Cup at the US Naval Academy and the nationals while attending FSU. Fred and his wife sailed their current vessel for the past twenty-six years on the Great Lakes.

William M. Simpson, Jr., Ph.D., PE

Associate Professor of Naval Architecture and Marine Engineering at the U.S. Coast Guard Academy, and is the president of East Haddam Naval Architecture, Inc. He is a graduate of the Coast Guard Academy, has a M.S. in naval architecture and marine engineering from M.I.T., and has a Ph.D. in aerospace engineering from the University of Maryland. He retired from active duty in the Coast Guard as a Captain in 1992 after 27 years of service. While on active duty, he served in various engineering and technical assignments including Commanding Officer of the Coast Guard Research and Development Center in Groton, CT. He served as Coast Guard Project Manager for Patrol Boat Acquisition and for Icebreaker Acquisition and as the Resident Inspector for the initial Island Class Patrol Boat construction. He is a member of ASNE and SNAME.

Capt Jack Ringelberg, MS, PE

Former President & CEO of JMS Naval Architects and Salvage Engineers, Mystic, CT 06355. President of Jamestown Marine Services, Jamestown RI. President of J MaRine Consultants, Panama City FL; State University of New York Maritime College, BE Marine Engineering. Webb Institute of Naval Architecture, BS Marine Eng, MS Naval Architecture. Captain, Engineering and Diving Officer, US Navy retired. Commanding officer, Navy Experimental Diving Unit. Chief Engineer, Aircraft Carrier, John F. Kennedy. Fellow and former officer of The Society of Naval Architects and Marine Engineers.

Development of DeepFlight Dragon

Graham Hawkes, Founder and Chief Technology Officer, DeepFlight
Dr. Kenneth Laws, Chief Scientist, DeepFlight

Introduction:

DeepFlight has developed a new submersible type, DeepFlight Dragon, to answer the market need for private, two-person undersea exploration craft that are relatively safe and simple to operate.

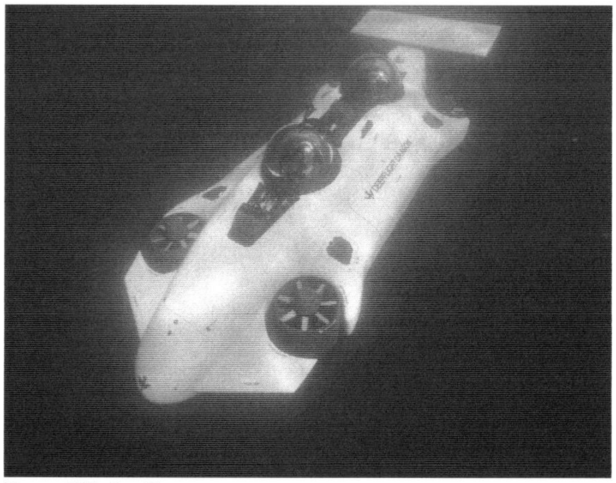

DeepFlight Dragon.

The Dragon continues to employ fixed (fail safe) positive buoyancy, with automatic surfacing that is unique to all DeepFlight craft. For ease of operation, the skill required for underwater flight or operation of traditional variable buoyancy systems was deemed a barrier to customer use, and the quadcopter was thought to be the optimum piloting solution for this newest craft.

Rationalization:

The innovative solution adopted therefore uses the same principles as electric quadcopters, whose challenges in manual flight are effectively solved by the embedded complexity of electronic sensors and automatic stabilization. This solution simplifies the man-machine interface by removing skill needed to pilot the sub. Essentially the concept is that the craft is flown by embedded smart computers and sensors, with pilot interface reduced to simple position or motion instructions.

The technology of deeply embedded control intelligence to command a machine is a powerful new way to create easy-to-use future vehicles. As a practical matter, such leap-ahead concepts are enabled by the convergence of enabling component technologies, such as cost-effective sensors, micro-controllers, and software, all able to interface with great fidelity and reliability with direct drive electric propulsion and powerful secondary batteries.

In the initial design phase, the Company anticipated how such leap-ahead technologies demonstrated in airborne craft could translate to undersea applications. Firstly, while forces would be similar in air and sea, velocity under sea is roughly $1/33^{rd}$ of that for airborne vehicles, drag forces would be similar in air and sea. Since reserve positive buoyancy for fail-safe operation need only be 4% of vehicle weight, power levels for maintaining vertical position would be similarly reduced. Thus the energy demand that limits practical duration of electric air craft would be solved subsea.

Design concept study results of Prototype Dragon:
With the relatively low speed and natural stability possible in the prototype Dragon (enabled by simple separation of center of buoyancy (CB and center of gravity (CG), it was hoped that this new craft could simplify airborne, complex control issues to the point of eliminating the necessity of automatic control stabilization. The rationale was to avoid any safety issues or reliance on automatic control systems or embedded intelligence going rogue. This effort was later shown not to be possible, necessitating and subsequent solutions for safety.

This company has some familiarity with developing new types of manned craft functioning outside of the industrial/regulatory safety net where traditional safety guidelines are design-based. For this new generation of craft, the company has needed to use more modern and adaptable "outcome-based" safety principles; in particular, a powerful tool, Failure Mode Effects Analysis (FMEA). Hence, the safety aspects of introducing embedded intelligence for stability control would be subject to FMEA.

Test setup for Bollard pull tests of prototype thruster.

Development:
The Basic Viability of this new craft depended on a power study. The Dragon project estimated energy consumption on the key assumption that the (proven) level of fixed positive buoyancy (4% of displacement) determined the required continuous hover (Bollard) vertical thrust and from there, the minimum submergence energy consumption based on achievable thruster efficiency. Accounting for additional loads for maneouvering and instrumentation, the conclusion was that the best results achieved so far for lithium Iron Phosphate

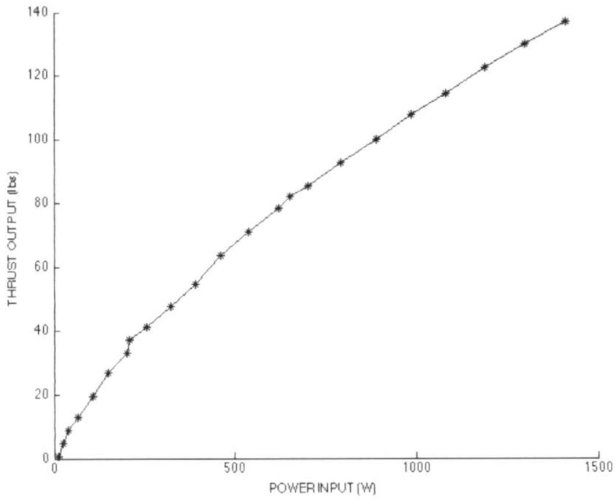

Thrust data obtained using Bollard test results for prototype thruster demonstrating better than 100 lbs per 900 W performance.

LiFePO4 batteries and propulsion efficiency fell about 30% short that which would be needed for a practical Dragon concept. However, that was considered to be within next stage developmental goals, primarily by improving thruster Froude efficiency and reducing mechanical drag losses. The goal was to achieve 100lbs Bollard pull for 1000 watts battery drain.

Thus the development risk for the concept boiled down to developing and achieving 100lbs Bollard thrust for 1000 Watts input for Dragon's vertical thruster. To eliminate risk early, the Company designed a prototype vertical thruster which was successfully tested December 2013. After achieving the design goals for battery drain at hover, the company green-lighted the Dragon prototype concept. The next issue (which would not be resolved until full scale testing) was the level of automatic stabilization needed (if any). As noted earlier, the Principle Designer's hope was that the craft would prove itself manually flyable.

First Sea Trials:

The prototype Dragon was first tested in [add month/year] for safe life support and other basic functions at company facility in Pt. Richmond California. Having passed this first round of tests, the Company moved the craft to its test site in Lake Tahoe, CA for open water trials. Here the challenges of pioneering a new type of submersible craft surfaced. Hawkes, as test pilot, found the craft easily controlled for surface taxi and dive to depth and safe return. However, once submerged, horizontal control was difficult.

DeepFlight team members Jay Tustin and Randal Fletcher prepare the Dragon for diving during initial open water trials at Lake Tahoe.

Several days were devoted to trim and other changes and, while the Dragon routinely proved to be robust and easily (manually) controlled on the surface and for descent and ascent, it continued to balk at being maneuvered in a horizontal plane. At times, the pilots, were able to complete a desired mission, but clearly not reliably.

The most significant problem was that in trying to move forwards or backwards with horizontal thrust, any induced pitch change caused the relatively large hover thrust to tilt and develop an unwanted horizontal vector. Usually the nose pitched down slightly on forwards vertical thrust, which induced a non-intuitive reverse vector thrust (opposite from helicopter or air drone), with the result that the Dragon seemed to stubbornly refuse to go where directed.

The solution would be to simply enable auto pitch (the sensors and circuits had already been included in the Dragon prototype design, but were not active). However, we first had to solve any critical safety effects indicated by worst case FMEA of relying on electronic stabilization.

The basic safety approach adopted was enabled by the fixed positive buoyancy design approach described previously. Because of this fixed positive buoyancy, failure mode effects of the attitude stabilization system can be mitigated by simply turning off the entire thrust sytem. Further, is was determined that limiting the stability control range by the pilots' vertical thrust setting provides an addition method to eliminate unwanted control by the pilot simply reducing vertical thrust setting. Thus there are two pathways to enable the pilot to disable stabilization and prevent any dangerous failure mode effects. Provided the Dragon is easily and safely surfaced and taxied back to base under manual control, any risky failure modes from pitch stabilization can be neutralized at any time at any depth.

The next few days were primarily devoted to demonstrating that the Dragon could be safely and easily recovered from any automatic control failure. In reality, positive buoyancy itself provides a basic safety net since any unwanted active failure mode of power or propulsion or control, as well as any auto-pilot, is easily ended by powering down and relying on natural buoyancy to surface the craft. As a final operating note, our internal safety guidelines for positively buoyant craft will warn against operation under any overhead hazard.

Stabilization:

With failure modes of automatic pitch stabilization solved by FMEA, Laws spent the next few days working to enable and tune Dragon's pitch stabilization. Stabilization control software was developed initially using simulation methods to estimate coefficients for a Proportional, Integral, Differential (PID) control algorithm. The Dragon employs a thrust control computer with on-board attitude sensors to independently control its two horizontal and four vertical thrusters. Once acceptable (PID) tuning was obtained in simulation, the algorithm was implemented on the vehicle's thrust control computer and tested in further trials in Lake Tahoe. Final tuning of

DeepFlight Dragon being launched at Port de Fontvieille, Monaco. (Our thanks to the port authorities for their support of operations.)

algorithm coefficients was accomplished updating values from within the cockpit via laptop computer.

The results were dramatic. After the changes, Dragon was docile and easily used, fully and easily controllable in all modes.

Vehicle Dynamics:
The worst stress test for auto pitch (with Hawkes as pilot and Laws in back cockpit with real time sensor data streaming on laptop), proved to be at maximum forwards speed where data showed an unexpectedly large nose-down, dynamic force threatening to overwhelm the limits of auto-pitch stabilization. Hawkes had planned that such nose down moment would be balanced by opposite pitch (tail down), dynamically induced by the large tail wing. Some minor changes were made on site, and proved to be beneficial, but relatively ineffectual. Hawkes approved the craft to continue to the next stage of open sea demonstrations, with a piloting restriction to avoid maximum forwards speed.

In September 2015, the Company shipped the Dragon to Monaco for open sea trials. With Hawkes as pilot, the prototype Dragon completed twenty eight dives in open water off Monaco, operating out of Port de Fontvieille. The Company accomplished most dives with guests in the back cockpit, but also ran a few dives, for continued software development, with Hawkes and Laws. The vehicle performed well in open sea trials in sea states typically running into maximum approved operations in sea state 3.

On return from Monaco in November 2015, the team worked at the Pt. Richmond facility to characterize the Dragon dynamics over a range of pre–set forward wing settings under high speed test runs and to dynamically balance or neutralize pitch changes at variable speed. Near neutral balance (the ideal) was only achieved with significant upwards lift from the nose wing. The nose section and wing were then re-designed and the body modified in winter 2015. The Company completed the prototype testing and upgraded the Dragon to pre-production status where it will be used for factory demonstrations and as mule for future development.

Certification issues:
The company is currently engaged in the process of obtaining certification from Germanischer Lloyd (DNV GL) while maintaining company confidentiality on our "Sea Glass13" composite pressure hulls. Our current development of the Dragon is a step forward in our strategy of making vehicles easier to operate while maintaining our primary emphasis on safety. The Company expects that, with DNV GL certification, we will open significant opportunities in expanding markets worldwide and continue our mission to open new frontiers for undersea exploration

and enable a new era of private exploration and adventure. The challenge remaining is largely within the USA because of ABS inability to provide certification services for composite pressure hulls. Fortunately USCG allows US Industry to innovate for private use under section 4 of NVIC 95-3.

Using Laser Scanning to Improve Operational Efficiency: Spool Metrology Case Study

Jason Gillham, CEO, 2G Robotics

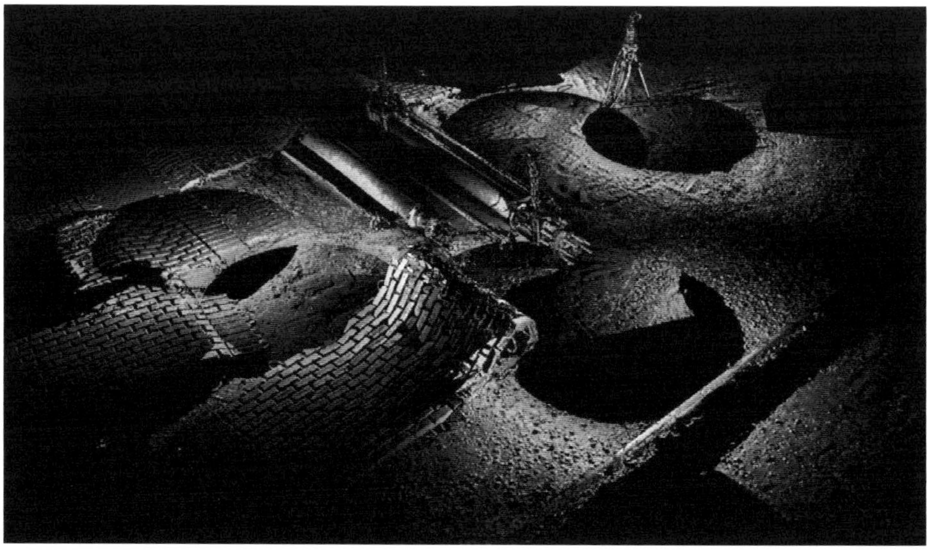

2G ROBOTICS

Spool Metrology using Laser Scanning

Background

- 2G Robotics' ULS-500 underwater laser scanner selected by Subsea 7 for a hybrid long baseline (LBL)/laser scanning solution
- Metrology performed in the North Sea

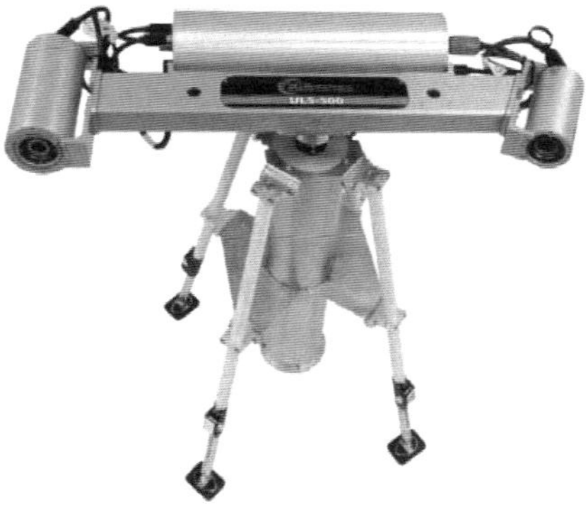

2G ROBOTICS

Spool Metrology using Laser Scanning

Objective

- Design five spool pieces to connect a newly installed bundle towhead to five previously laid spool pieces

Methodology

- Acoustic LBL used to obtain towhead and spool piece positions and headings
- Laser scanning used to obtain required accuracy for dimensional control survey of towhead structure with flanges

2G ROBOTICS

Spool Metrology using Laser Scanning

Operation

- Two saddle mounted transponders with control spheres deployed by WROV onto the largest of the five spools to position centre flange within the LBL acoustic array
- ULS-500 performed 360° scans at eight locations around the spools.
- Laser data used to accurately position each flange within the LBL array

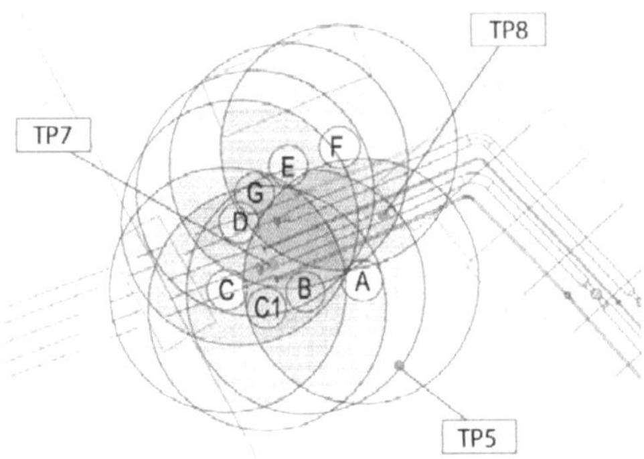

2G ROBOTICS

Spool Metrology using Laser Scanning

Results

- Metrology took **less than a quarter of the time** of a traditional LBL approach
- Laser scanning approach generated a complete 3D visualization comprised of **14.6 million points**
- Entire data set captured in only **3.5 hours**
- Data captured in real-time and processed while offshore

2G ROBOTICS

Spool Metrology using Laser Scanning

"The end results were very impressive and the detail within the point cloud is fantastic. The **operational time savings will be motivation to use this system again**."
-Pieter Jansen, Survey and Subsea Data Manager

2G ROBOTICS

New Technology

New Long-range Laser System

- Range: 15m+
- Points per line: 1024
- Sample Rate 33 profiles/sec
- Integrated stills imagery

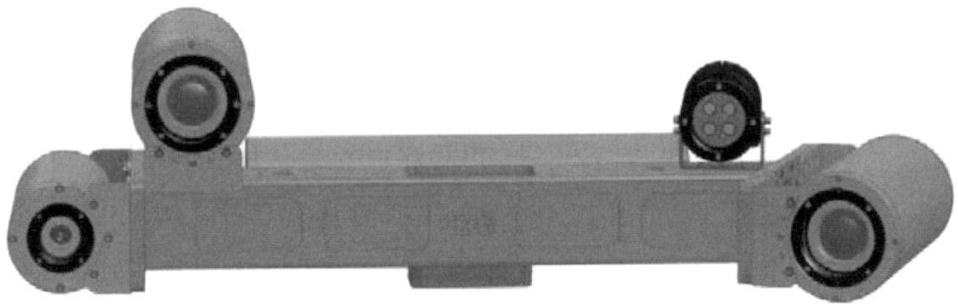

2G Robotics

Sonar 101 for First Responders

Ruben Morales

The Need

- Searching for evidence and drowning victims are just two of a growing list of tasks that First Responders are required to carry out. We are witnessing a greater demand for waterborne evidence recovery (i.e San Bernardino), parasitic device detection, and underwater infrastructure inspections.

The Challenge

- There are many challenges associated with the use of sonar for Search and Recovery Missions some of which include:
  - Geography
  - System portability
  - Training
  - Basic understanding of sonar operations
  - Ease of use

What is Sonar?

- Sonar (originally an acronym for SOund Navigation And Ranging) is a technique that uses sound propagation to navigate, communicate with or detect objects on or under the surface of the water.

What Sonar is not?

- Sonar is not a magic tool that guarantees location of underwater objects.

Types of Sonar Systems available

- Multibeam
- Mechanically Scanning
- Side Scan Sonar
- Tracking / Beacons

Multibeam Sonar

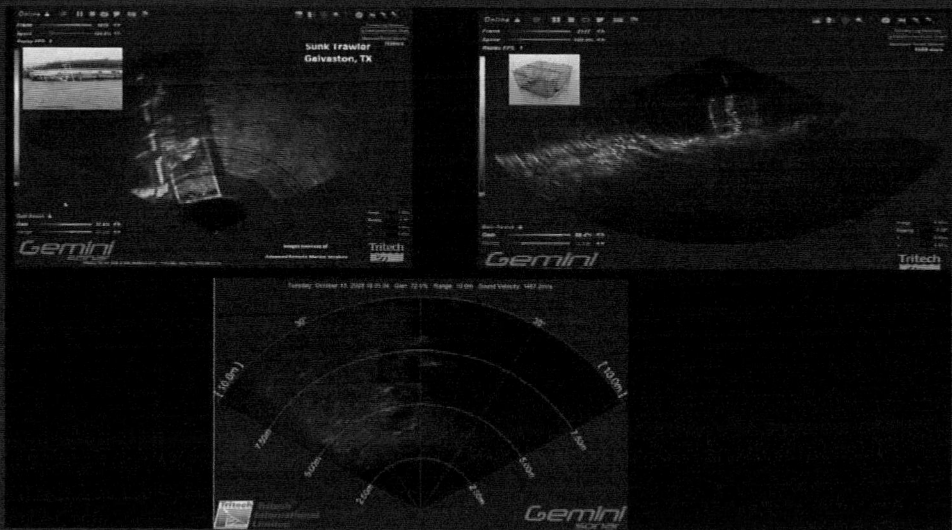

Mechanically Scanning Sonar

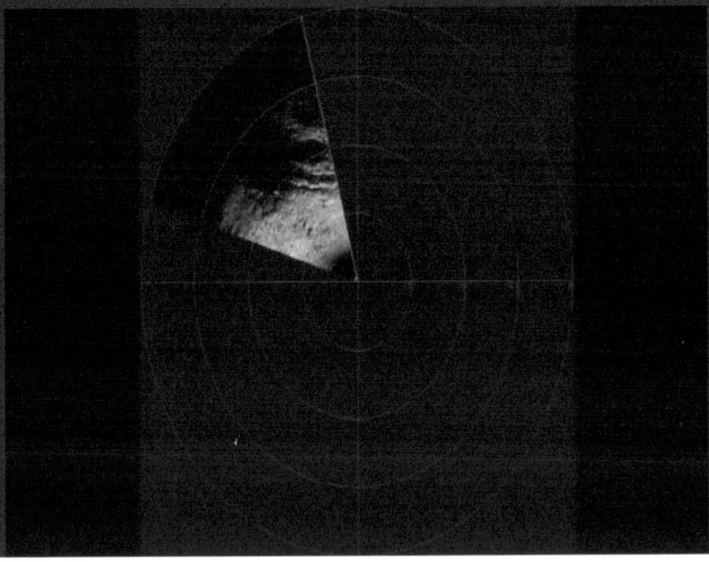

Side Scan Sonar

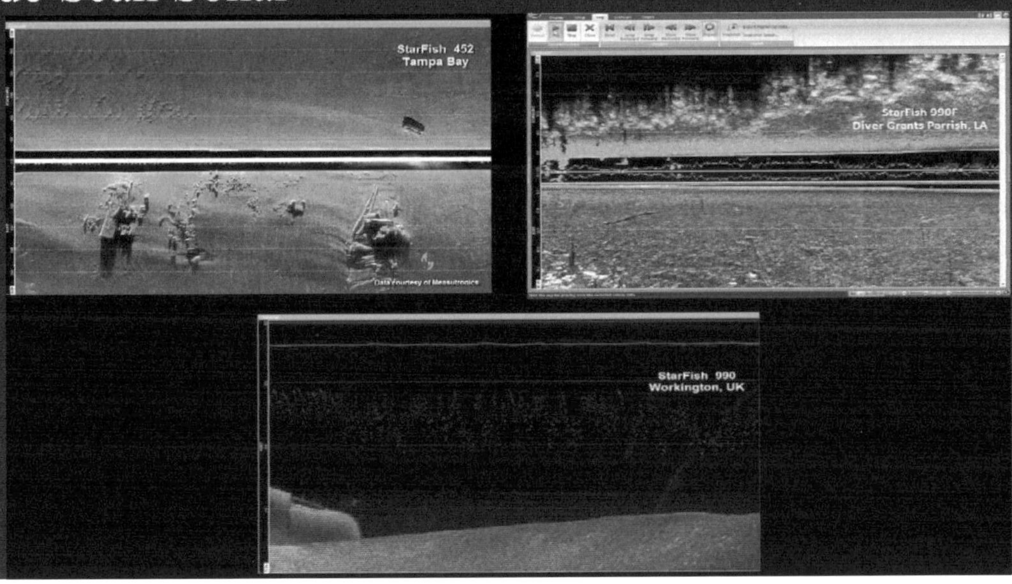

Tracking / Beacons

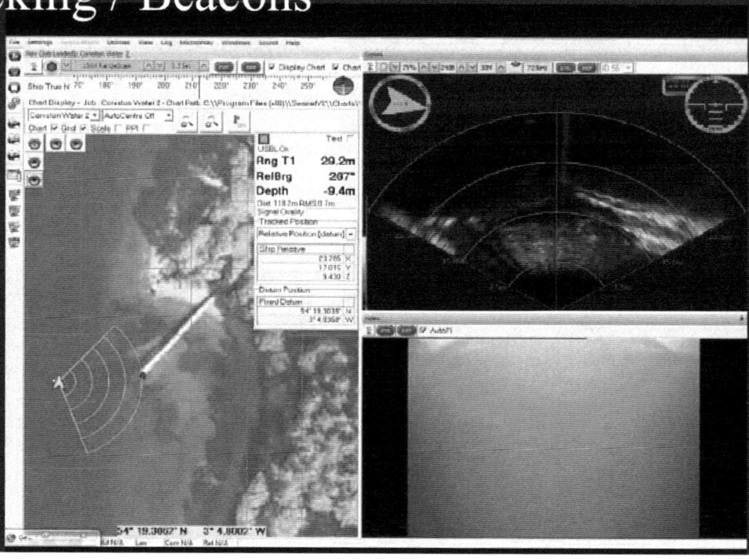

Questions?

Development of Methods to Eliminate Oil Pollution Threats in Sunken Vessels

Aaron Glanville
Gahanna Lincoln High School
Gahanna, Ohio

Sunken Vessels

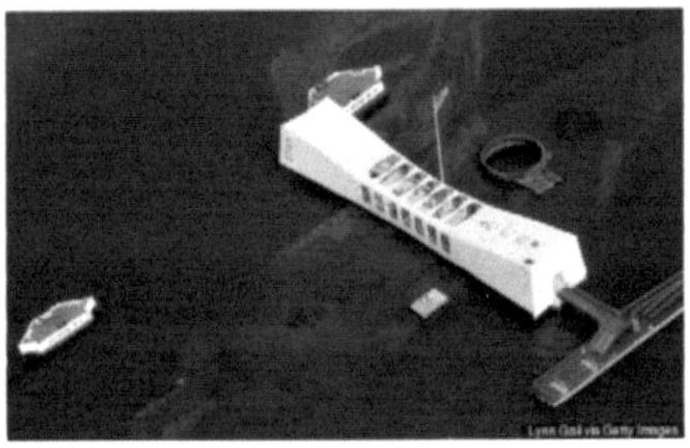

- Oil Tankers
- Bunker Fuel
- 8,569 sunken ships, 1,583 tankers, and 6,986 non-tanker vessels
- 2.5 to 20 million metric tons of oil

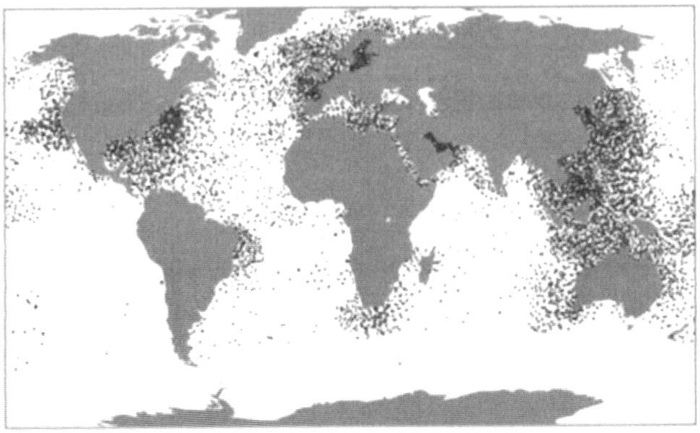

Environmental Impact

- Artificial Reefs
- One quarter of marine life
- Sensitive

Economical Impact

- Impoverished Pacific Islands
- Fishing and Tourism

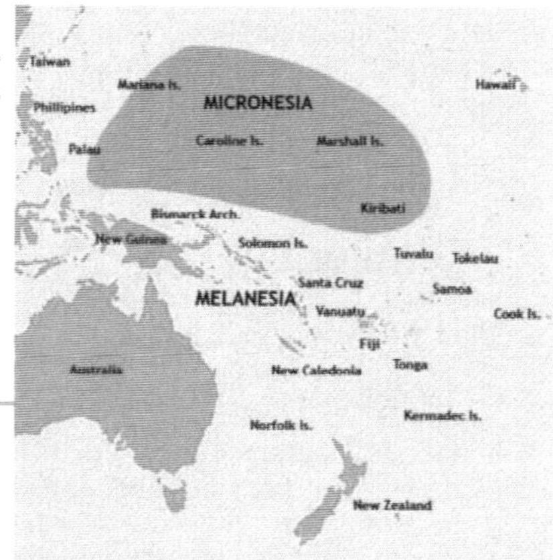

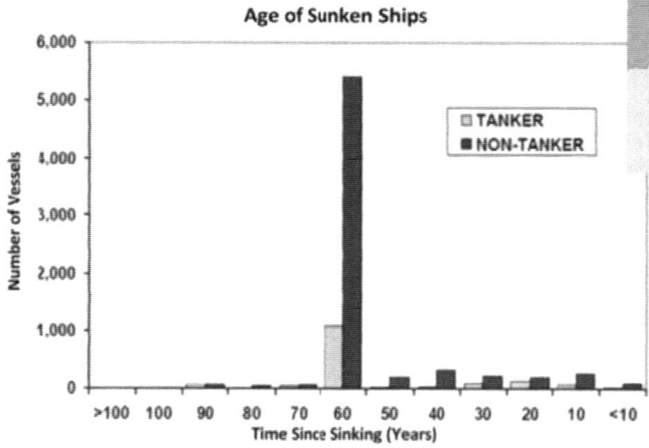

Purpose

To construct a cost effective tool package for an underwater remotely operated vehicle [ROV] to extract oil from a sunken ship to prevent oil pollution to the surrounding aquatic ecosystem.

Orthographies S.W.O.R.D. 1.0

Side

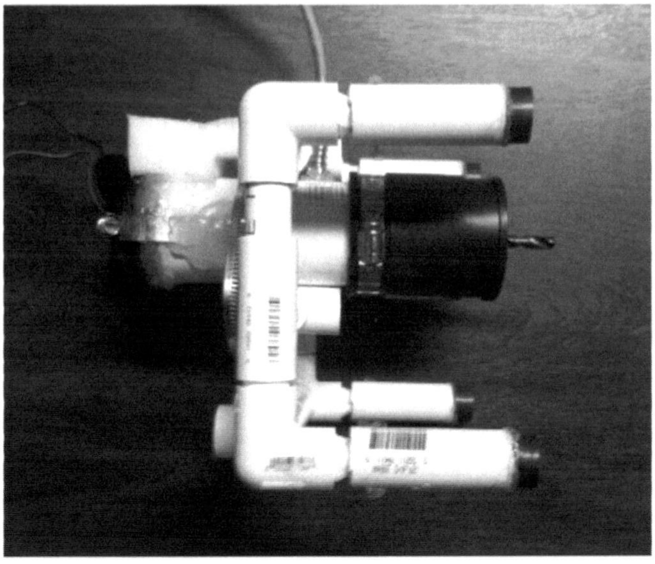

Orthographies S.W.O.R.D. 1.0

Top

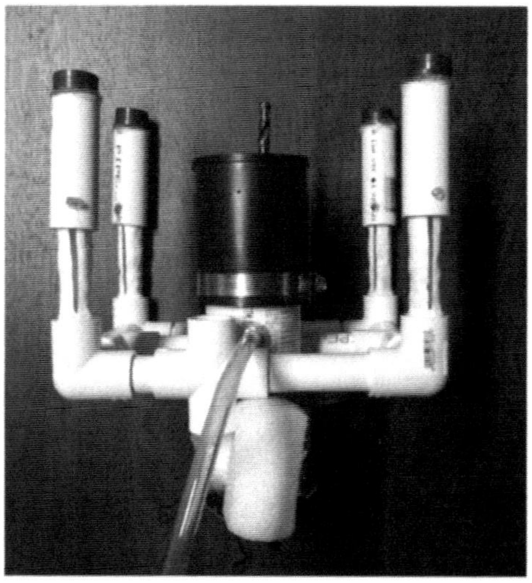

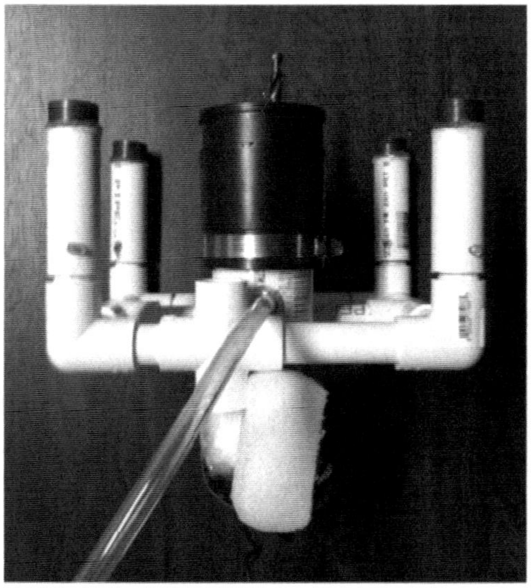

Orthographies S.W.O.R.D. 1.0

Front

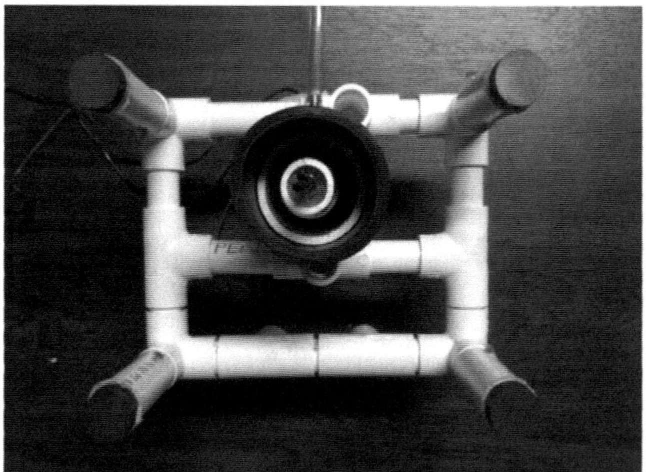

Construction

Basic Frame

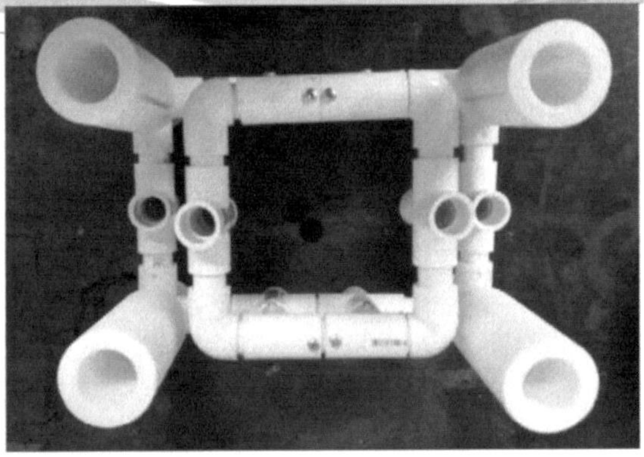

Top View

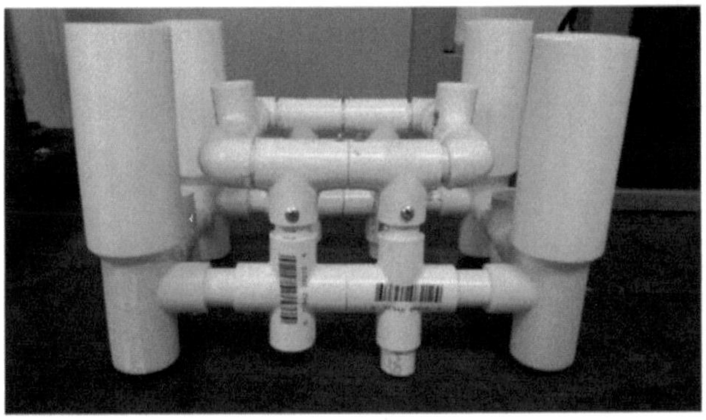

Side View

Construction

Three Dimensional Modeling

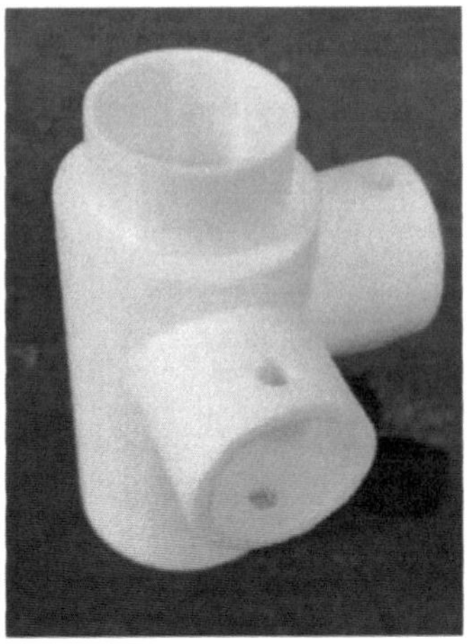

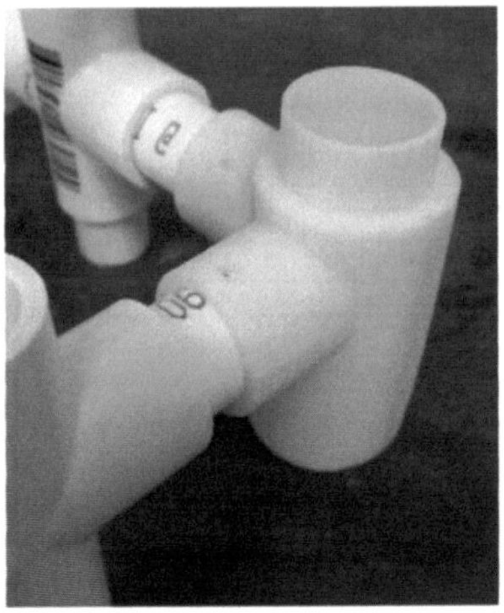

Construction

Three Dimensional Modeling

Inner Tube

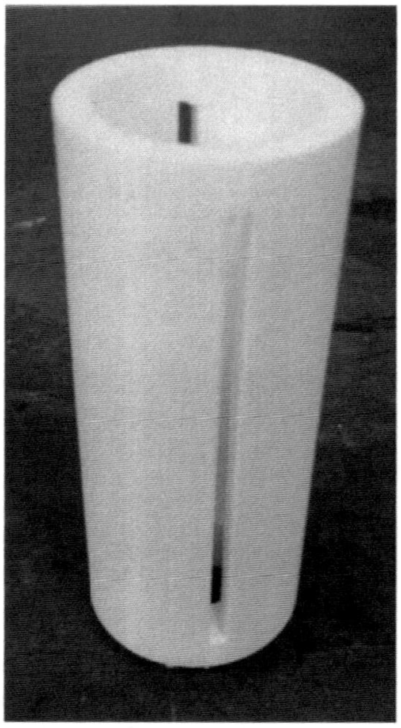

Outer Tube

Construction

Three Dimensional Modeling

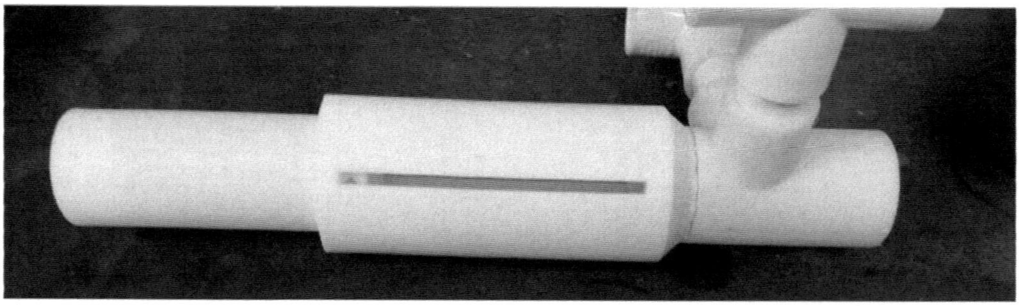

Construction

Magnetic Locking System (MLS)

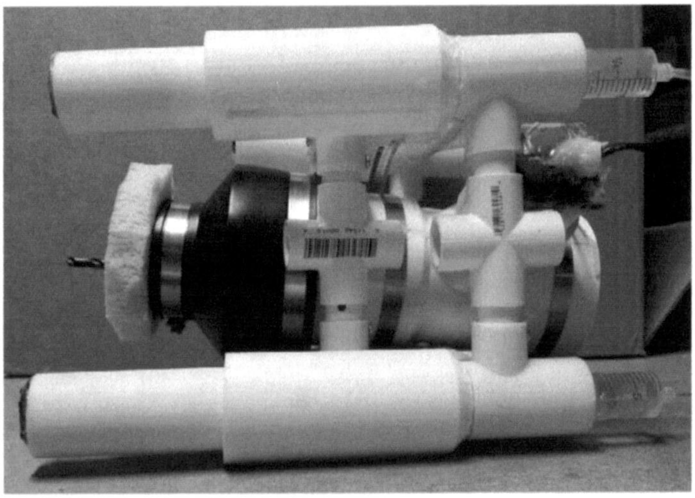

Construction

Magnetic Locking System (MLS)

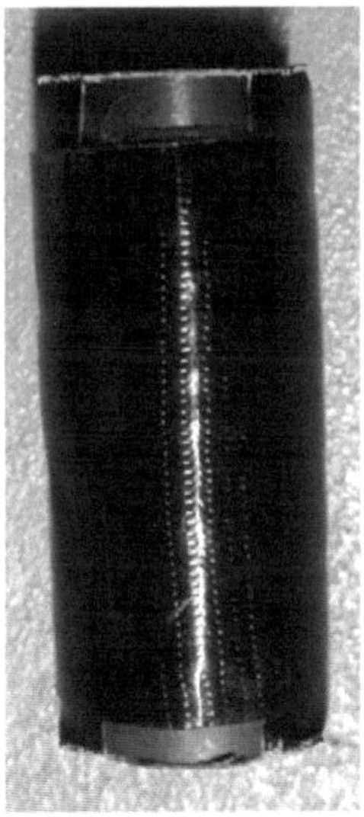

Construction

Magnetic Locking System (MLS)

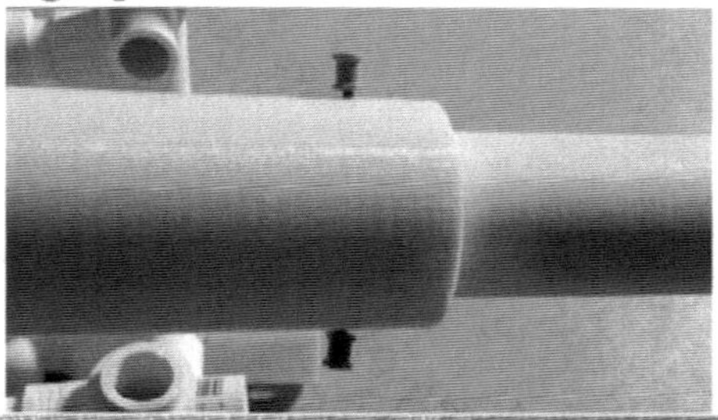

Construction

Magnetic Locking System (MLS)

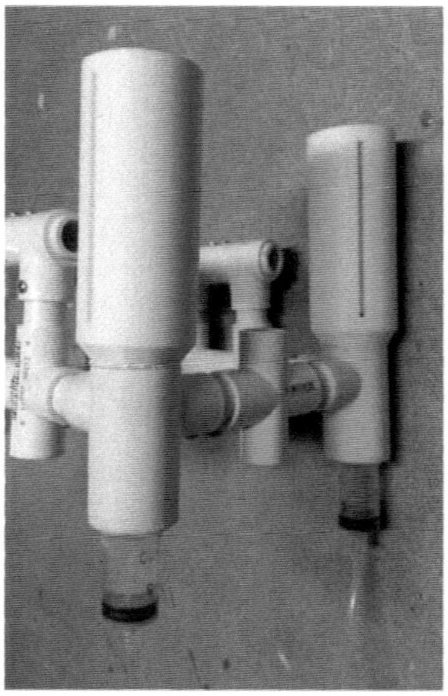

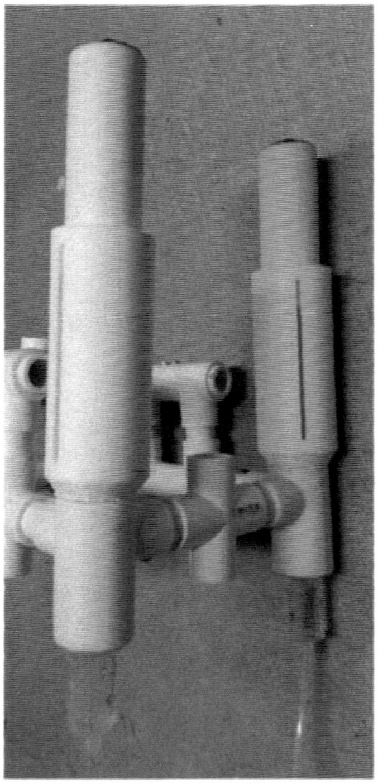

Construction

Waterproof Drill

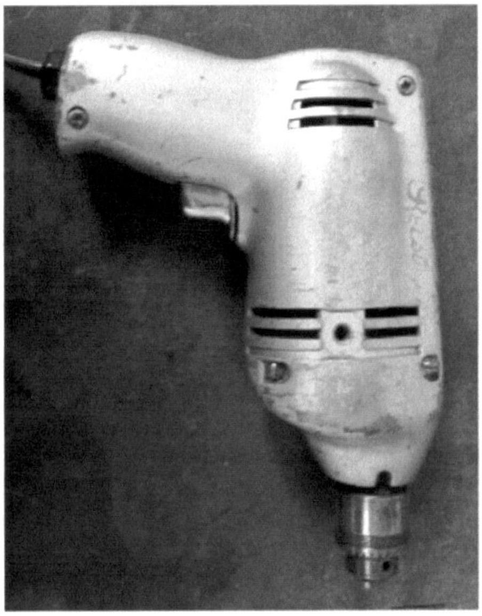

Construction

Waterproof Drill

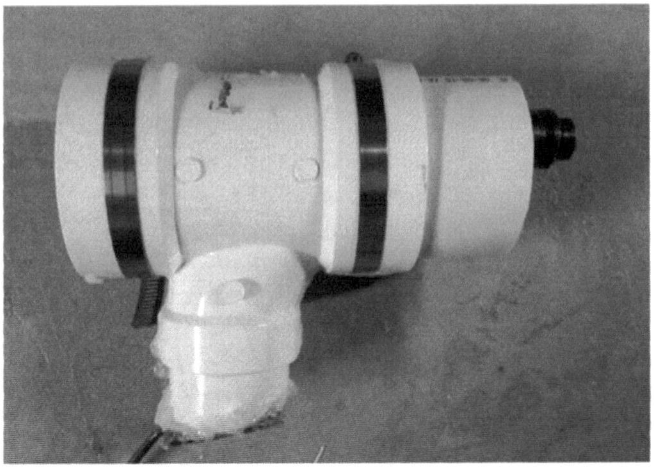

Construction

Waterproof Drill

Construction

Waterproof Drill

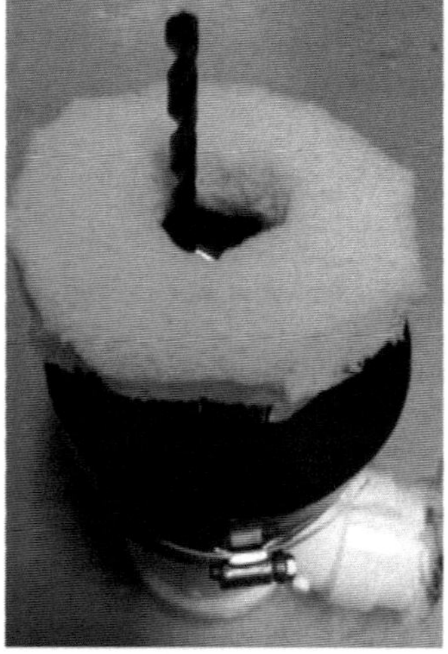

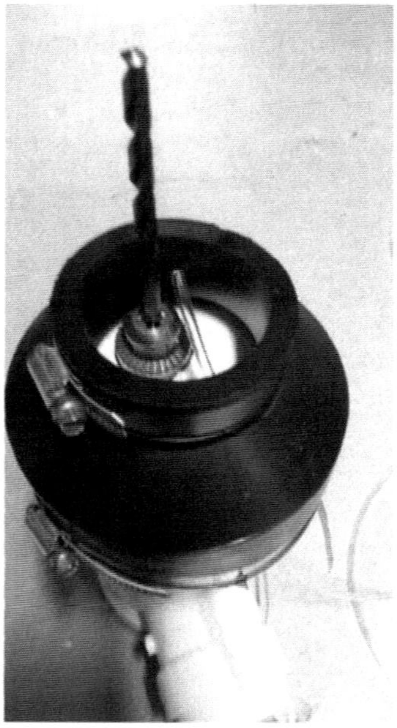

Construction

Laser Pointer Aiming System (LPAS)

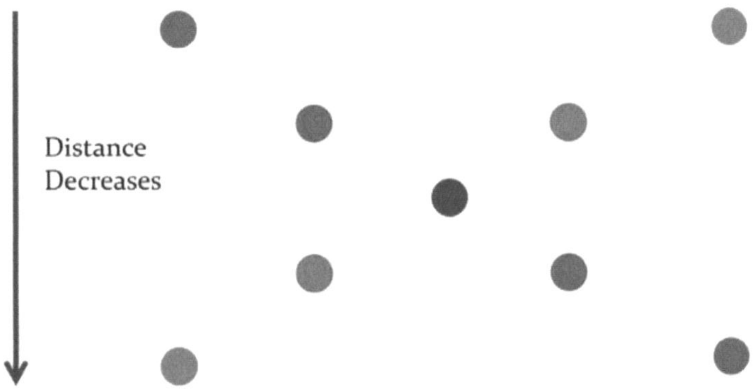

Construction

Laser Pointer Aiming System (LPAS)

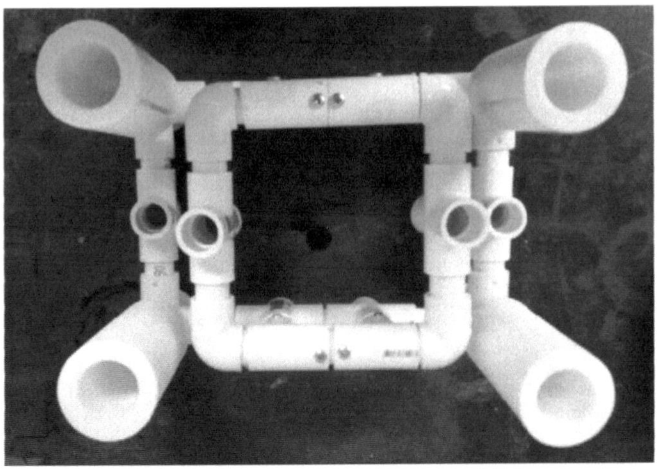

Construction

Laser Pointer Aiming System (LPAS)

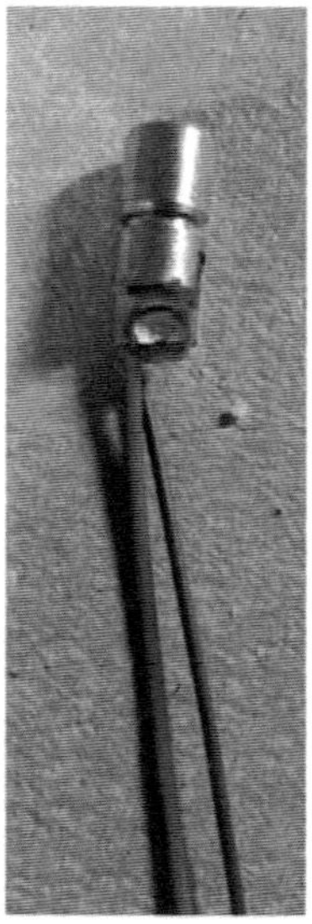

Construction

Laser Pointer Aiming System (LPAS)

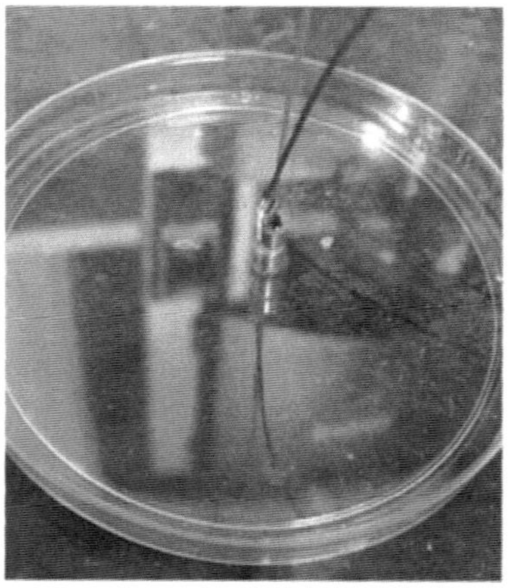

Construction

Remotely Operated Vehicle (ROV)

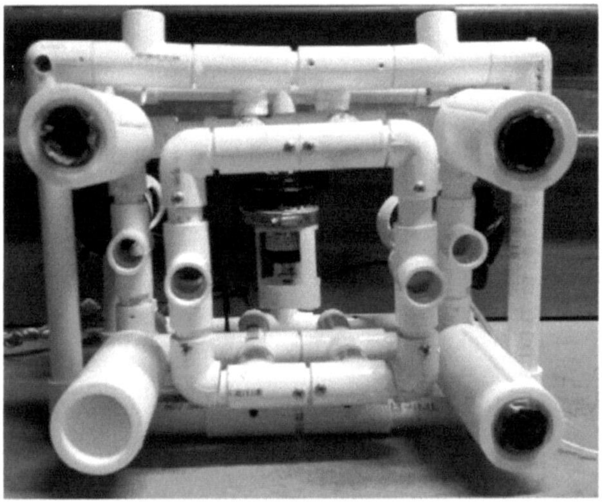

Construction

Remotely Operated Vehicle (ROV)

Testing

Laser Pointer Aiming System (LPAS)

Original

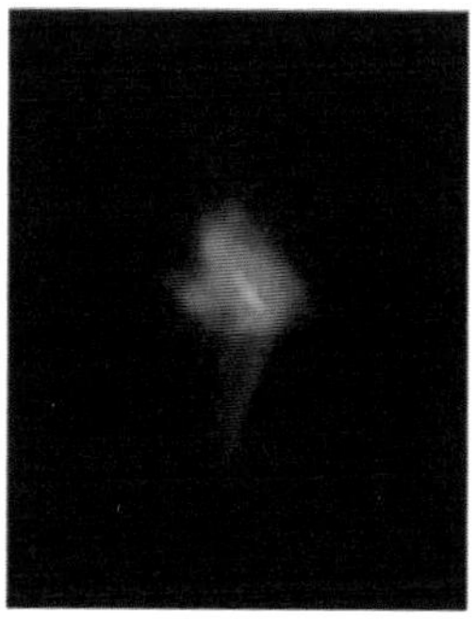

Pipet

Petri Dish

Testing

Drill Bits- Force/Time

Testing

Drill Bits- Force

Range is 0.046875 Inches

Range is 1.35 Newtons

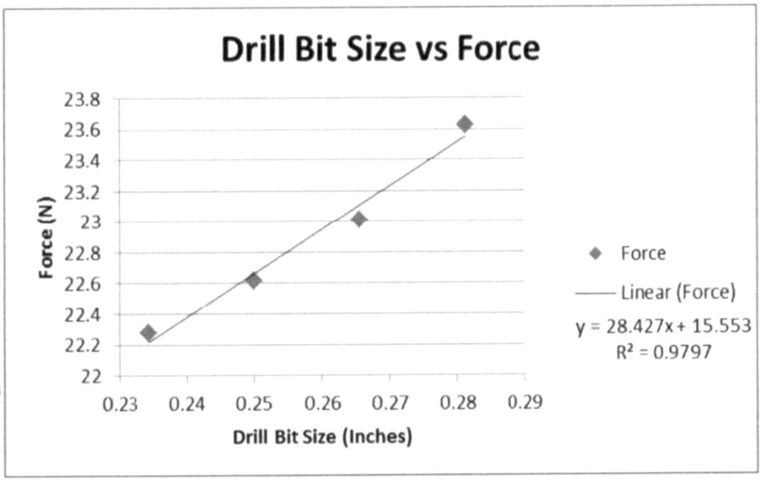

Testing

Drill Bits- Time

Range is .35 seconds

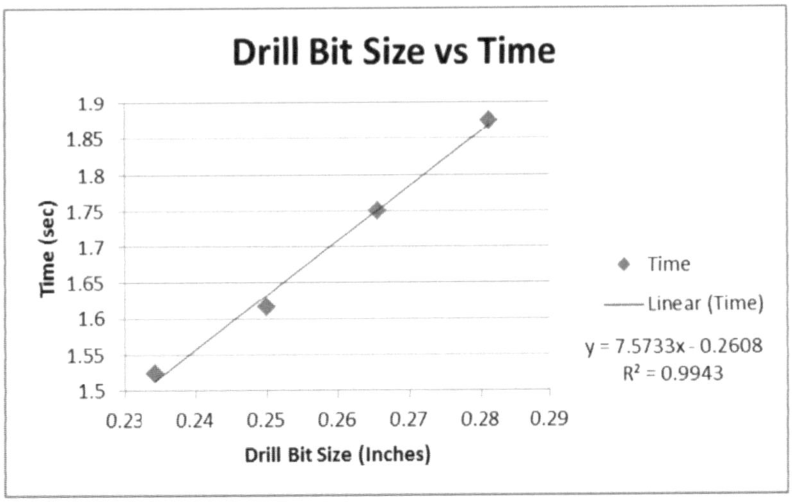

Testing

Hydraulics- Drop in Force

AveForce x (25/AveForce) = 25 N

AveTime x (25/AveForce) = AveTime @ 25N

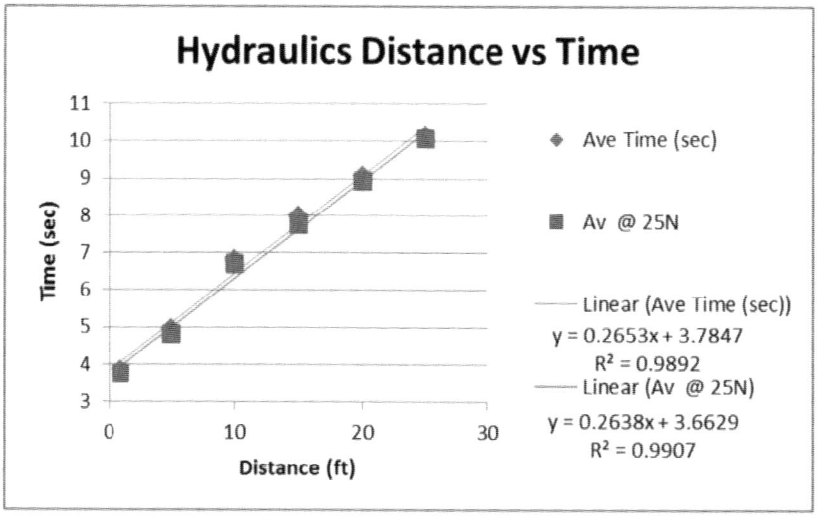

Testing

Hydraulics- Drop in Force

Newton's Second Law

$$F = M \times A$$

$$F = \frac{M \times (V_f - V_i)}{T}$$

$$\downarrow F = \frac{M \times (V_f - V_i)}{\uparrow T}$$

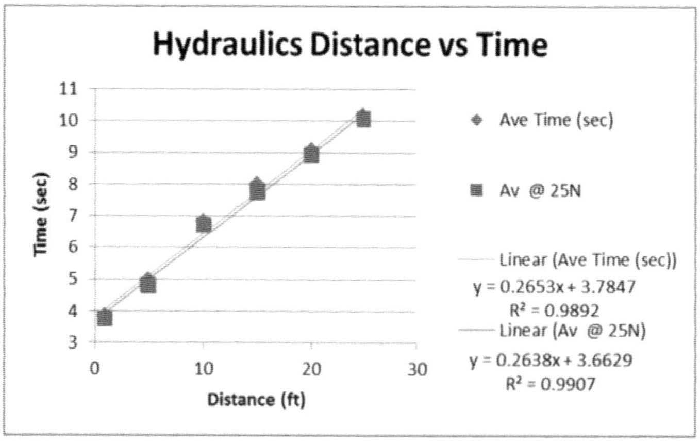

Testing

Hydraulics- Max Force

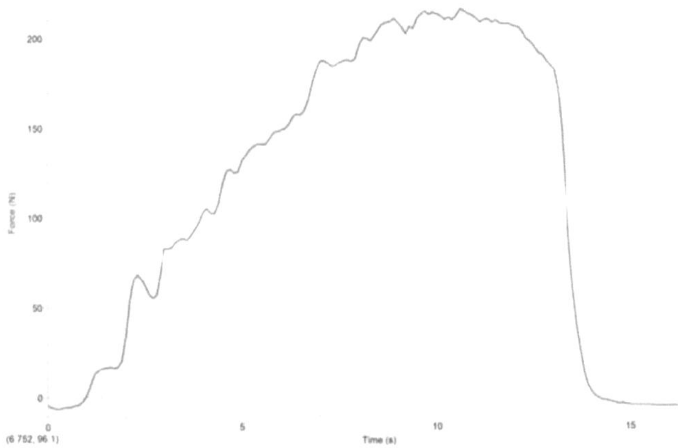

Average peak at 214 N

Testing

Attachment Strength- MLS Arm

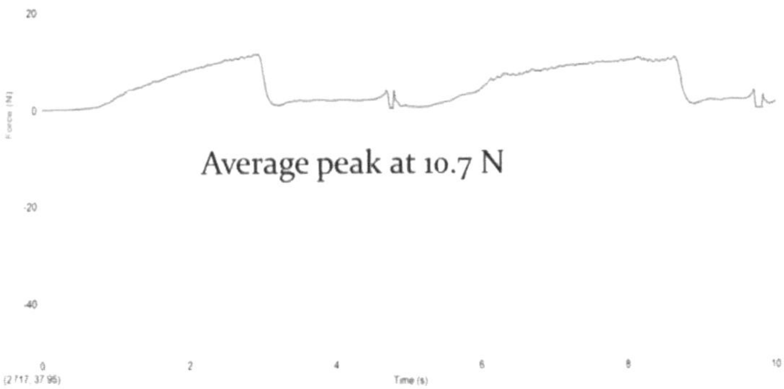

Average peak at 10.7 N

Previous Testing

Mock Extraction

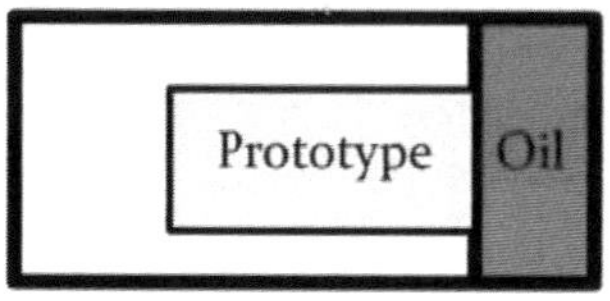

Conclusions

- Able to devise a method to waterproof a drill
- MLS functioned as designed
- LPAS: Diode's light intensity was greatly decreased
- Further testing is needed to determine the prototype's overall capabilities

Future Improvements

Electromagnetism

LPAS

- -Stronger laser pointers
- -Glass instead of plastic

Further Testing

- -Mock Extraction
- -While attached to an ROV
- -Force to remove prototype

Acknowledgements

Mr. Donelson

Mr. Bruns

Mrs. Rice

Dr. Languis

Classmates

References

- Amato, E. (2007). Development of European guidelines for potentially polluting shipwrecks. *DEEPP Project*. Retrieved September, 2014, from http://ec.europa.eu/echo/files/civil_protection/civil/marin/pdfdocs/deepppilotproject.pdf
- Bunker fuel and maritime fuel oil standards. (2012, November 20). Retrieved September, 2014, from http://svmshippingblog.wordpress.com/2012/11/20/bunker-fuel-and-maritime-fuel-oil-standards/
- Etkin, .. S., Rooji, H. V., & McCay, D. F. (n.d.). Risk Assessment Modeling Approach for the Prioritization Of Oil Removal Operations from Sunken Wrecks. *Risk Assessment Modeling Approach for the Prioritization Of Oil Removal Operations from Sunken Wrecks*. Retrieved September, 2014.
- Giant slick now major threat to Spanish wildlife. (2002, November 21). *US Newswire*. Retrieved September, 2014, from http://www.prnewswire.com/
- Hamer, M. (2002, December 14). Salvage hope for leaking oil. *New Scientist*. doi:GALE|A95679017
- Hamer, M. (2010, September 7). Why wartime wrecks are slicking time bombs. *New Scientist*. Retrieved September, 2014, from http://www.newscientist.com/article/mg20727761.600-why-wartime-wrecks-are-slicking-time-bombs.html?full=true#.VC6WNGddW4p
- Knowlton, N. (n.d.). Corals and Coral Reefs. Retrieved September, 2014, from http://ocean.si.edu/corals-and-coral-reefs
- Miyaji, Y. (2013, August 19). Photographer sees deterioration in Japanese ships sunk in 'second Pearl Harbor' - AJW by The Asahi Shimbun. Retrieved September, 2014, from http://ajw.asahi.com/article/behind_news/social_affairs/AJ201308190079
- Monfilis, R. (n.d.). Sunken WWII shipwrecks of the Pacific and East Asia: The need for regional collaboration to address the potential marine pollution threat. Retrieved September, 2014, from http://www.sciencedirect.com/science/article/pii/S0964569106000901
- Morton, E. (n.d.). The Eerie Ghost Ships at the Bottom of Chuuk Lagoon. Retrieved September, 2014, from http://www.slate.com/blogs/atlas_obscura/2013/11/21/chuuk_truk_lagoon_is_filled_with_the_remains_of_japanese_world_war_ii_ships.html
- Oil slick moving closer to national park; WWF leading efforts to save wildlife caught in spill. (2014, November 22). *US Newswire*. Retrieved September, 2014, from http://www.prnewswire.com/
- Oil tanker leak [Abstract]. (2000). *Clean Water Report*. doi:GALE|A65240707
- Oil tanker spill statistics. (2013). Retrieved September, 2014, from http://www.itopf.com/fileadmin/data/Documents/Company_Lit/OilSpillstats_2013.pdf
- Potentially Polluting Wrecks in U.S. Waters. (n.d.). Retrieved September, 2014, from http://sanctuaries.noaa.gov/protect/ppw/
- Rogers, P. (2002, May 22). Crew to try to remove oil from sunken tanker off California coast. *San Jose Mercury News*. Retrieved September, 2014, from http://www.mercurynews.com/
- Rogowska, J., & Namiesnik, J. (2012, September 11). Environmental Risk Assessment of WWII Shipwreck Pollution. Retrieved September, 2014, from http://link.springer.com/chapter/10.1007/978-94-007-4942-9_16#page-1
- Roylance, F. D. (2011, July 08). World War II: Shipwrecks could threaten U.S. coast. Retrieved September, 2014, from http://articles.baltimoresun.com/2011-07-08/news/bs-md-shipwreck-oil-20110708_1_fuel-oil-fuel-bunkers-crude-oil
- Shipwrecks from the Second World War could cause catastrophic damage to the environment, scientists have warned. (2003, May). *History Today*. doi:GALE|A100879517
- Sunken WWIIships pose an environmental threat. (2011). *Mooresville Tribune*, 4A.
- Wortel, O. (2006, November 8). Chuuk Lagoon Facing ?Imminent Threat? from WWII Ships. Retrieved September, 2014, from http://www.stpns.net/view_article.html?articleId=11543153147611337

Questions?

Contact Information:
Aaron Glanville
glanvilleaaron@gmail.com

The Argo

-Sank: 1937
-Discovered: August, 2015
-Light oils
-Estimated 100,000-200,000 gallons

FlexiClean

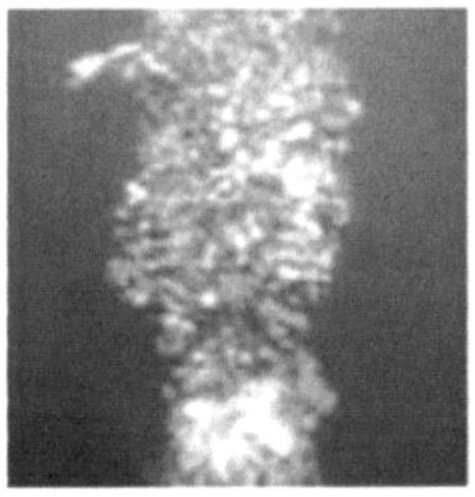

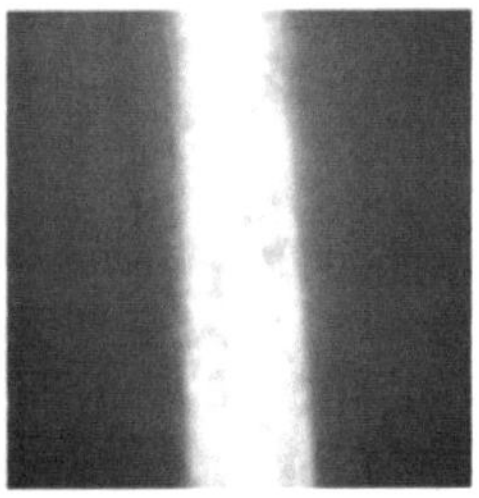

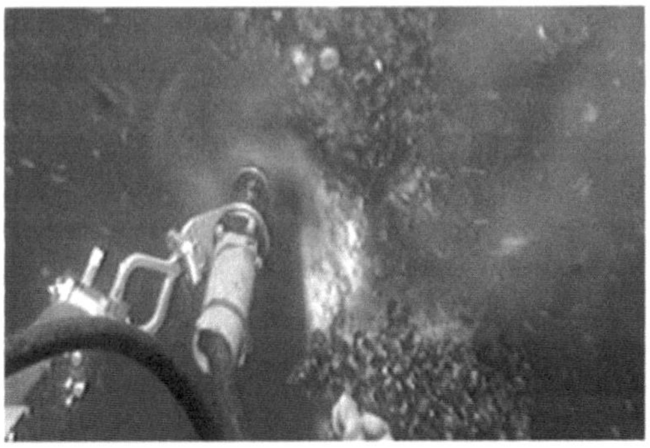

Current Solutions

- Example: USS Mississinewa (2001)
- $11 million
- Oil salvage (1.8 million gallons)
- Sold to Singapore (0.50 per gallon)
- 15 divers
- 10,000 minutes
- 80 to 112 feet depth

Moskito

MIKO MARINE AS

Norwegian Company

Released in the spring, 2015

Competitive Advantage

Cost Effective

- Impoverished nations

ROV Platform

- Does not require divers
- No need for training
- Divers can not be underwater for long periods of time

Easily Deployed

Underwater Technology for Undergraduates: Innovations and Ideas

CDR Andy Gish, PhD, PE
US Naval Academy

Underwater Intervention '16
New Orleans, LA

1

Presentation Outline

- Background and Motivation
- Ocean Engineering at USNA
  - Lab and field facilities
- Specific Examples:
  - Scale model salvage project
  - Hydrophone construction and testing
  - Underwater tool for ROV
  - SmartPerch AUV
- Discussion

2

Background and Motivation

- Underwater Technology is part of many undergrad engineering programs
  - Traditional engineering courses do not address unique underwater challenges
  - Shortage of good textbooks on the subject
  - Shortage of interested and qualified faculty
- Project-based learning (PBL) is effective and popular with students
- Existing PBL resource examples:
  - LEGO Mindstorm robotics
  - SeaPerch ROV
  - MATE ROV Competition
- **<u>CHALLENGE</u>: How to most effectively use PBL to teach underwater technology principles (design, construction, and operation)?**

3

United States Naval Academy Ocean Engineering

- 4000 undergraduate students, 0 graduate students
- 25 different majors (10 engineering)
- Faculty: 50% tenure-track civilian, 50% military officers
- Very rigorous academic/military/physical regimen
- 60-70 Ocean Engineering majors per class
  - 14 required core engineering courses
  - 4 engineering electives (senior year)
    - Underwater track: Underwater Work Systems and Life Support Systems
  - Senior Capstone Design Project (2 semesters)
  - Opportunities for independent research projects (1 or 2 semesters)

4

United States Naval Academy
Ocean Engineering

- 4000 undergraduate students, 0 graduate students
- 25 different majors (10 engineering)
- Faculty: 50% tenure-track civilian, 50% military officers
- Very rigorous academic/military/physical regimen
- 60-70 Ocean Engineering majors per class
  - 14 required core engineering courses
  - 4 engineering electives (senior year)
    - Underwater track: Underwater Work Systems and Life Support Systems
  - Senior Capstone Design Project (2 semesters)
  - Opportunities for independent research projects (1 or 2 semesters)

5

United States Naval Academy
Lab Facilities

- **380 foot towing tank:**

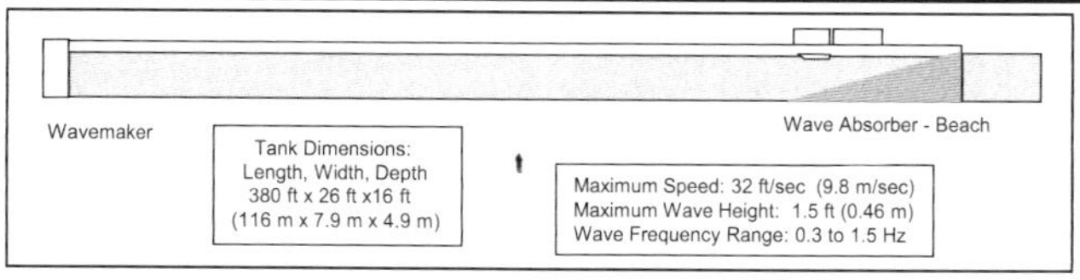

- **120 foot towing tank:**

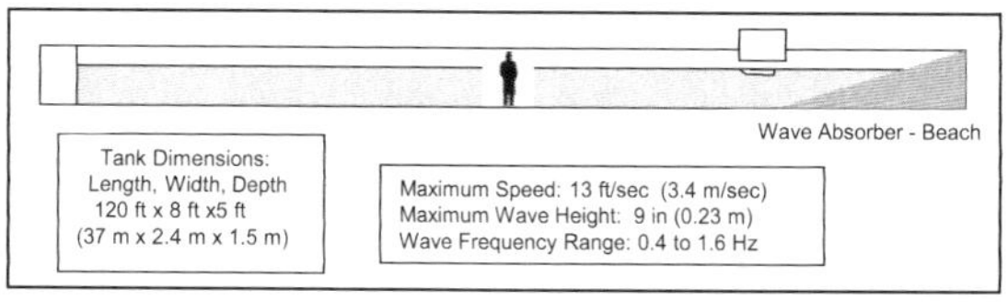

- **Circulating water channel:** max. speed 25 ft/sec
  - test section 16"x16"x5'
- **Coastal engineering basin:** max. depth 1 foot
  - used to study wave effects on beaches and structures
- **Ballast tank:** depth=4 feet
  - used for static tests and ballasting vehicles

6

United States Naval Academy Field Facilities

- Pier/seawall access to Severn River and College Creek
- Small boat/Yard Patrol (YP) craft access to river and Chesapeake Bay

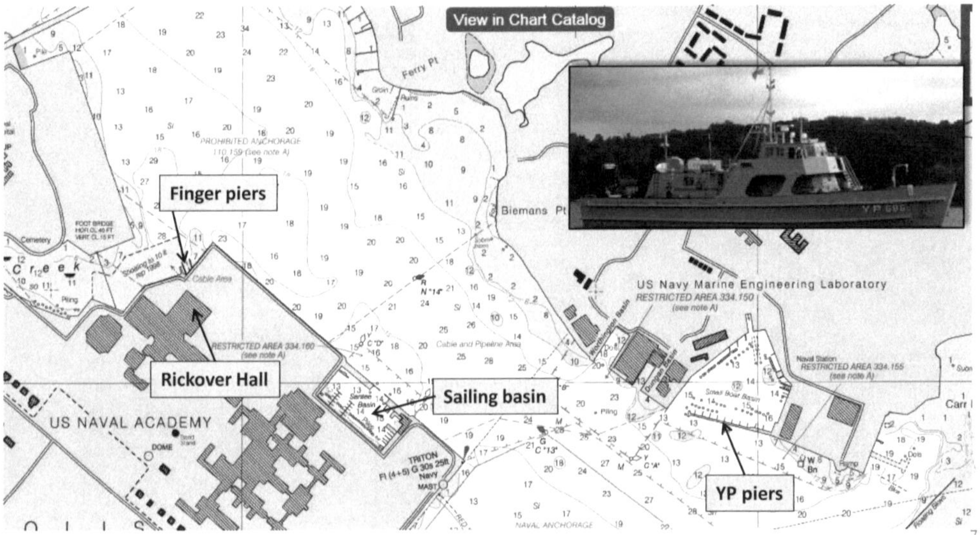

7

Scale-Model Salvage Project

- **Objective**: Gain experience and appreciation for difficulties in planning and executing a submerged salvage operation with ROVs
- **Focus**: Underwater Salvage and ROV Operations
- **Scenario:**
  - Two model ship hulls sunk in 380 ft tow tank (depth=11 ft)

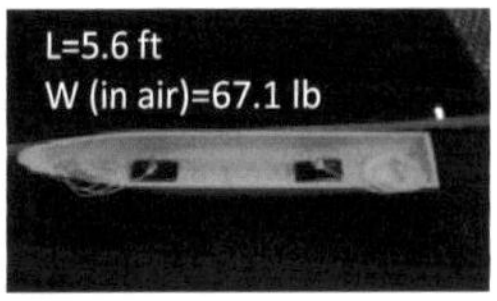

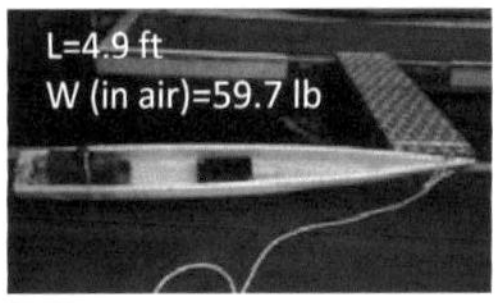

- Available Resources:
  - VideoRay Pro4 ROV (w/ Micron SeaSprite sonar)
  - VideoRay Pro3 ROV
  - Aquabotix HydroView Sport ROV
  - 2 diver lift bags (50 lb and 100 lb cap.)
  - Shop air supply (30 psi) and hose
  - Rigging hardware (line, shackles, carabiners, etc)

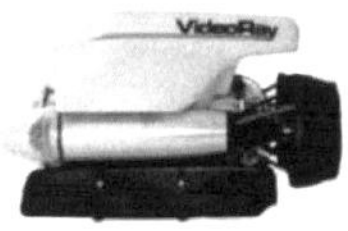

8

Scale-Model Salvage Project

- **Assignment**:
  1. Survey site with ROVs
  2. Perform basic salvage calculations (CG, CB, lifting force)
  3. Develop rigging plan to attach lift bag(s) and raise vessels
  4. Execute the salvage plan within a given time limit
  5. Provide video footage of the operation

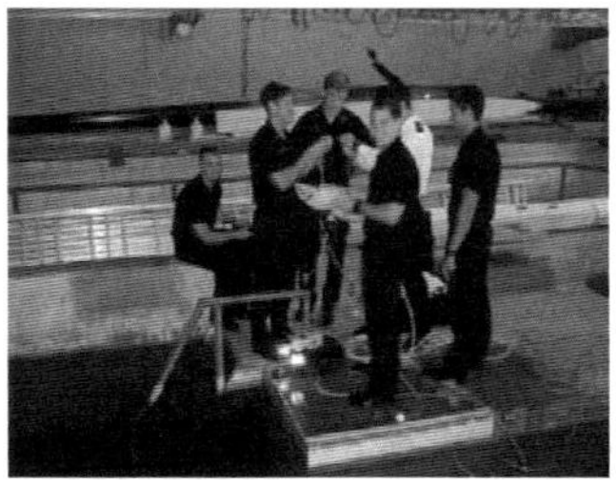

- **Lessons Learned:**
  - Simpler is better
  - Loose lines and hoses always get tangled
  - Planning and run-throughs are critical when you have limited on-site time
  - Tether drag/pull makes precise maneuvering very difficult
  - Teamwork and project management

9

Scale-Model Salvage Project-Video

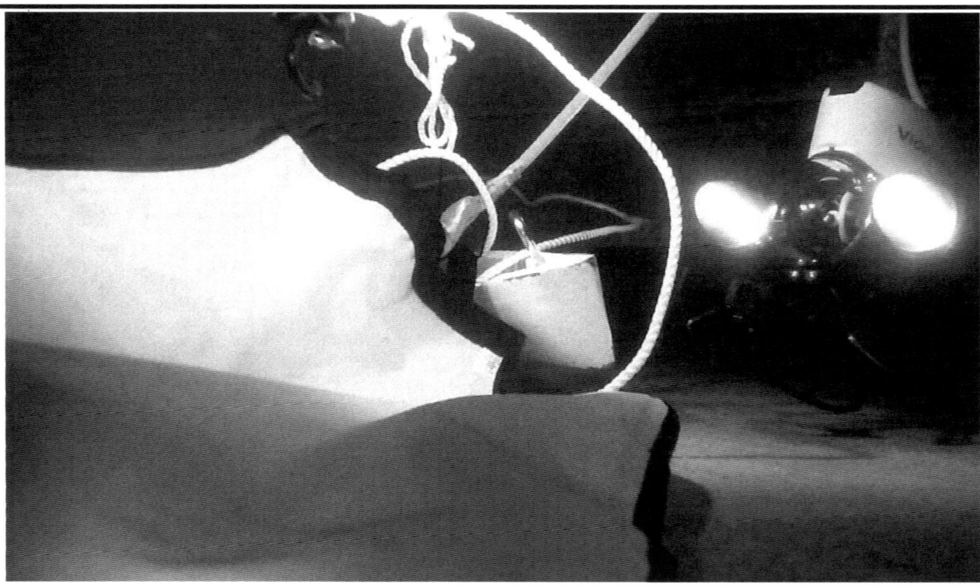

10

Scale-Model Salvage Project-Video

11

Hydrophone Construction and Testing

- **Objective**: Construct simple, inexpensive omnidirectional hydrophones, and use them to illustrate basic underwater acoustic principles
- **Focus**: Acoustics, fabrication and waterproofing techniques

- Reference: Kevin Hardy, "How to Build a Hydrophone", available online: http://www.dosits.org/files/dosits/hydrophone_instruc_w_image.pdf
- **Assignment:**
  1. Follow instructions to build a working hydrophone
     - Involves basic electronics, soldering, oil compensation, waterproof cable seals
  2. Measure max. detection range for different frequency acoustic sources
  3. Conduct 2-way acoustic ranging demonstration
  4. Observe ambient sounds in the local waters

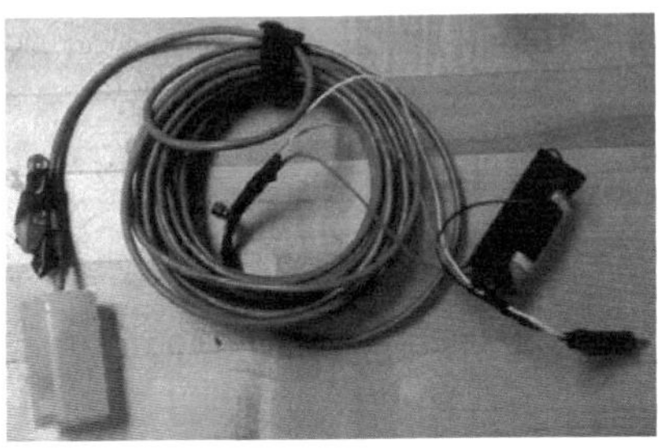

- **Lessons Learned:**
  - WOW! You really can hear things underwater!
  - Relationship between freq. and attenuation
  - Accomplishment of making something that works

12

Underwater Tool Design

- **Objective:** Design a socket wrench to mount on one of the available ROVs, capable of tightening or loosening a range of bolt/nut sizes
- **Focus**: Underwater tool design principles
- **Assignment:**
  1. Develop a conceptual design for a bi-directional socket wrench, operable from the surface, to mount on the Videoray Pro4 ROV
  2. Sketch the tool and its connection to the ROV
  3. Describe the tool's operation and method of providing reaction torque
  4. (FUTURE) Build and test the tool
- **Lessons Learned:**
  - Keep it simple
  - Counter-torque is critical for rotary tools
  - Major hurdles between concept design and detailed design/build

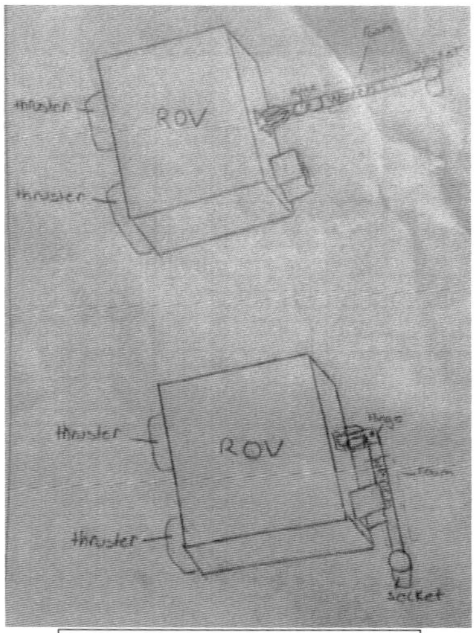

Student design, using ROV thrusters for torque

13

SmartPerch AUV

- **Objective:** Utilize a low-cost, user-friendly AUV platform to demonstrate and experiment with underwater vehicle concepts
- **Focus**: Hydrodynamics, control systems, power systems, navigation, and communications

- Start with SeaPerch kit (plus 1 extra motor)
- AUV basic features:
  - Mbed microcontroller
  - Pelican case pressure hull (tested to 15 ft)
  - MPX5050DP pressure sensor
  - MPU9150 9-DOF compass/gyro/accelerometer
  - XBee wifi comms (via floating antenna)
  - Micro SD card data logging
  - 12v and 5v Li-ion batteries

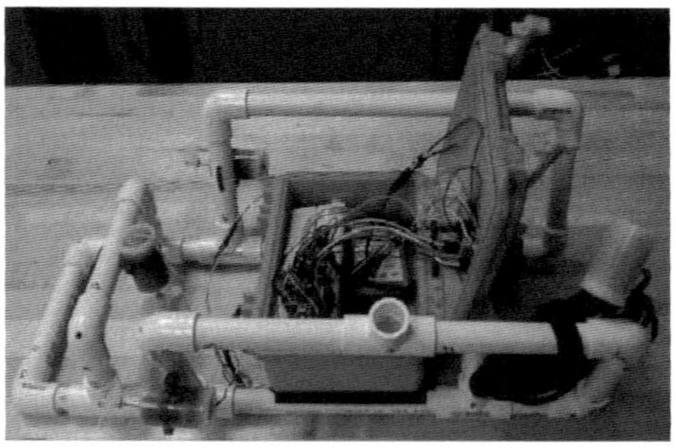

- Work in progress – Not yet implemented into course

14

SmartPerch AUV

- Future expansion plans:
  - Temperature sensor
  - Hydrophone(s)
  - Camera
  - Acoustic comms
  - Optical (laser) comms
  - Battery recharge
  - Advanced control and autonomy (MOOS-IvP)
  - Other sensors???

15

Student Feedback

- Overall response : overwhelmingly positive!
- Most students like open-ended labs, with unknown outcomes
- Some complaints about "failures" of activities
- Specific student comments:
  - "I enjoyed that a lot of the class has been hands-on. I am more of a visual learner so it helped me grasp a lot of the concepts better."
  - "The most effective learning experience was the salvage project. We got to work together as a team to plan out our operations and then could see the difficulties of putting that plan into action. It greatly contributed to my understanding of the challenges posed by the underwater environment."
  - "The salvage team project taught me that no matter how much you plan for different possible scenarios, something else will always go wrong that you haven't considered."

16

Discussion

- Questions?
- Suggestions to improve or expand these examples?
- Other project ideas?

Contact information:
CDR Andy Gish
U.S. Naval Academy
410-293-6585
gish@usna.edu

17

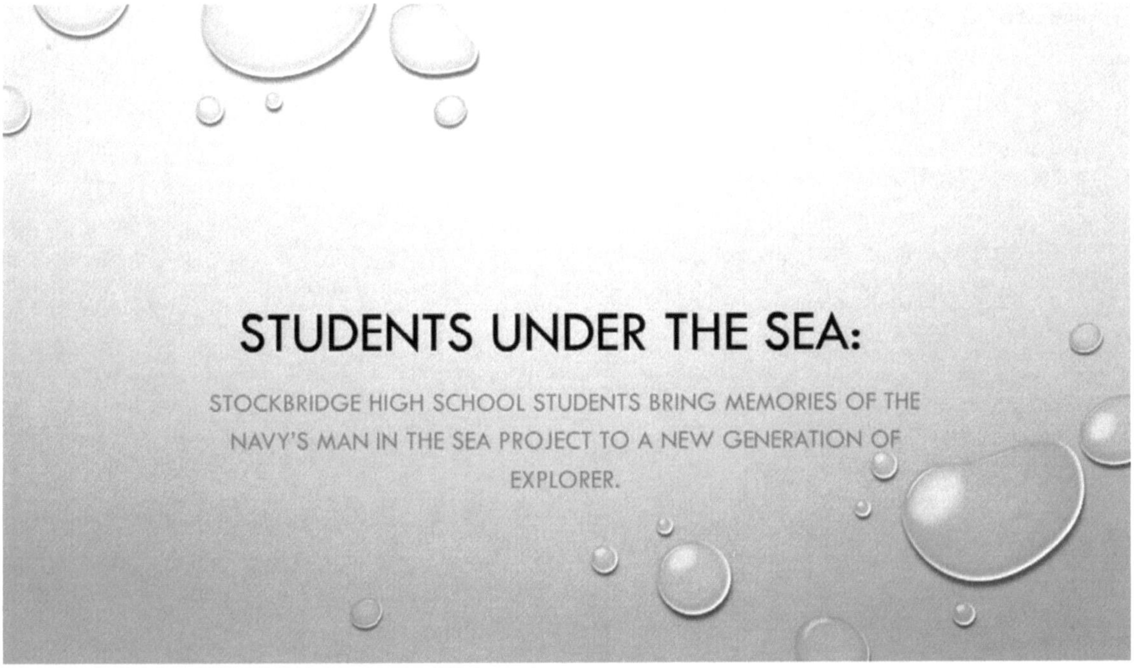
STUDENTS UNDER THE SEA:
STOCKBRIDGE HIGH SCHOOL STUDENTS BRING MEMORIES OF THE NAVY'S MAN IN THE SEA PROJECT TO A NEW GENERATION OF EXPLORER.

INSPIRATION
U.S.NAVY
SEALAB
U.S. NAVY
50
1965 2015
SEALAB
SEALAB
AMERICA'S FORGOTTEN QUEST TO LIVE AND WORK ON THE OCEAN FLOOR
BEN HELLWARTH

PLANNING THE ADVENTURE

- GRANT WRITING AND FUNDING
- SCUBA CERTIFICATION
- CLASSROOM PREPARATION
- ENGAGING YOUNGER STUDENTS

PADI SCUBA CERTIFICATION

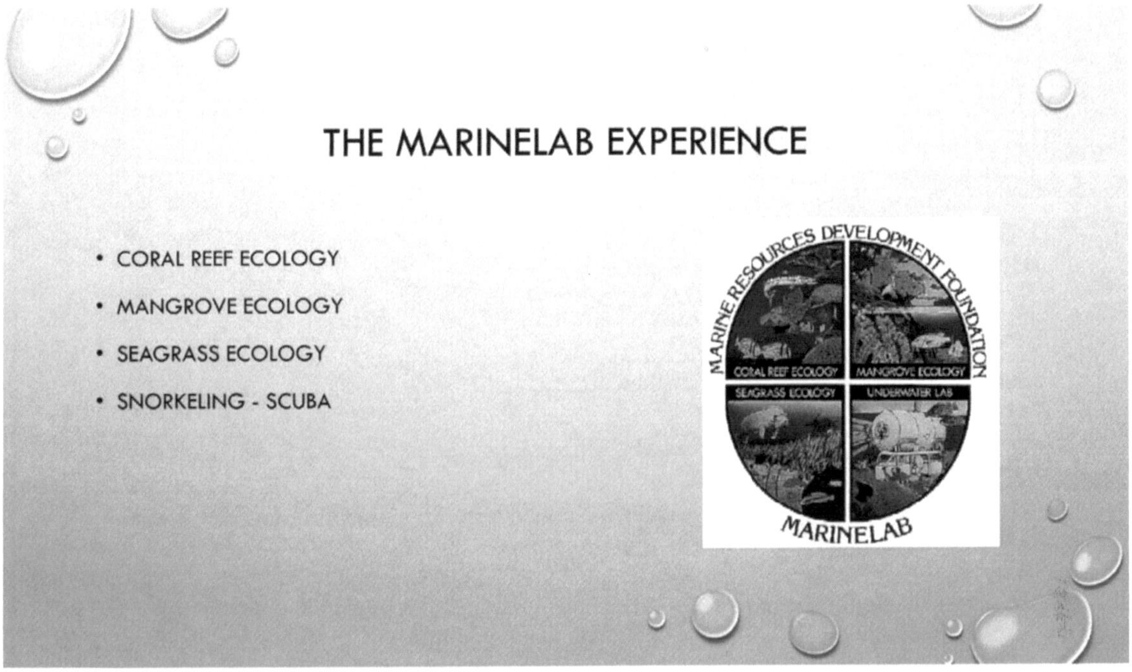
THE MARINELAB EXPERIENCE
• CORAL REEF ECOLOGY
• MANGROVE ECOLOGY
• SEAGRASS ECOLOGY
• SNORKELING - SCUBA
MARINE RESOURCES DEVELOPMENT FOUNDATION
CORAL REEF ECOLOGY
MANGROVE ECOLOGY
SEAGRASS ECOLOGY
UNDERWATER LAB
MARINELAB

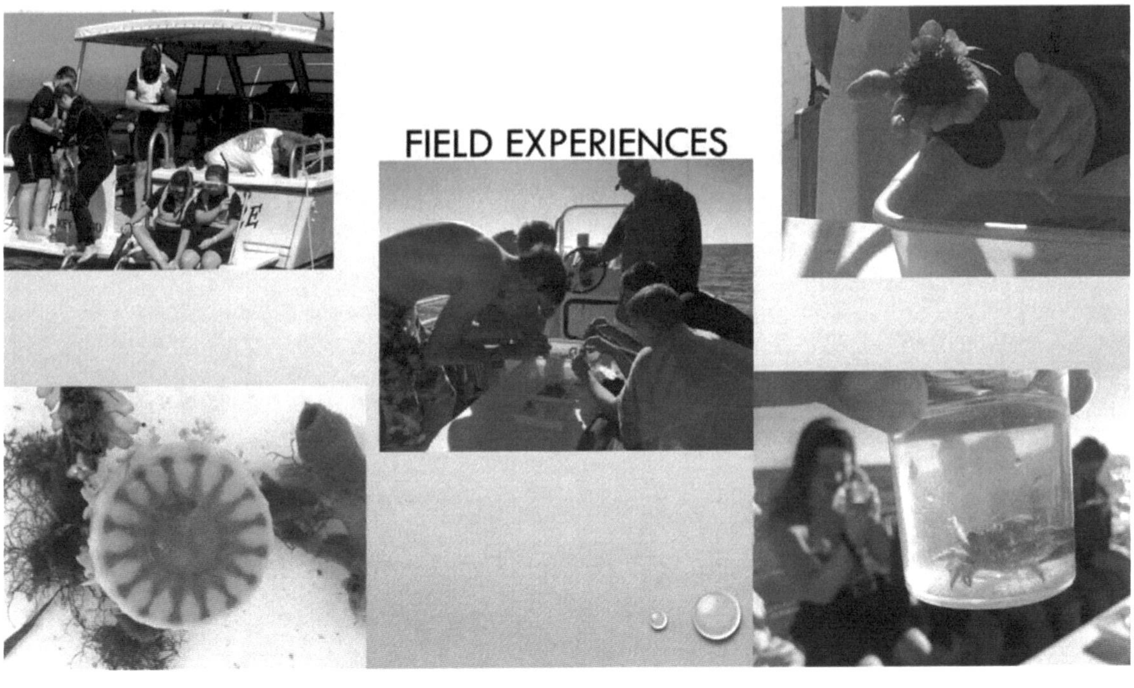
FIELD EXPERIENCES

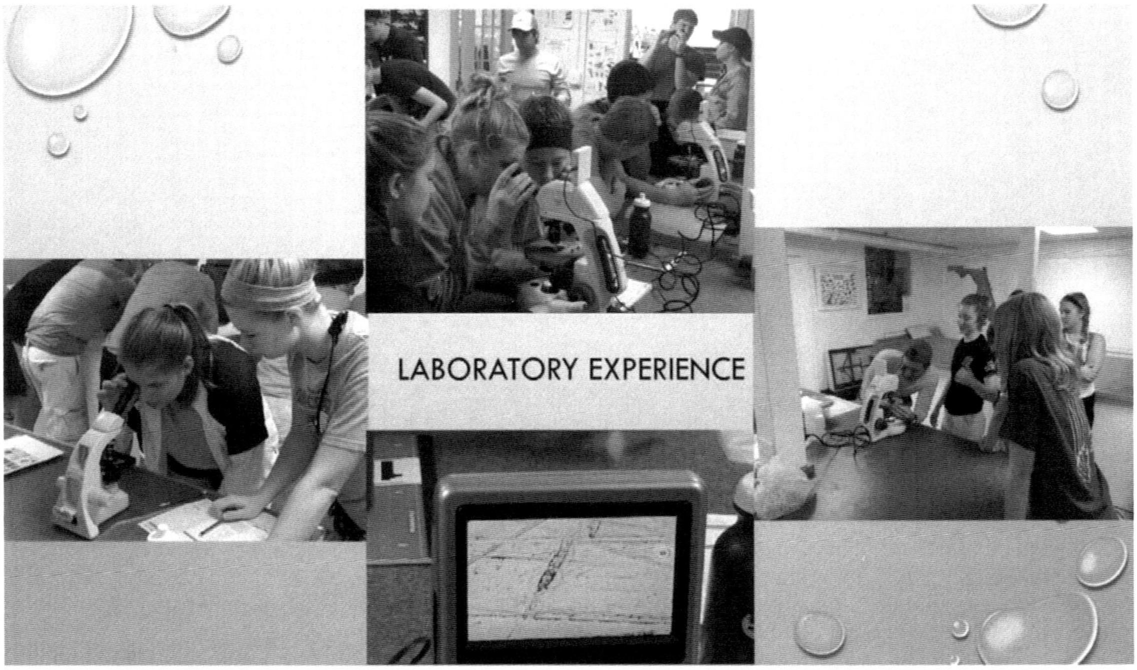
LABORATORY EXPERIENCE

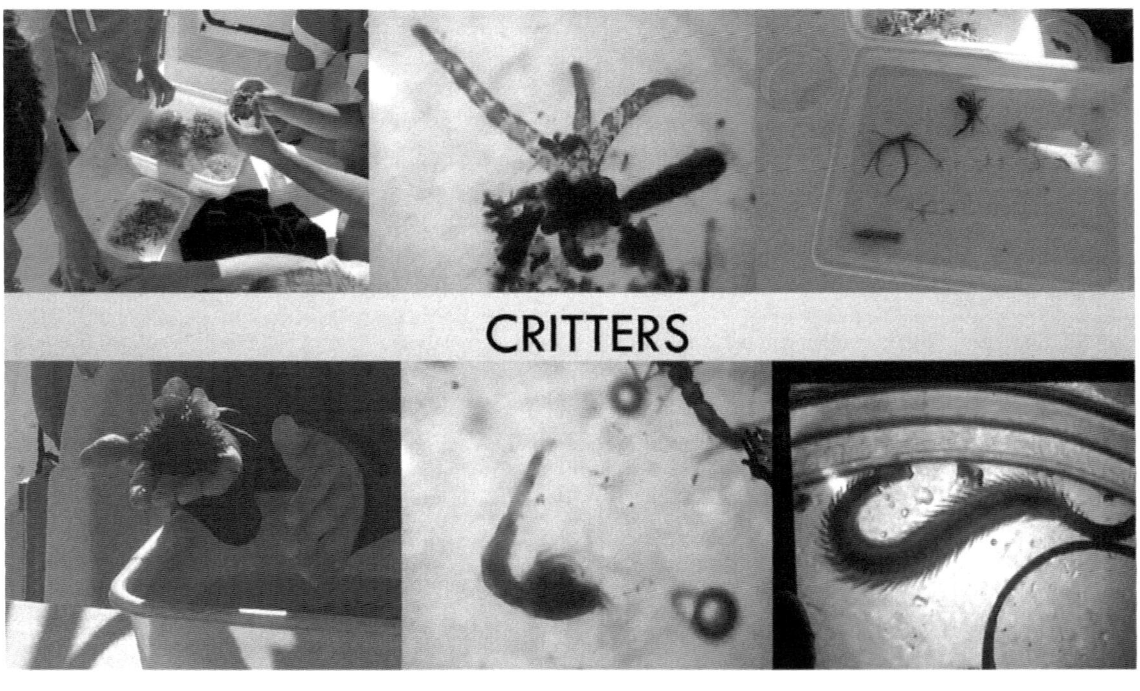
CRITTERS

AQUANAUT TRAINING

HISTORY

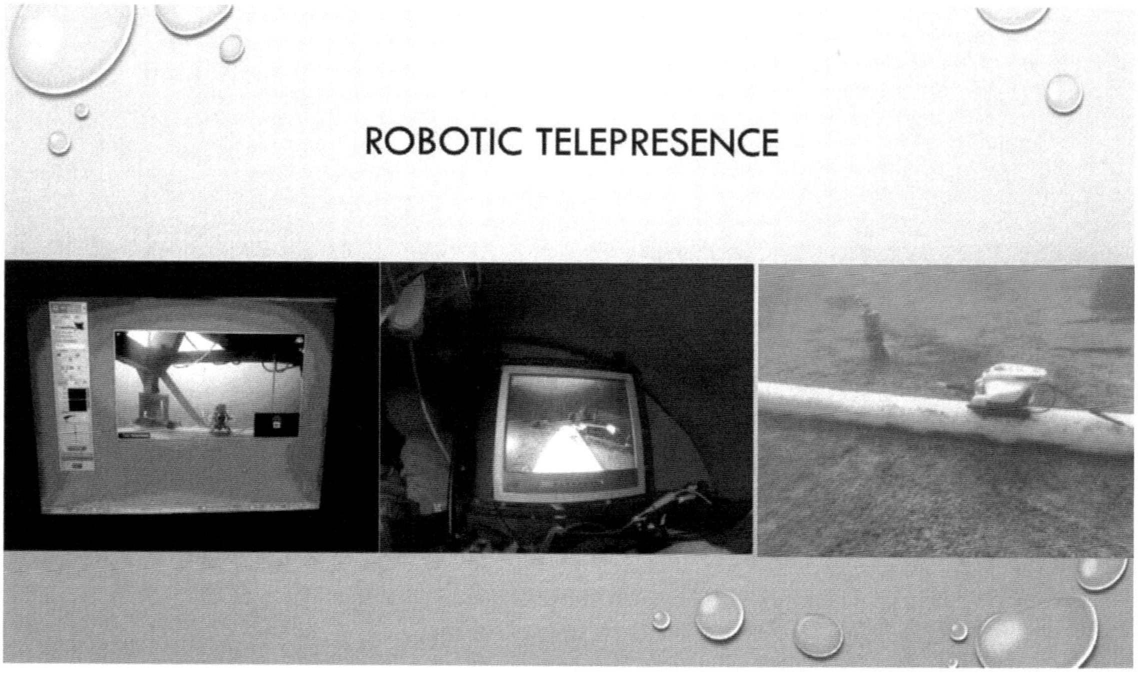
ROBOTIC TELEPRESENCE

TESTING THE CAMERA SYSTEM

ATLANTIS: THE LOST ISLAND

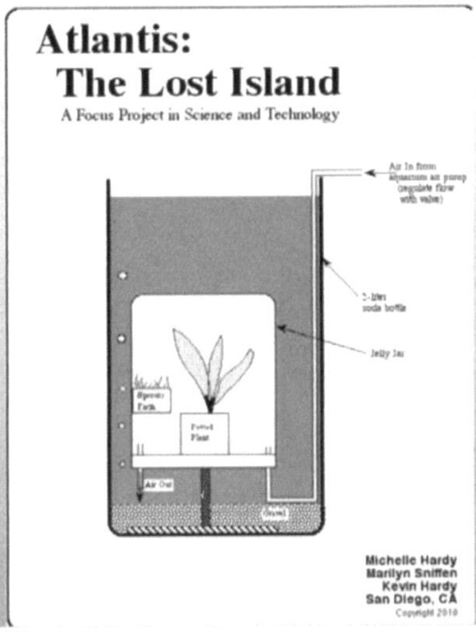

ENGAGING YOUNGER STUDENTS

OUTREACH

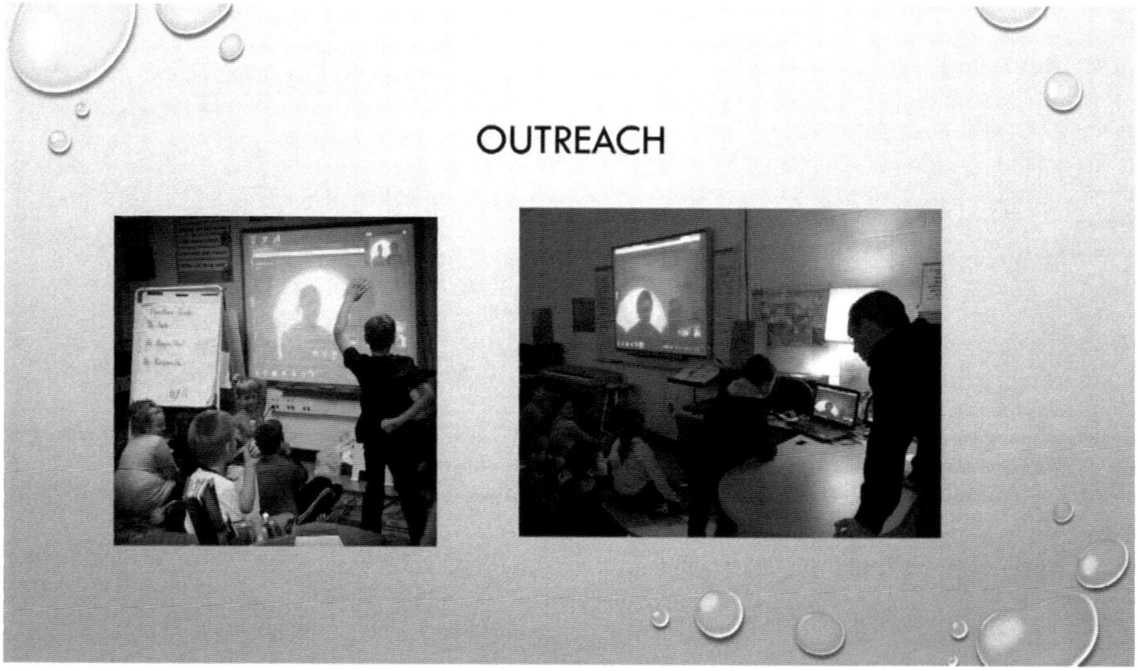
OUTREACH

MEDIA

OFF GASSING

LESSONS LEARNED

STEAM…CONNECTED LESSONS

San Francisco International OCEAN Film Festival

THUNDER BAY INTERNATIONAL FILM FESTIVAL

#WATERIS

San Francisco International Ocean Film Festival
5th Annual Student Film Competition
FREE to submit
Deadline for Submission
January 11, 2016

FUTURE PLANS

OceanGate FOUNDATION

Inspiration Through Exploration®

OceanGate Foundation is a not-for-profit educational outreach organization.

Our Open The Oceans program ignites curiosity, passion and lifelong learning by diving into the deep.

Our goal is to create STEM-engaged students using our access to submersible technology, interesting expeditions, and amazing content.

opentheoceans.org

OceanGate

OceanGate, Inc. is a privately held company that operates manned submersibles for industry, research and exploration.

OceanGate manned submersibles provide access for deep-sea technology testing, site survey and inspection, research and data collection and media production.

OceanGate is committed to opening the oceans and ensuring the responsible use of marine resources.

oceangate.com

Where are the last unexplored places?
outer
inner
@opentheoceans
OPEN THE OCEANS

OPEN THE OCEANS
Inspiration Through Exploration®
@opentheoceans
OPEN THE OCEANS

@opentheoceans
OPEN THE OCEANS

MAST
MAST

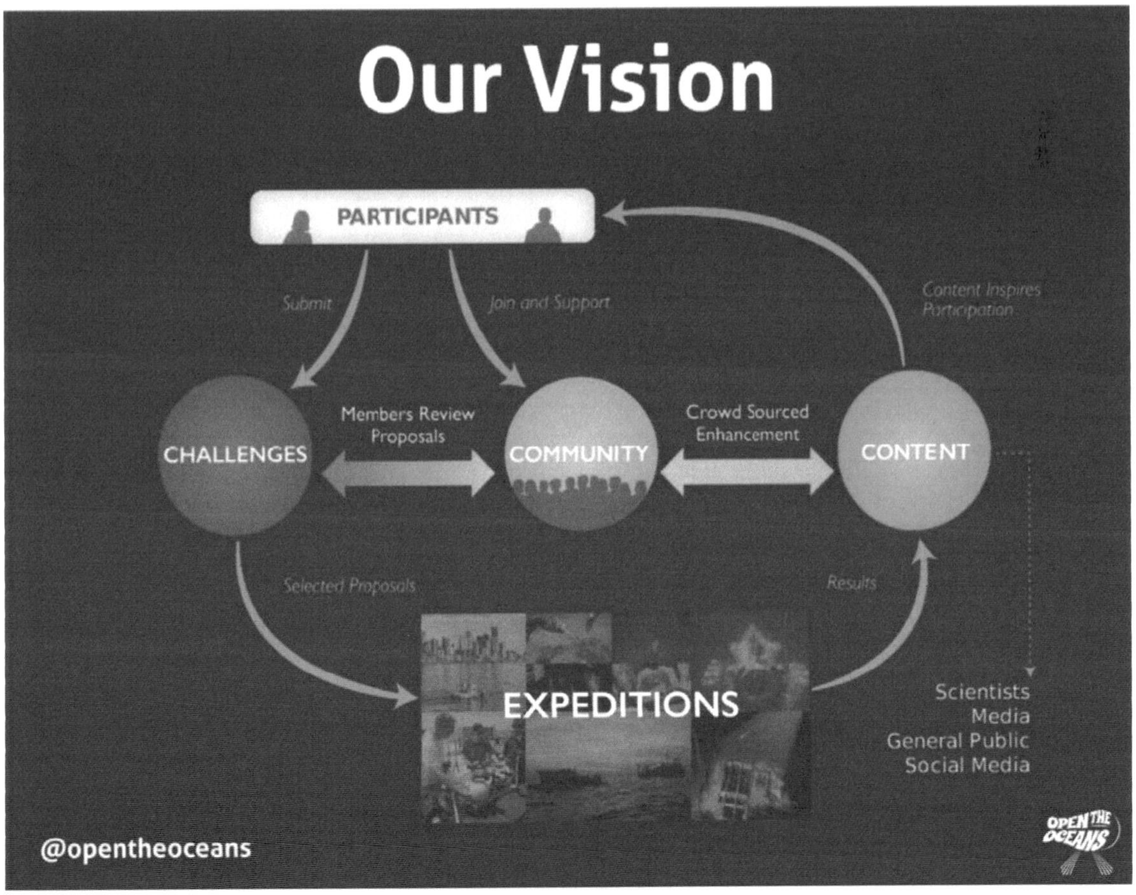
Our Vision
PARTICIPANTS
Submit
Join and Support
Content Inspires Participation
CHALLENGES
Members Review Proposals
COMMUNITY
Crowd Sourced Enhancement
CONTENT
Selected Proposals
Results
EXPEDITIONS
Scientists
Media
General Public
Social Media
@opentheoceans
OPEN THE OCEANS

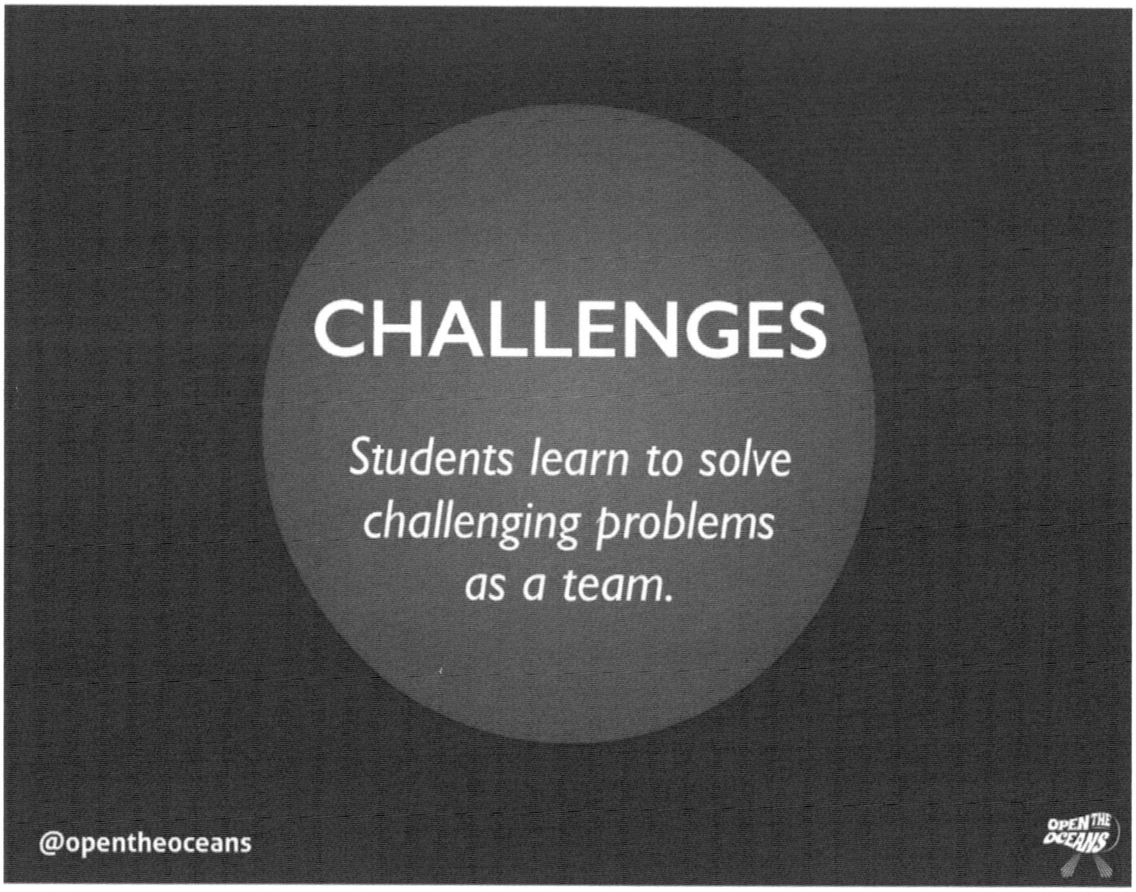
CHALLENGES
Students learn to solve challenging problems as a team.
@opentheoceans
OPEN THE OCEANS

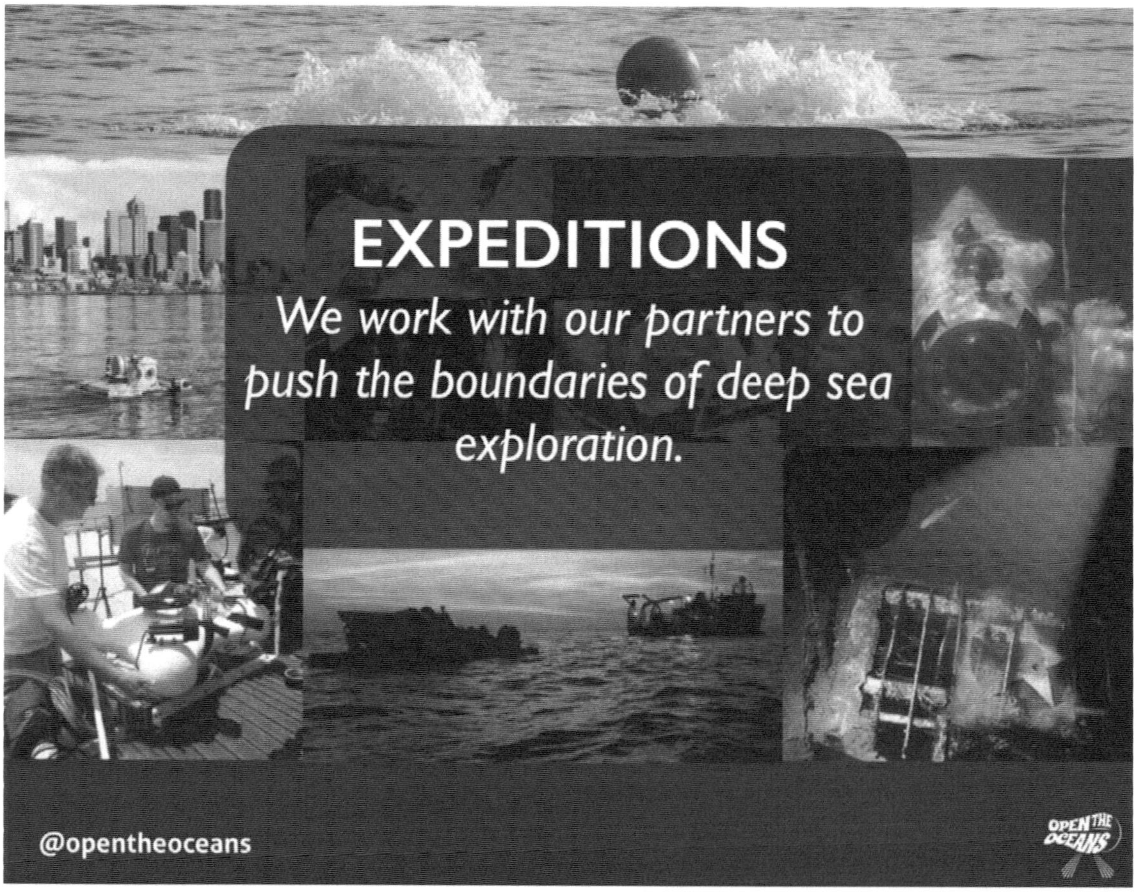
EXPEDITIONS
We work with our partners to push the boundaries of deep sea exploration.
@opentheoceans
OPEN THE OCEANS

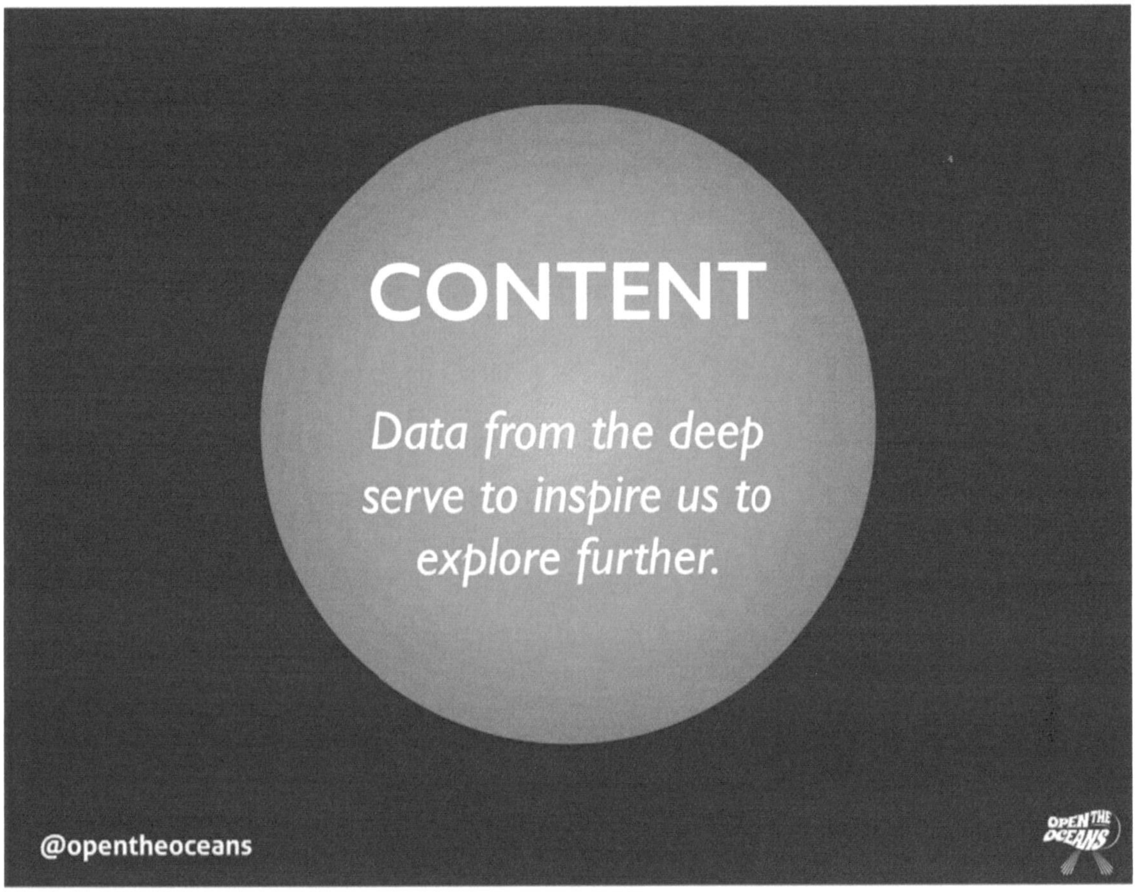
CONTENT
Data from the deep serve to inspire us to explore further.
@opentheoceans

COMMUNITY
We envision a community inspired and passionate about deep sea exploration
@opentheoceans

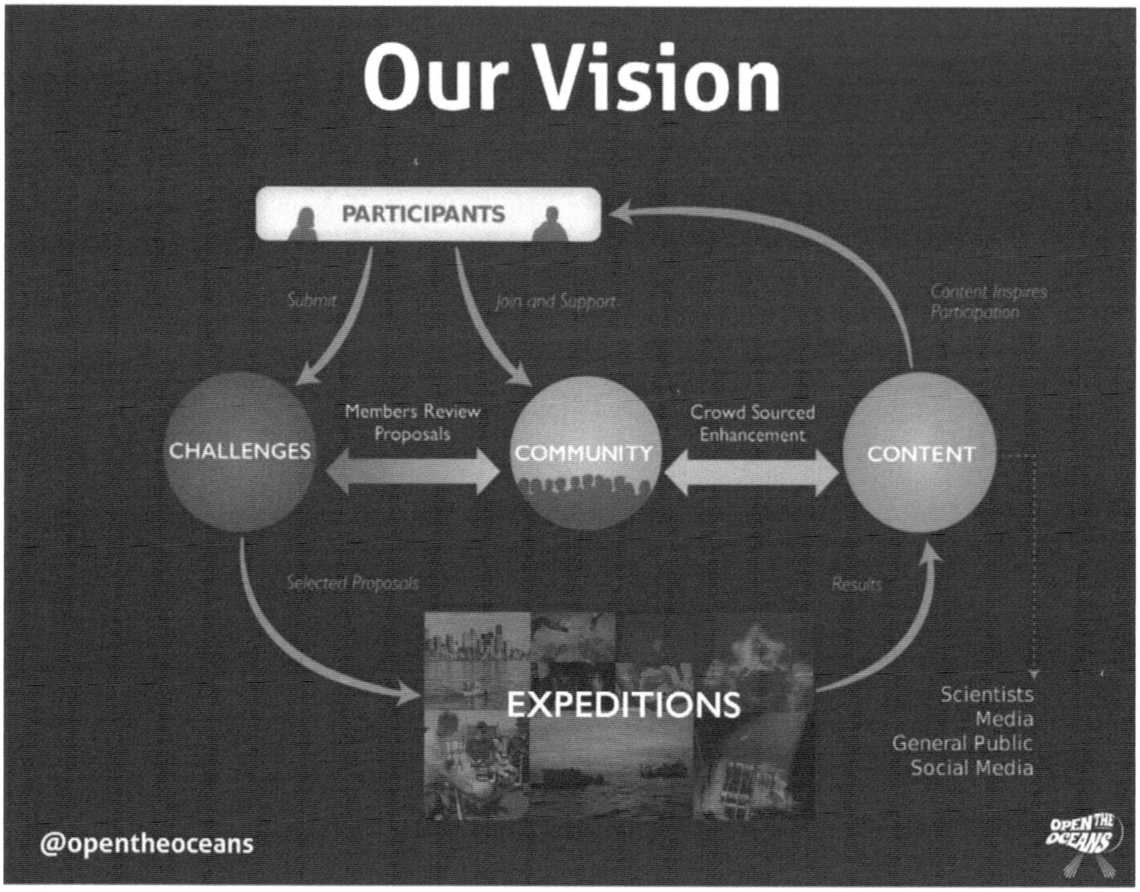
Our Vision
PARTICIPANTS
Submit
Join and Support
Content Inspires Participation
CHALLENGES
Members Review Proposals
COMMUNITY
Crowd Sourced Enhancement
CONTENT
Selected Proposals
Results
EXPEDITIONS
Scientists
Media
General Public
Social Media
@opentheoceans
OPEN THE OCEANS

Institutional Partners
Teachers and Researchers
OceanGate
NOAA
NATIONAL OCEANIC AND ATMOSPHERIC ADMINISTRATION
U.S. DEPARTMENT OF COMMERCE
NATIONAL MARINE SANCTUARIES
SEATTLE CENTRAL COLLEGE
One of the Seattle Colleges
W
UNIVERSITY of WASHINGTON
UNIVERSITY OF Rhode Island
@opentheoceans
OPEN THE OCEANS

@opentheoceans
OPEN THE OCEANS

Open The Oceans Team 1
@opentheoceans
OPEN THE OCEANS

Open The Oceans Challenge 2
2014
Winners

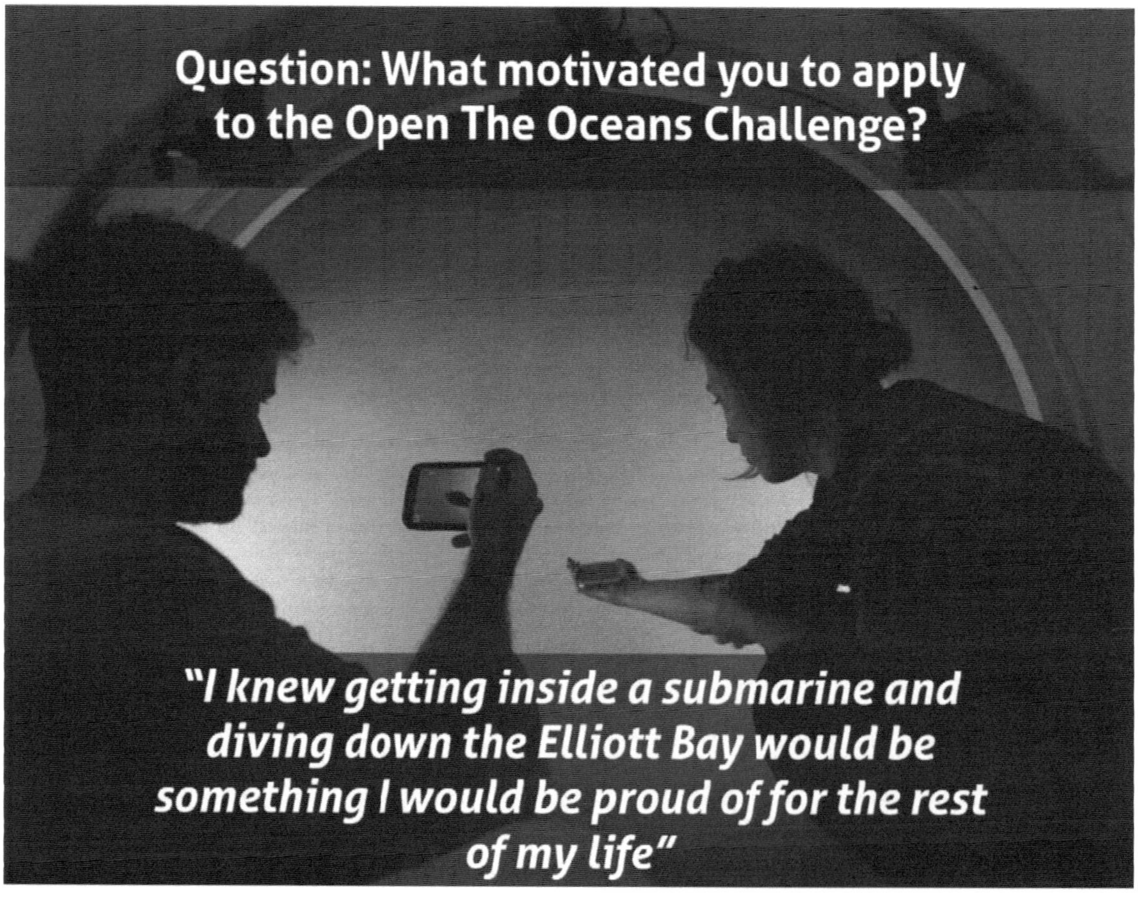
Question: What motivated you to apply to the Open The Oceans Challenge?
"I knew getting inside a submarine and diving down the Elliott Bay would be something I would be proud of for the rest of my life"

OTO Results: 3D Model

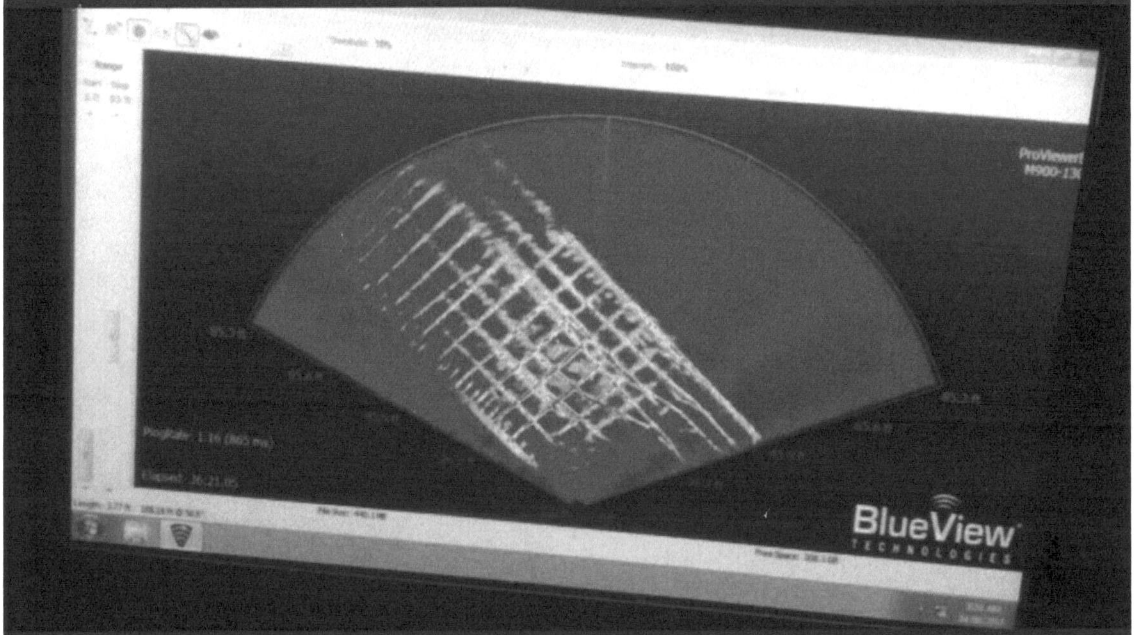

@opentheoceans

OTO Results: Colonization

- Preferential colonization by anemones was discovered on the wreck.
- It is unclear whether there is a systematic pattern to this or whether it is related to patchiness
- Still and video images can be correlated to location on the wreck using time stamps and sonar footage.

@opentheoceans

OTO Results: Lessons

- Students are incredibly motivated to participate in submersible expeditions.
- Students bring real insights to the field component.
- Learning what *doesn't* work is as important as what does.
- Having time and guidance to work up the data is an important aspect the program.

@opentheoceans

NSF IUSE Proposal

Inspiring the Next Generation of Ocean Technologists (INGOT)

A program developed in partnership with Seattle Central College to build on what we have done so far.

@opentheoceans

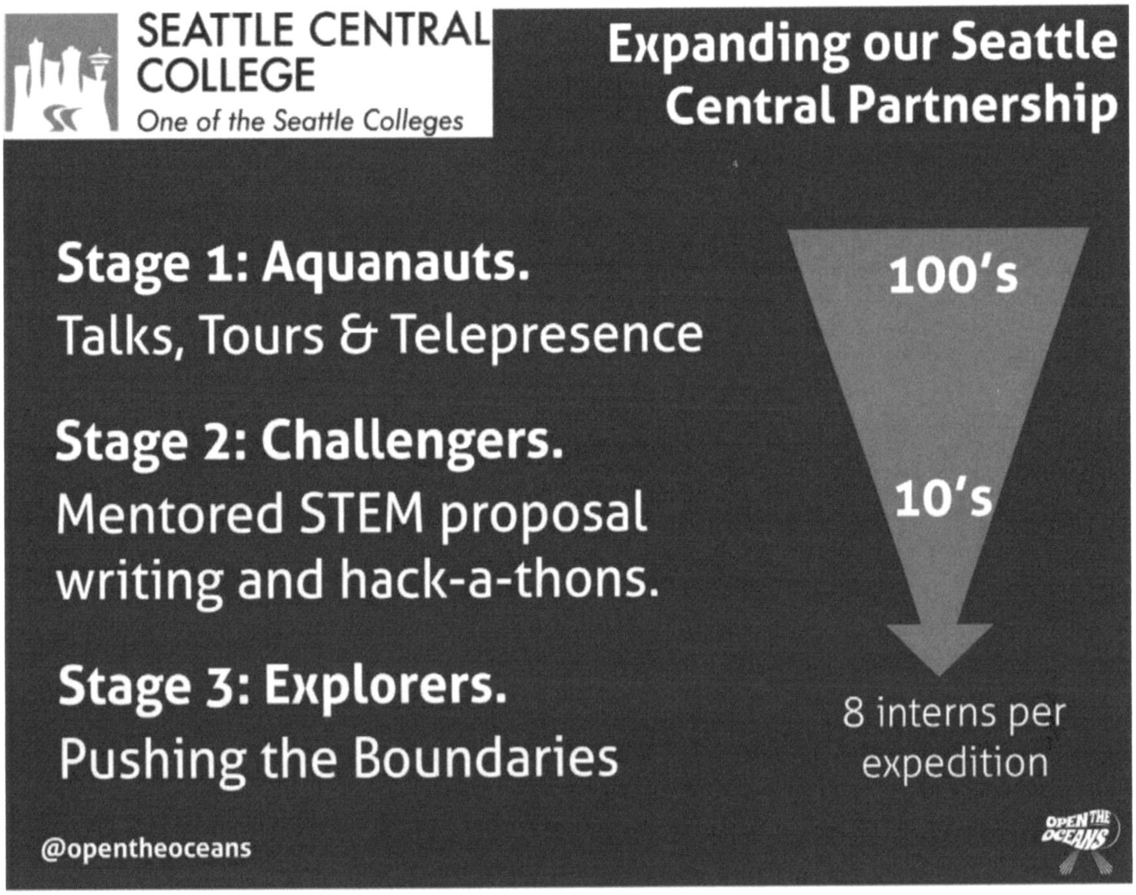
SEATTLE CENTRAL COLLEGE
One of the Seattle Colleges
Expanding our Seattle Central Partnership
Stage 1: Aquanauts.
Talks, Tours & Telepresence
Stage 2: Challengers.
Mentored STEM proposal writing and hack-a-thons.
Stage 3: Explorers.
Pushing the Boundaries
100's
10's
8 interns per expedition
@opentheoceans
OPEN THE OCEANS

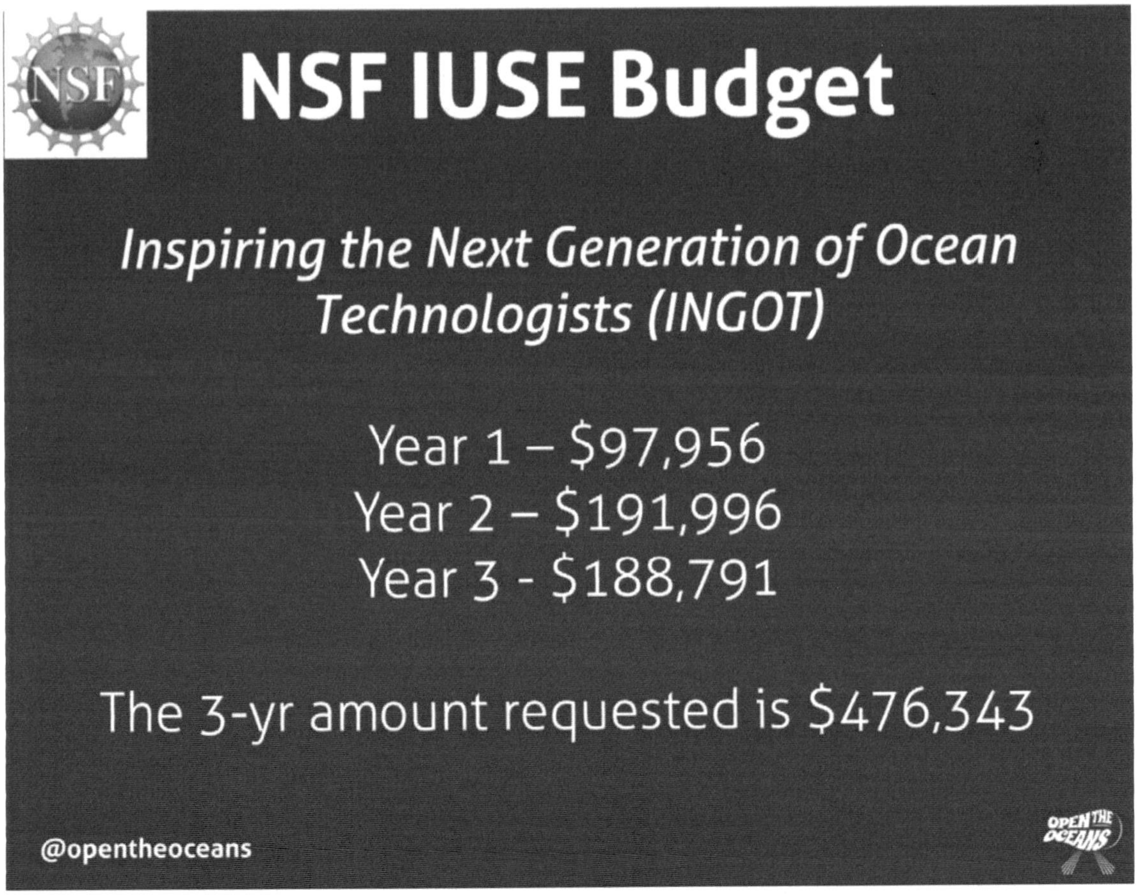
NSF
NSF IUSE Budget
Inspiring the Next Generation of Ocean Technologists (INGOT)
Year 1 – $97,956
Year 2 – $191,996
Year 3 - $188,791
The 3-yr amount requested is $476,343
@opentheoceans
OPEN THE OCEANS

OPEN THE OCEANS
Interested in Learning More?
cpsarason@oceangatefoundation.org
@opentheoceans
OPEN THE OCEANS

Development of Student's Soft Skills in Preparation for a Career on the Water

ALPENA COMMUNITY COLLEGE

Presented by:
David Cummins
cumminsd@alpenacc.edu
989-358-7224
and
RC Laugal
laugalr@alpenacc.edu
989-358-7417

Thanks to:
OCEANEERING
www.oceaneering.com
National Marine Sanctuaries
NOAA
MATE

Development of Student's Soft Skills in Preparation for a Career on the Water

What employers are looking for:

- On time
- Take the initiative
- Is productive
- Works well with others
- Ability to communication

ALPENA COMMUNITY COLLEGE

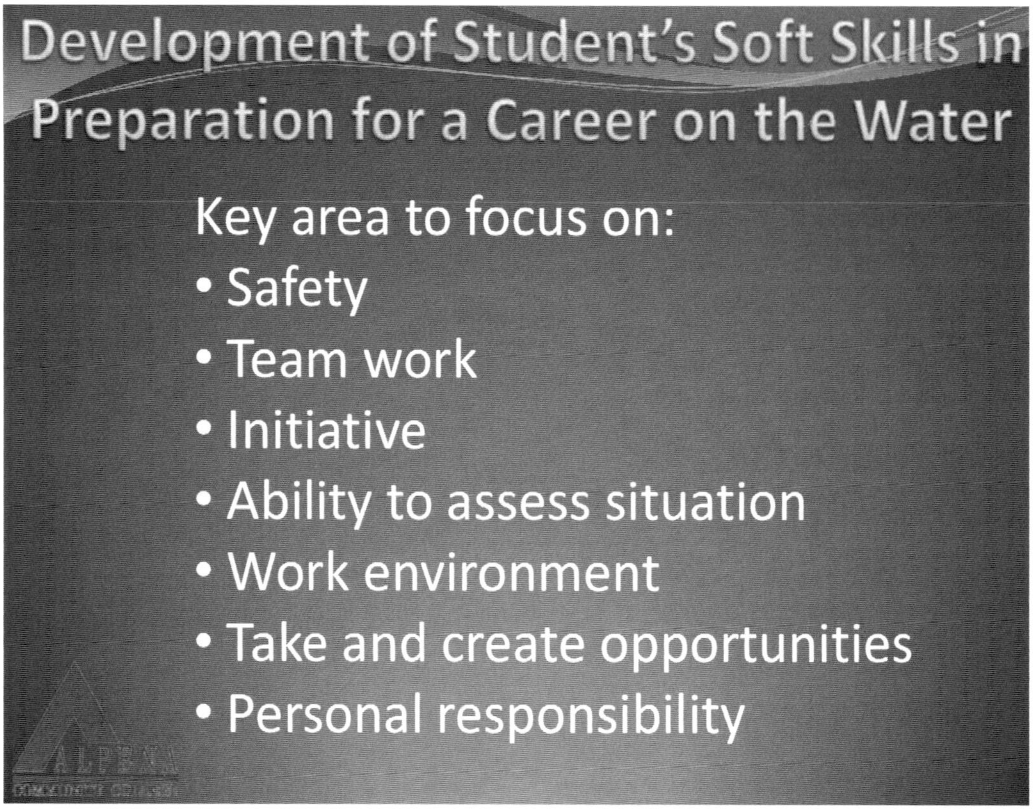
Development of Student's Soft Skills in Preparation for a Career on the Water
Key area to focus on:
• Safety
• Team work
• Initiative
• Ability to assess situation
• Work environment
• Take and create opportunities
• Personal responsibility

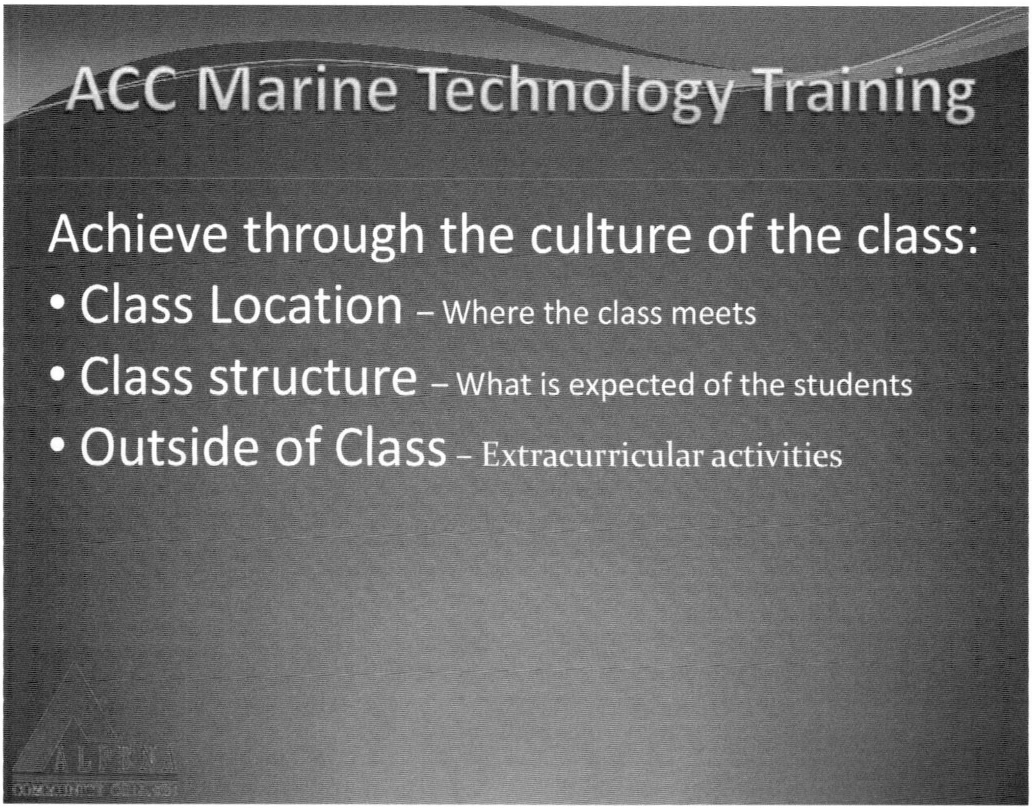
ACC Marine Technology Training
Achieve through the culture of the class:
• Class Location – Where the class meets
• Class structure – What is expected of the students
• Outside of Class – Extracurricular activities

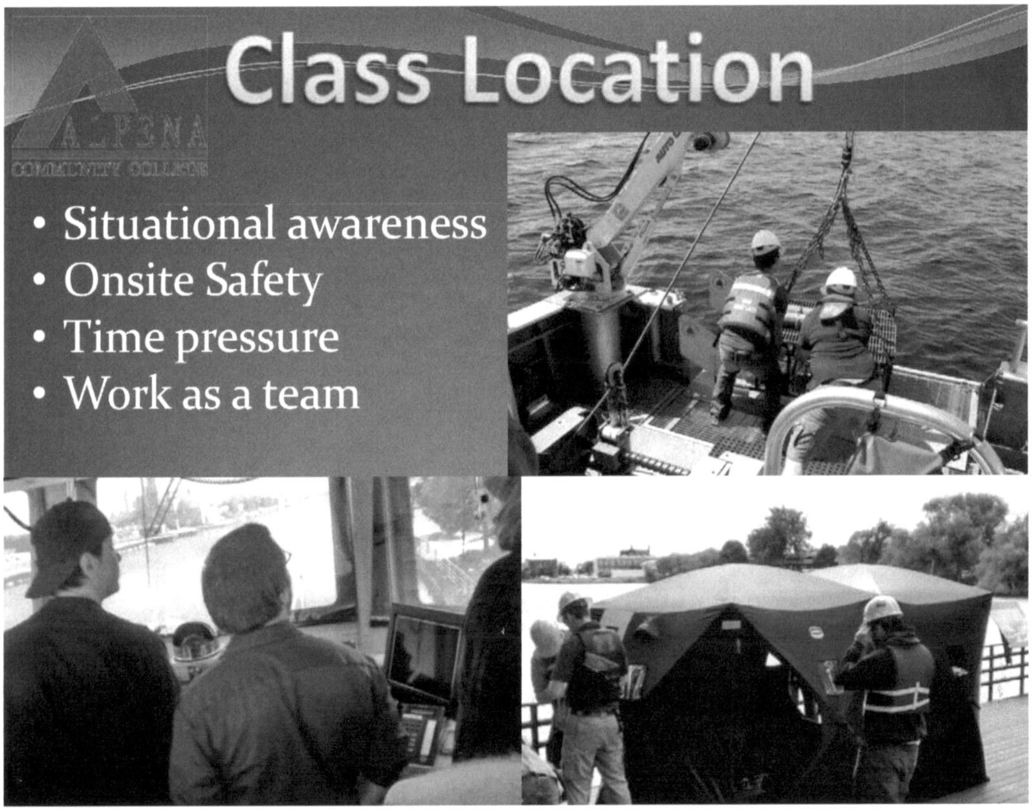
Class Location
• Situational awareness
• Onsite Safety
• Time pressure
• Work as a team

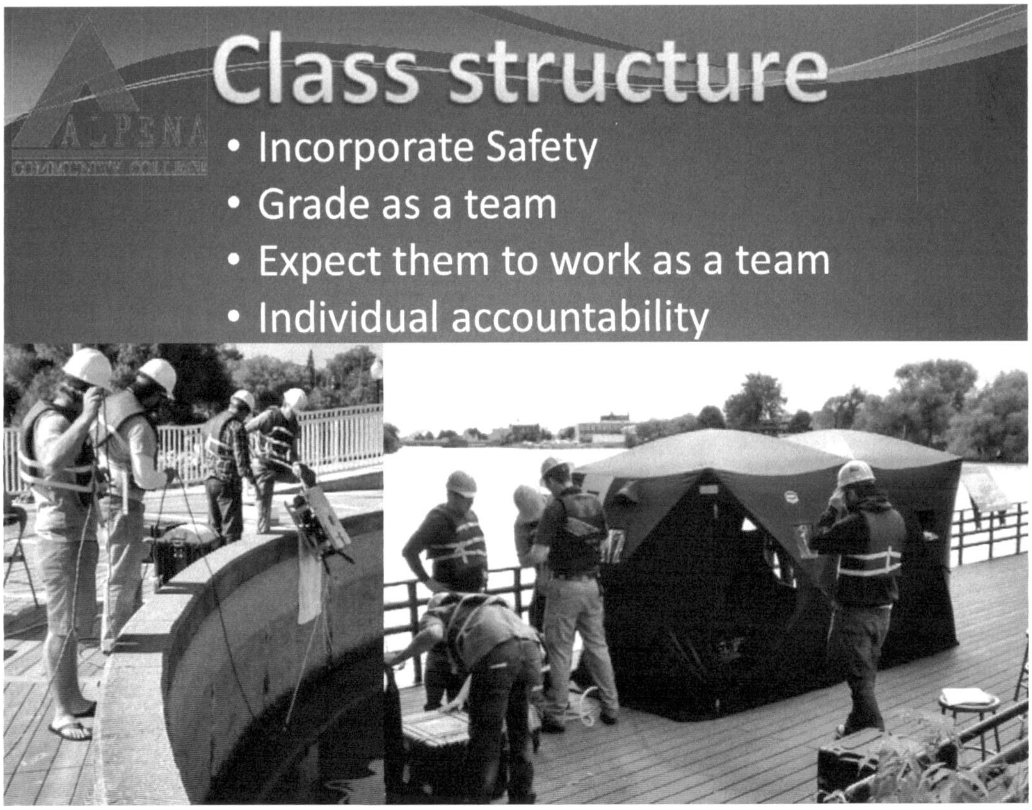
Class structure
• Incorporate Safety
• Grade as a team
• Expect them to work as a team
• Individual accountability

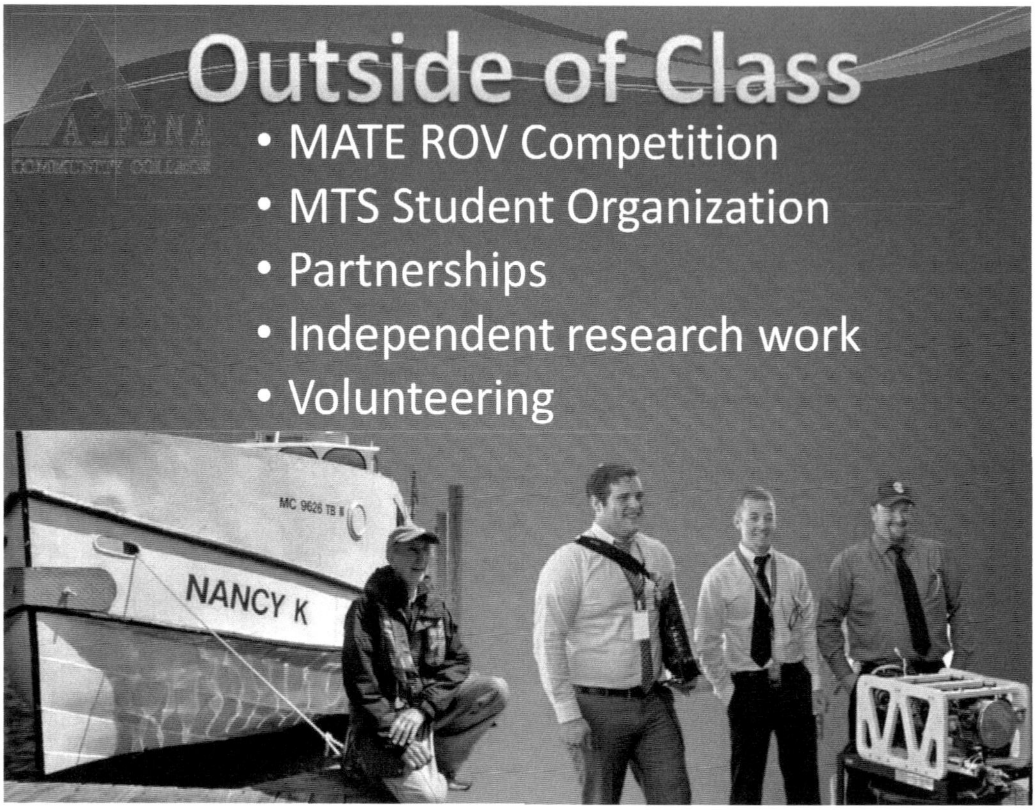
Outside of Class
• MATE ROV Competition
• MTS Student Organization
• Partnerships
• Independent research work
• Volunteering
NANCY K
MC 9626 TB

Group Discussion
OCEANEERING
Presented by:
David Cummins
cumminsd@alpenacc.edu
989-358-7224
and
RC Laugal
laugalr@alpenacc.edu
989-358-7417
MATE
NOAA
NATIONAL MARINE SANCTUARIES
THUNDER BAY
ALPENA
COMMUNITY COLLEGE
OCEANEERING
www.oceaneering.com
Marine Tech Program

Fred Donelson - UI 2016

Getting ROVs Into the Curriculum

Our World is changing ...

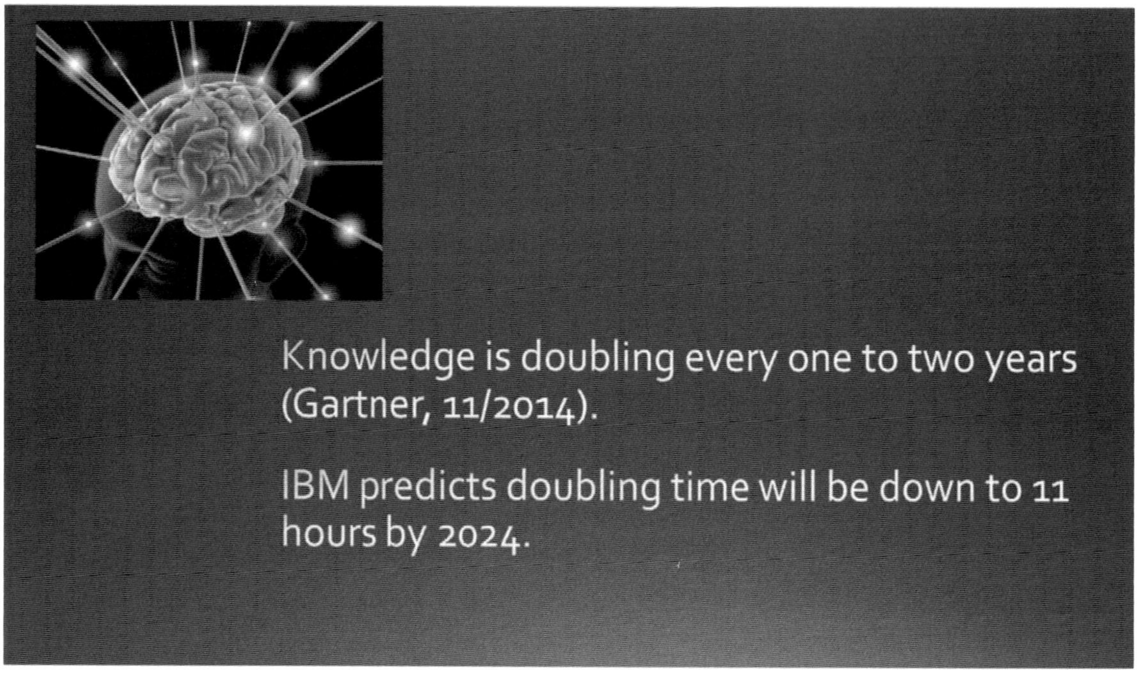
Knowledge is doubling every one to two years (Gartner, 11/2014).
IBM predicts doubling time will be down to 11 hours by 2024.

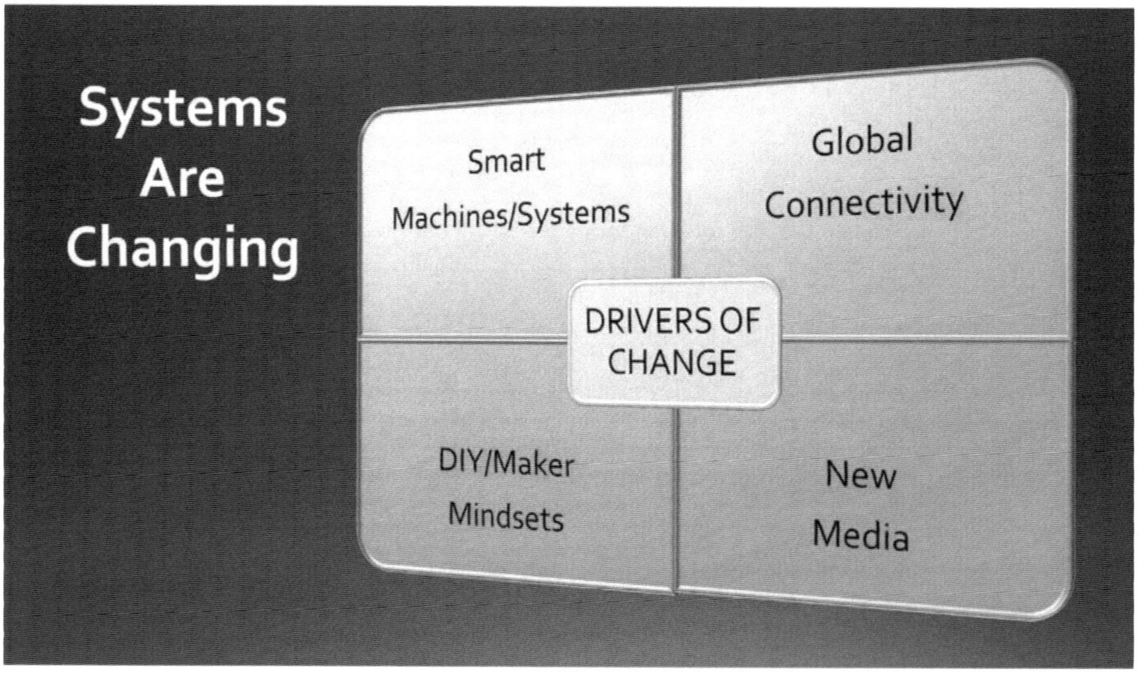
Systems Are Changing
Smart Machines/Systems
Global Connectivity
DRIVERS OF CHANGE
DIY/Maker Mindsets
New Media

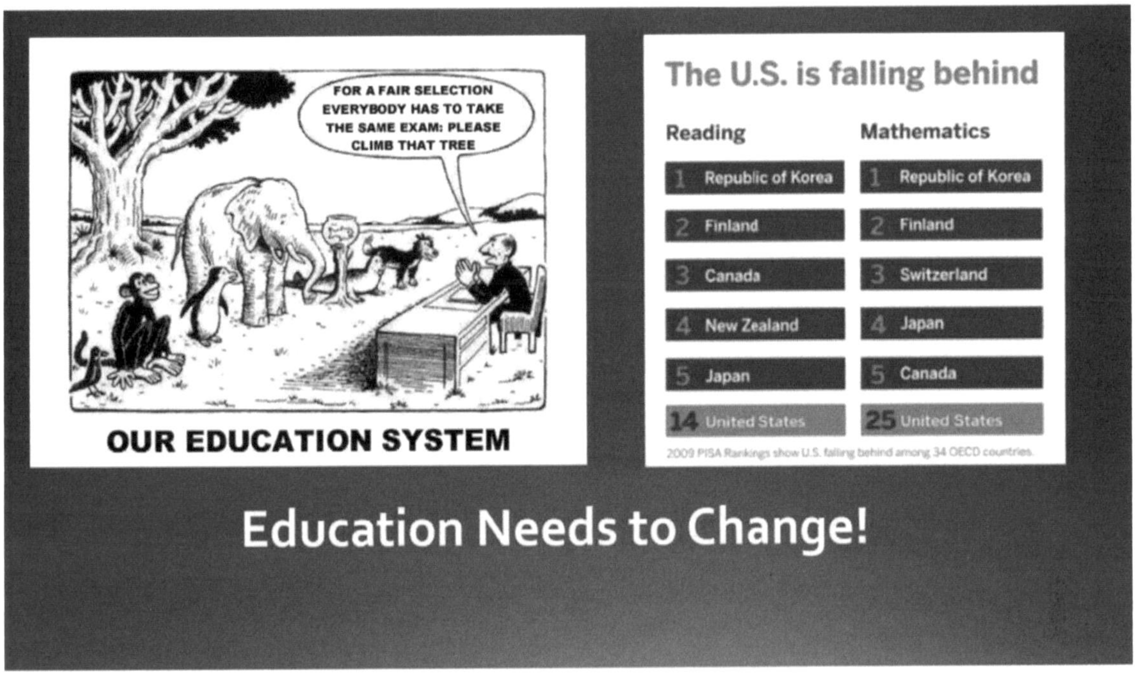
FOR A FAIR SELECTION EVERYBODY HAS TO TAKE THE SAME EXAM: PLEASE CLIMB THAT TREE
OUR EDUCATION SYSTEM
The U.S. is falling behind
Reading
1 Republic of Korea
2 Finland
3 Canada
4 New Zealand
5 Japan
14 United States
Mathematics
1 Republic of Korea
2 Finland
3 Switzerland
4 Japan
5 Canada
25 United States
2009 PISA Rankings show U.S. falling behind among 34 OECD countries.
Education Needs to Change!

"I don't love studying.
I hate studying.
I like learning. Learning is beautiful."
- Natalie Portman

Gahanna Lincoln High School
Science Academy

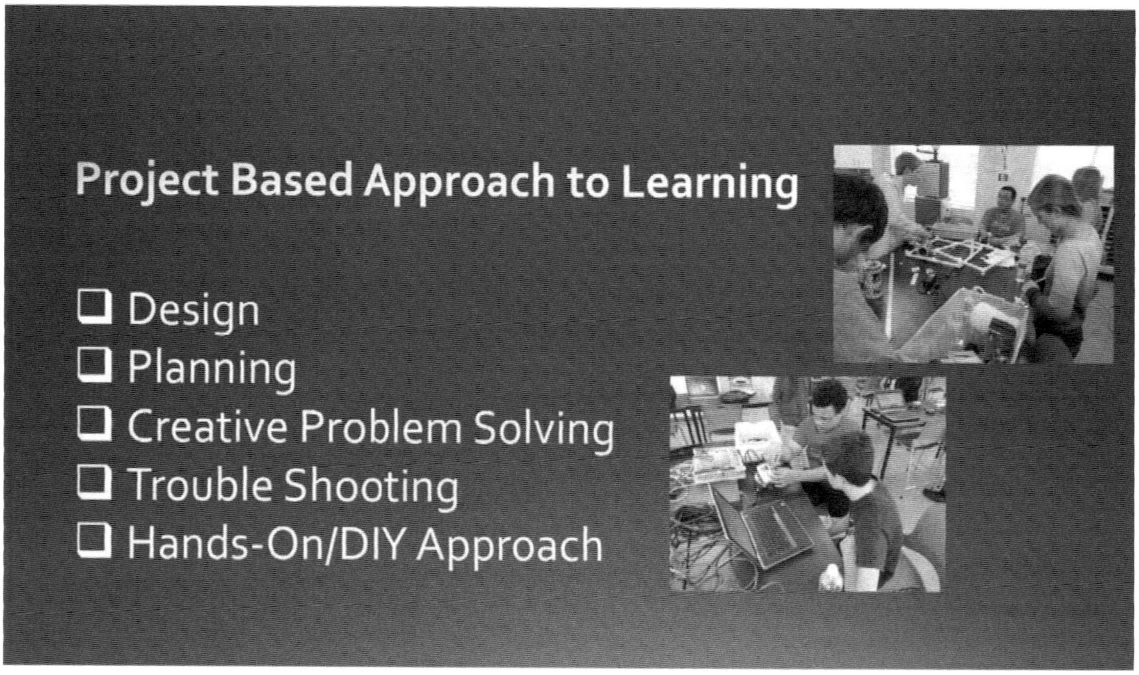
Project Based Approach to Learning
❑ Design
❑ Planning
❑ Creative Problem Solving
❑ Trouble Shooting
❑ Hands-On/DIY Approach

The China Project
中国项目
Zhōngguó xiàngmù

Rita's

One Roadblock Continues:
TESTING ...
NOT ENOUGH TIME ...

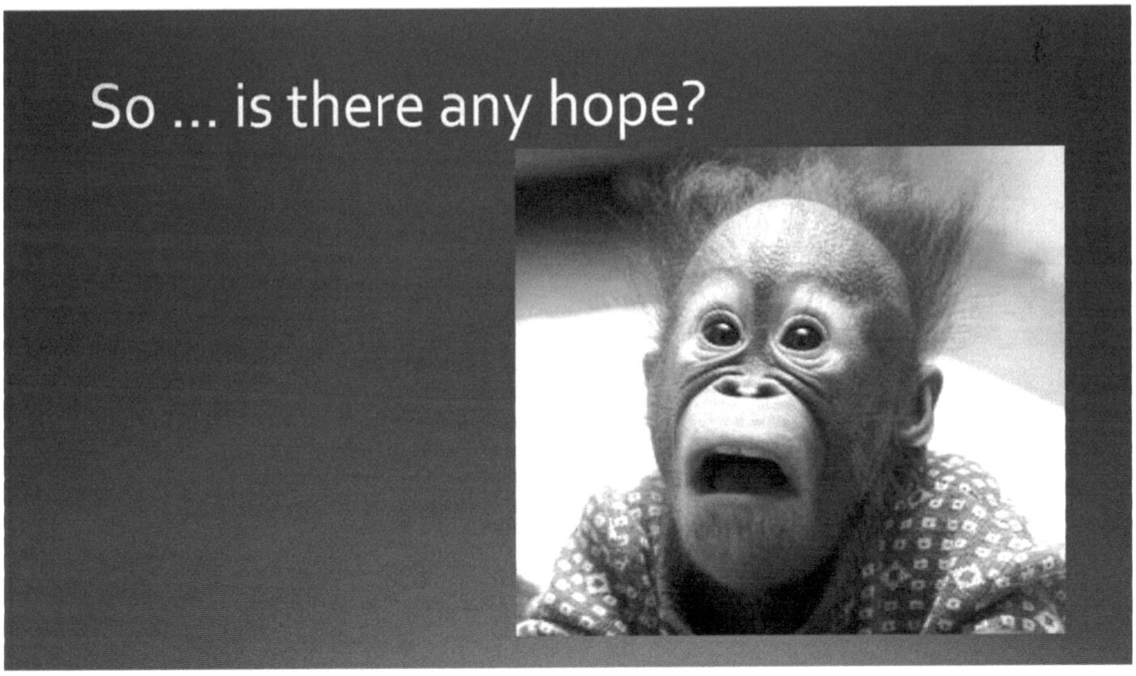
So ... is there any hope?

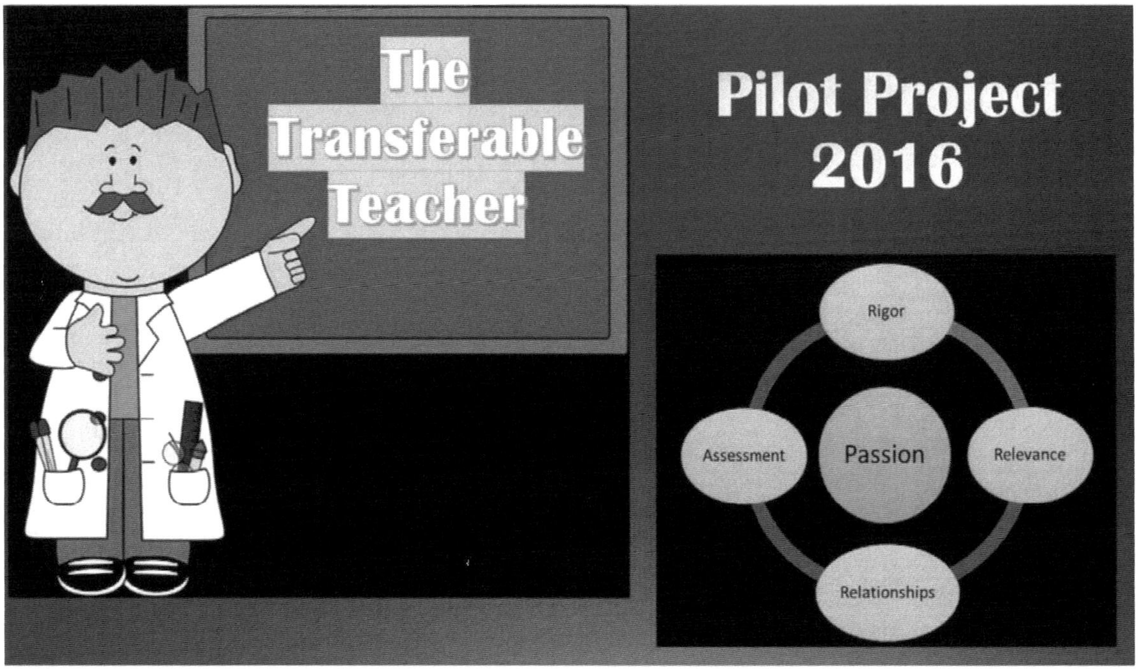

Goal:
Create a semester curriculum that can be subdivided into small projects or lessons that fit into core science classes or other normal classes.

Delivery Materials

- Outline of 8 Units (One Semester, 2 lessons/week, 40 minute lessons)
- Science Objectives for each unit
- Power Point Presentation for each unit
- Videos for skills needed for each unit
- Worksheet with problems for each unit
- Final Project Suggestions and helps for setting up missions, judges

Orthographic/Isometric Sketching

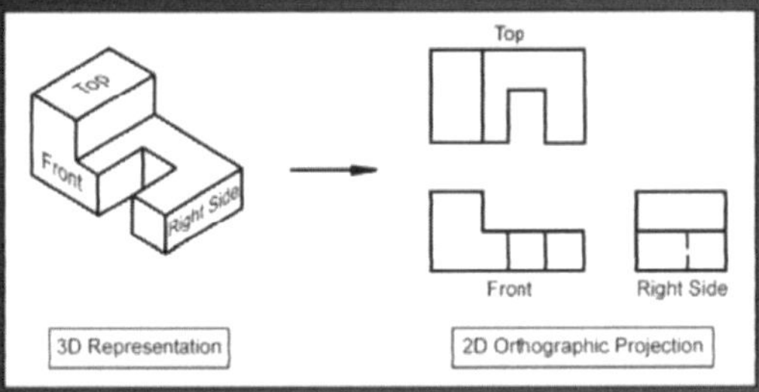

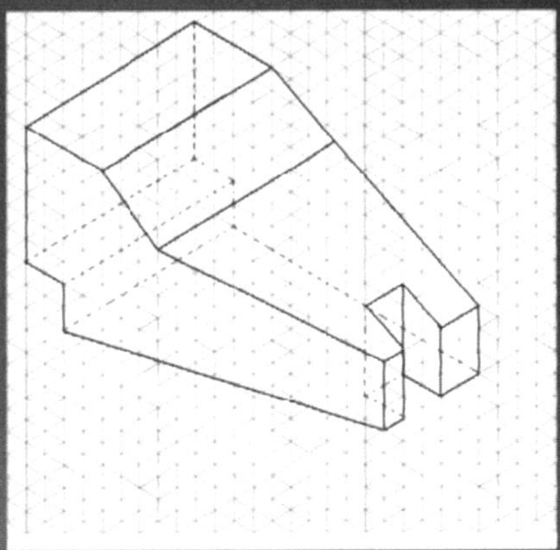

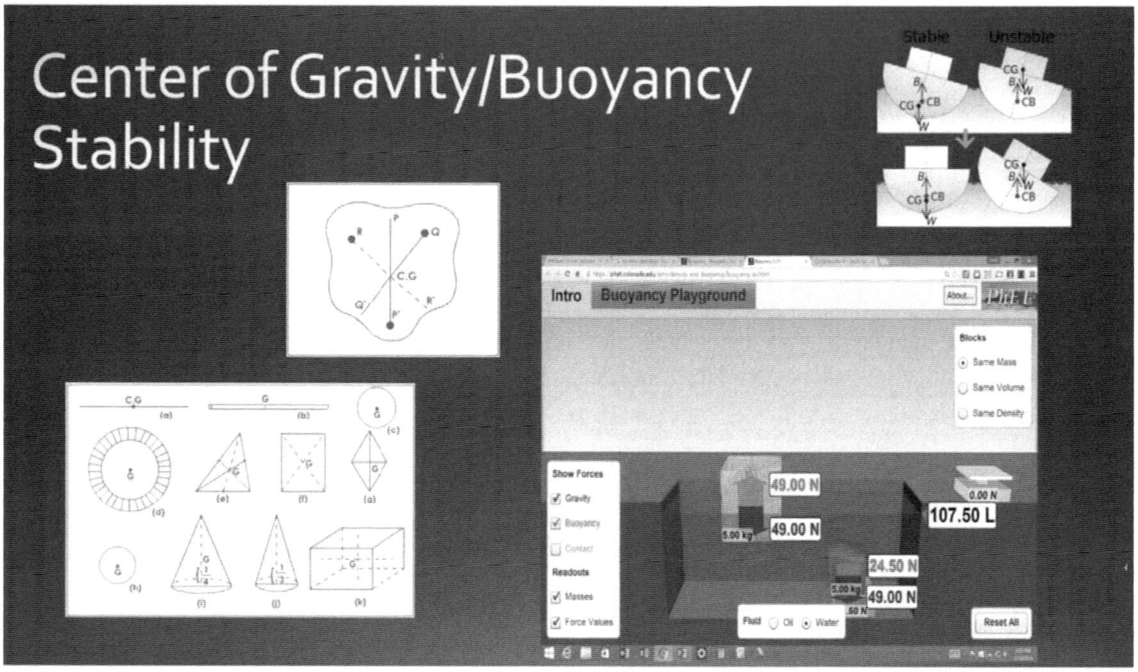

Center of Gravity/Buoyancy Stability
Stable
Unstable
Intro
Buoyancy Playground
Blocks
Same Mass
Same Volume
Same Density
Show Forces
Gravity
Buoyancy
Contact
Readouts
Masses
Force Values
49.00 N
49.00 N
24.50 N
49.00 N
107.50 L
Fluid
Oil
Water
Reset All

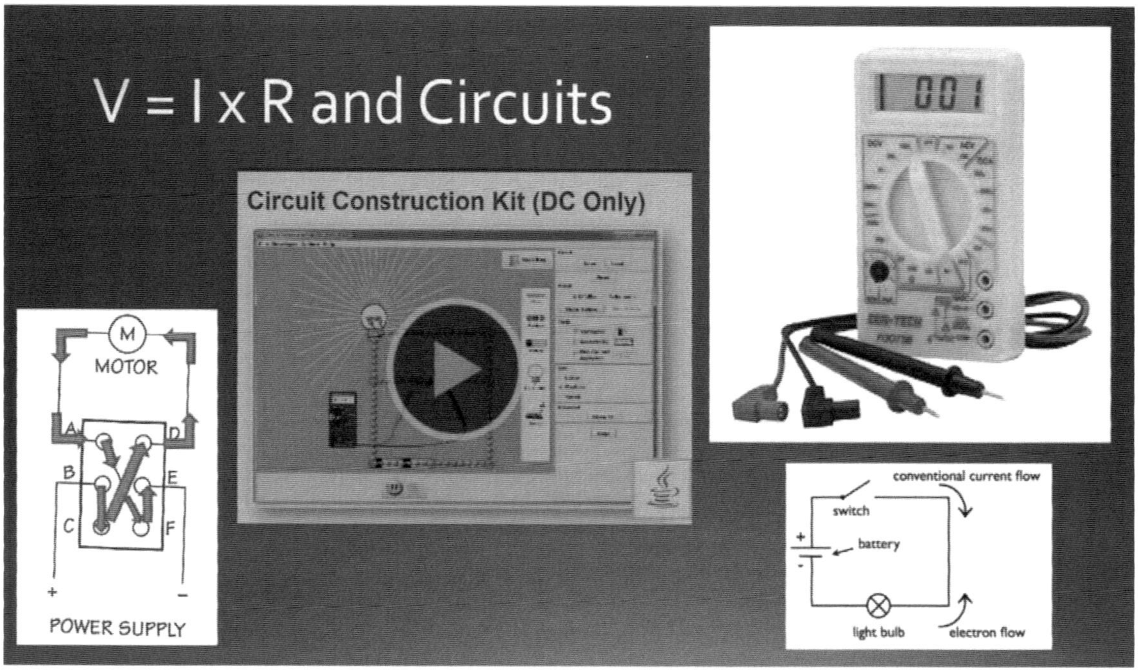

V = I x R and Circuits
Circuit Construction Kit (DC Only)
MOTOR
POWER SUPPLY
conventional current flow
switch
battery
light bulb
electron flow

Motors, Waterproofing, and Loads

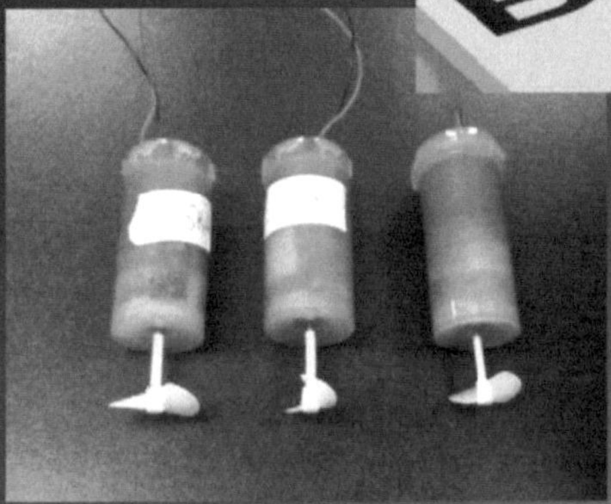

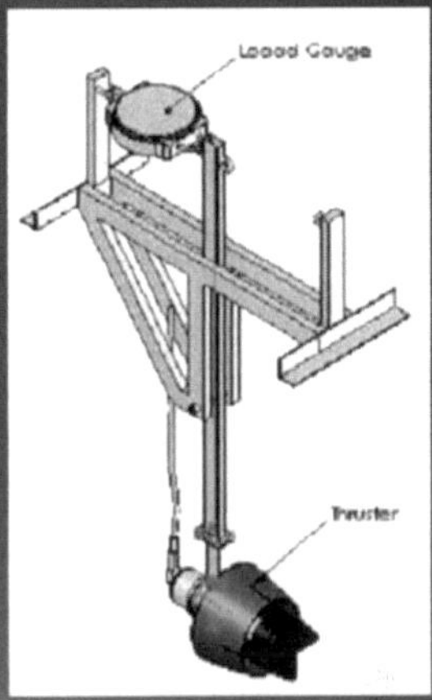

Tethers and Resistance

Wire gauge	*Ohms/1000 ft (approx.)*
20	10
18	6
16	4
14	2.5
12	1.5

Tool Packages: Sensors

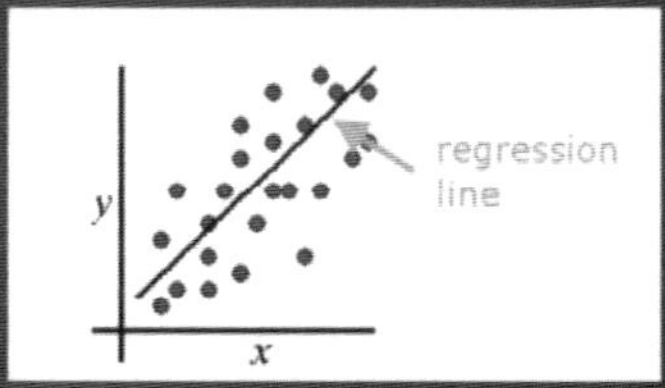

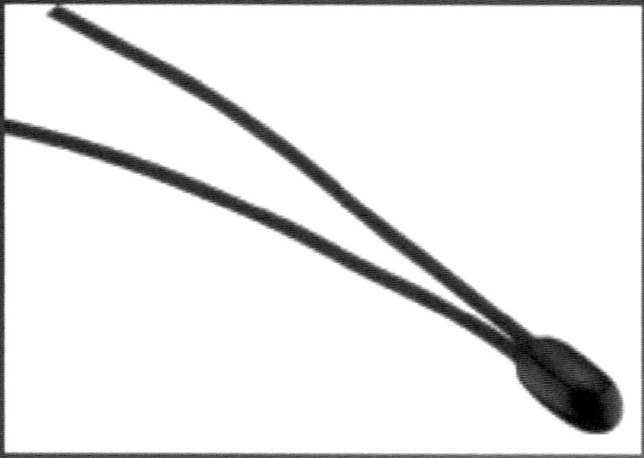

Tool Packages: Grippers/Hydraulics

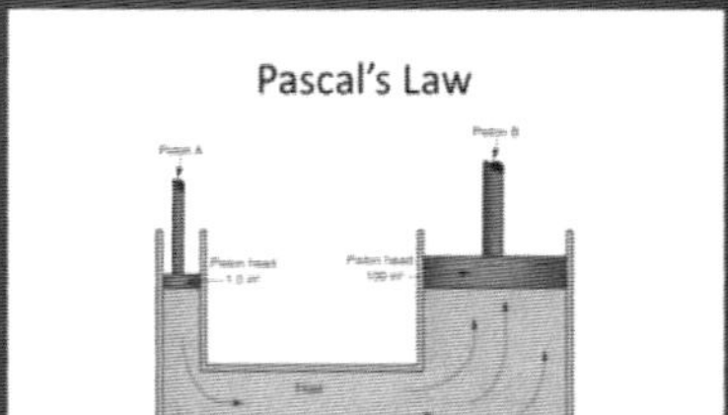

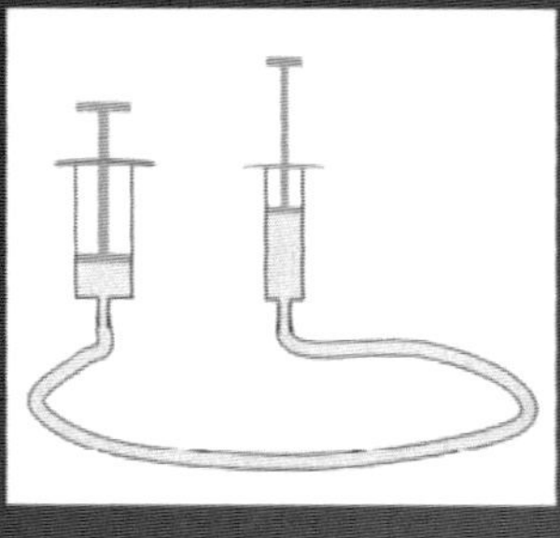

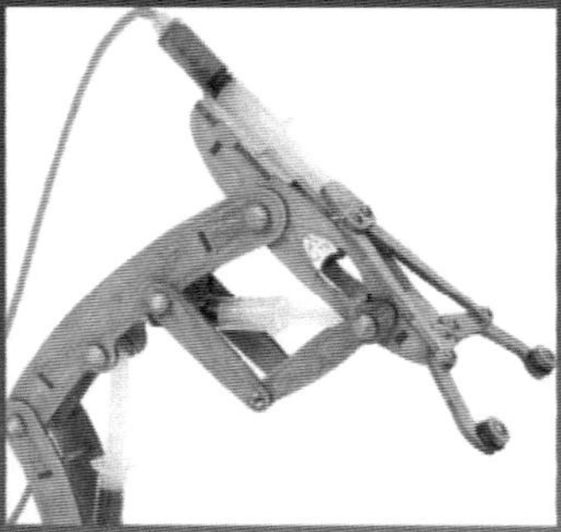

Projects and Teamwork: Constraints, Deadlines, Gantt Charts

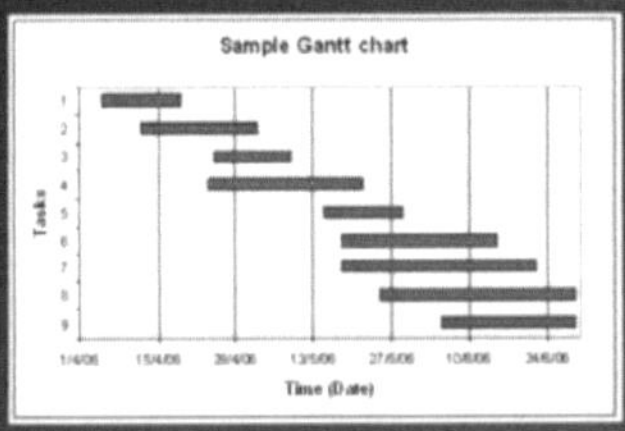

gantter
FREE project scheduling

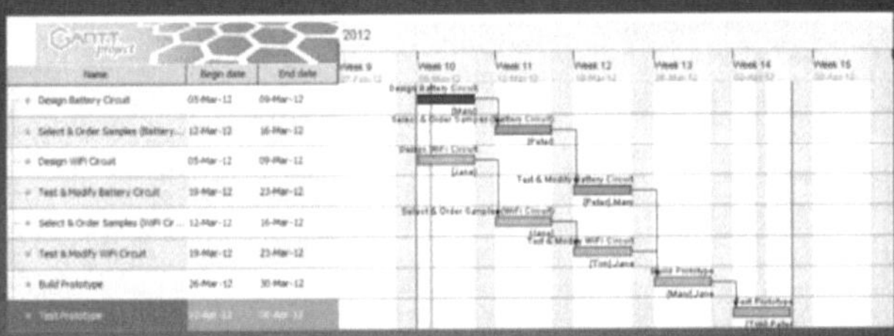

Fred Donelson
fdonelson@gmail.com

Twitter : @mrdglhs

Facebook: Fred Donelson

Short Term ROV Training Solutions for the Marine Industry

Hans Van Sumeren
Northwestern Michigan College
Underwater Intervention 2016

Northwestern Michigan College

- Founded in 1951 - First Community College in the State of Michigan
- Four campuses in Traverse City, 4000 full time students
- First Community College to offer BS degrees in Michigan
- GLMA -1 of 6 federally authorized maritime academies
- Aviation – manned and unmanned training

Strategic Directions

- Global preparation
- National/International leadership
- Networked delivery
- Lifelong relationships
- New credentials

Trends – Higher Education

From	To
Time	Competency
Place	Connect anywhere
Knowledge transfer	Application/synthesis
Reactive	Proactive
Long cycle time	Short cycle time
Public good	Private good
Access	Success

ROV training specifics

- Content built on industry feedback
- 5-15 person cohorts
- M-F, 9 hours/day, ~10 weeks
- Homework and examinations
- 2/3 applied laboratory and hands on training
- Campus housing and local short term options
- Flexible offering
- VA tuition support
- $10,000 in tuition and fees

Multiple Training Environments

- Great Lakes
- NMC Harbor
- Research vessels
- Rivers
- Under ice
- Indoor training facility

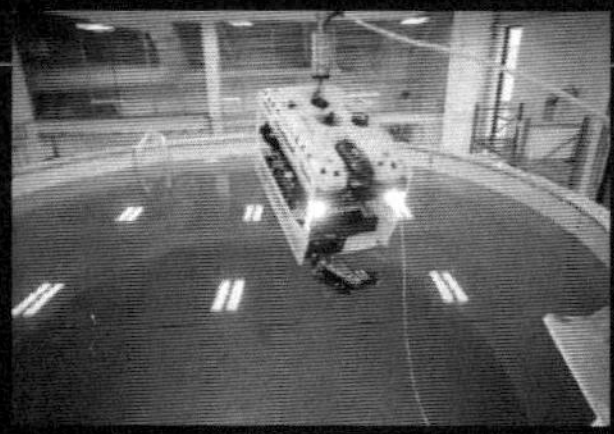

ROV Systems and operations

- Multiple platforms
- Fiber and copper systems
- Integrated USBL
- Manipulator training
  - 5 function hydraulic
  - Single function sticks
- Multiple scanning sonars
- ROV integrated MBES
- Greensea Automation and Control
- CP Probe

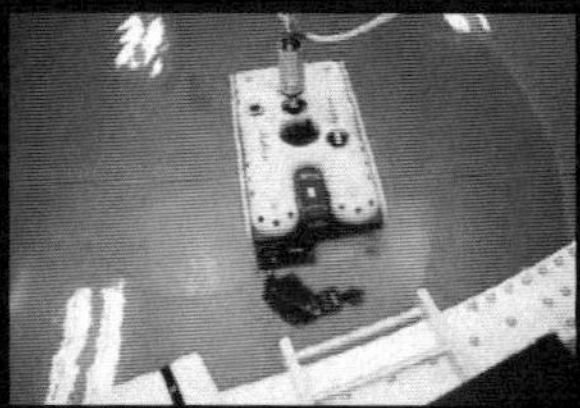

Emphasis on Safety, Communication, Teamwork and Procedure

- PEC Core Compliance
- First Aid/AED/CPR
- JSA
- Vessel safety
- Crane and hoist
- Extreme weather

Electronics, Fiber, Terminations, connectors

- Safety
- Circuit Theory and Analysis
- Circuit Construction and Testing
- Troubleshooting
- Measure and Analyze

Fluid Power and Pump Systems

- Fluid Power Safety
- Basic Hydraulic/Fluid Power Systems
- Sources of Power and Fluids
- Fluid Storage and Distribution
- Actuators and Pressure Control Valves
- Directional and Proportional Control Valves
- Troubleshooting

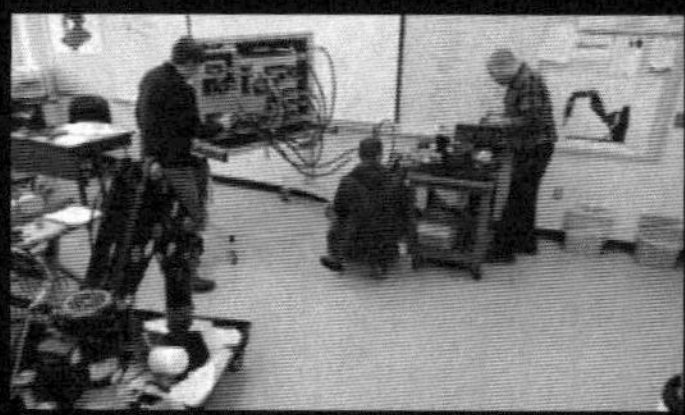

10 week program specifics

- PEC Core Compliance
- First Aid/AED/CPR
- Basic/intermediate Electronics and Hydraulics
- Advanced training in tether terminations, fiber optics, electronics troubleshooting
- Advanced hydraulics troubleshooting and manipulator training
- ROV systems and operations – multiple environments
- ROV Acoustics

Multiple Training/Education Opportunities

- 10 week ROV Training Program
- Associates in Marine Technology
- Bachelors of Science in Marine Technology
- Credentials of Value (CSWA, PMIA, IFPS)
- Professional Development Opportunities
  - Hydraulics
  - Acoustics
  - Data Processing

Questions?

Northwestern Michigan College

Hans Van Sumeren
Great Lakes Water Studies Institute
hvansumeren@nmc.edu
231-995-1793

UNDERWATER INTERVENTION 2016

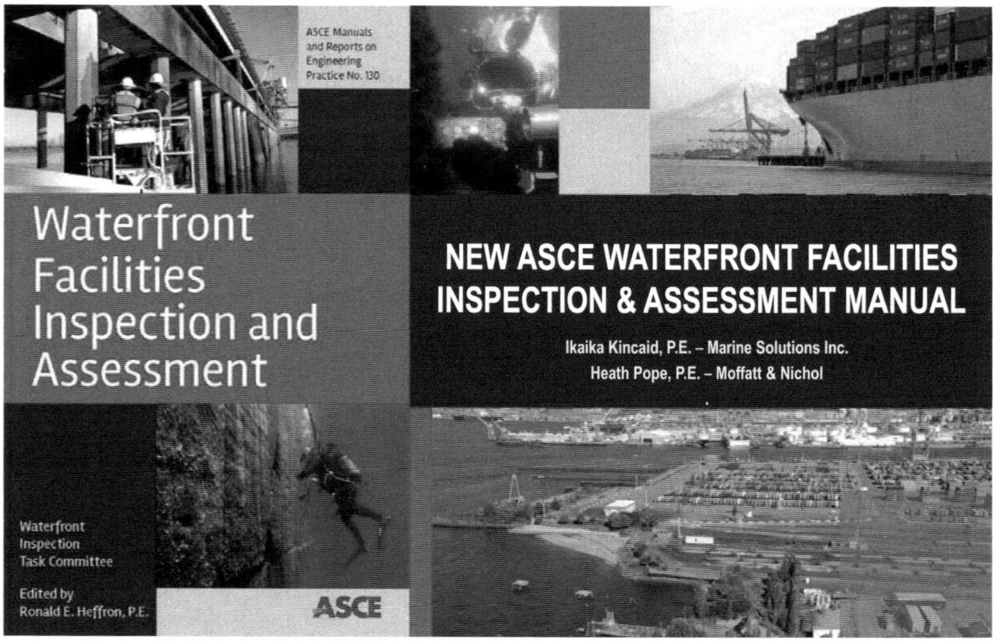

Agenda

- Need and Purpose
- Scope
- The Team
- New Topics Introduced
- Cutting-edge and Controversial Topics Introduced
- Content of the Guidelines

NEW ASCE WATERFRONT FACILITIES INSPECTION & ASSESSMENT MANUAL | UI 2016 | 2

Need and Purpose

- "Underwater Investigations Standard Practice Manual" (ASCE Manual 101)
  - Published in 2001
  - Written by engineers, for engineers
  - Provides guidance on many structure types, from piers & wharves to bridges, dams and tunnels
- The Need for a Waterfront Facilities focused manual became evident
  - Inclusive of the entire structure and fixed appurtenances
  - New concepts and technical approaches needed

Need and Purpose

"Waterfront Facilities Inspection & Assessment Manual"

- Covers inspection of the entire asset
- Provides guidance to Owners, as well as "executing" engineers

Scope

- Comprehensive treatment of waterfront facilities:
  - Piers/jetties
  - Wharves/quays
  - Bulkheads/quaywalls
  - Wave screens
  - Marinas
  - Boat ramps
  - Floating structures
  - Buoys
  - Slope protection

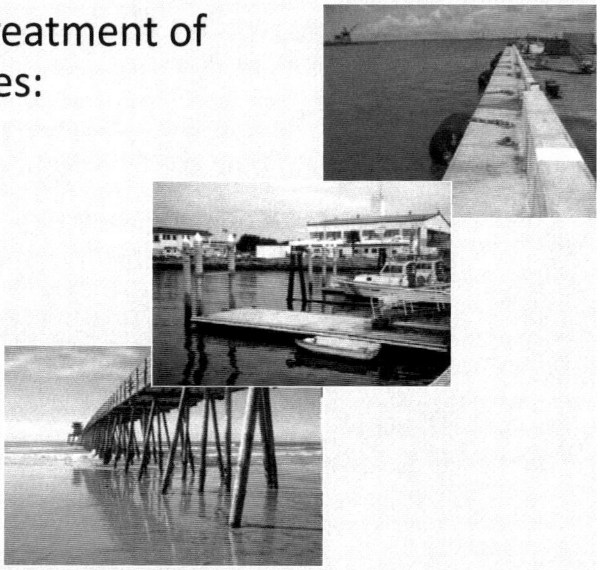

NEW ASCE WATERFRONT FACILITIES INSPECTION & ASSESSMENT MANUAL | UI 2016 | 5

Scope

- Guidance provided for:
  - Structural components
    - Above water and underwater
  - Fixed utilities
  - Equipment
  - Mooring hardware
  - Topside paving and drainage
  - Safety features
  - Appurtenances
  - Excluded: container cranes; material offloading/conveyance equipment

NEW ASCE WATERFRONT FACILITIES INSPECTION & ASSESSMENT MANUAL | UI 2016 | 6

Scope

- Working knowledge of waterfront structures is assumed
- Focuses on "what, when, why, and where"
- Detailed "how to" guidance NOT provided

The Team

- Members:
  - Ron Heffron, Chairman
  - Noah Elwood, Secretary
  - Terry Browne
  - Andrew Cairns
  - Sean Chapman
  - Steve Curtis
  - John Daley
  - Frank Davidson
  - Bill Bruin
  - Elizabeth Burkhart
  - Anna Dix
  - Joshua Johnson
  - Bryan Jones
  - Ikaika Kincaid
  - Shawn Lindmark
  - Matthew Martinez
  - Todd Mitchell
  - Bruce Ostbo
  - Ralph Petereit
  - Heath Pope
  - Kirk Riden
  - Charlie Roberts
  - Paul Roberts
  - Craig Sams
  - Alberto Sanchez
  - Shelley Sommerfeld
  - Tom Spencer
  - Warren Stewart
  - Erling Vegsund
- Blue Ribbon Panel Reviewers:
  - Lee Barco
  - Richard Jenkins
  - Angel Lim
  - William Stahlman
  - Philip Vitale

The Team

Representing:

- Port Authorities
- U.S. Navy
- Consulting Engineers
- Academia

NEW ASCE WATERFRONT FACILITIES INSPECTION & ASSESSMENT MANUAL | UI 2016 | 9

New Topics Introduced

- Seven Inspection Types Remain from Manual 101:
  - Routine Inspection
  - Structural Repair or Upgrade Inspection
  - New Construction Inspection
  - Baseline Inspection
  - Special Inspection
  - Repair Construction Inspection
  - Post-Event Inspection

- Eighth Inspection Type Introduced:
  - Due Diligence Inspection

NEW ASCE WATERFRONT FACILITIES INSPECTION & ASSESSMENT MANUAL | UI 2016 | 10

New Topics Introduced

- Service Life Modeling
- Definition of element-level ratings, with sketches
- Mooring and berthing system condition inspections and rating scheme
- Addition of utility system condition inspections and rating scheme
- Addition of coating system defect definitions
- Addition of load isolators and bearing defect definitions
- Addition of a comprehensive appendix on specialized inspection techniques

NEW ASCE WATERFRONT FACILITIES INSPECTION & ASSESSMENT MANUAL | UI 2016 | 11

New Topics Introduced

- Extensive coverage of "Special Considerations" for specific structure and system types
  - Pile-supported waterfront structures
  - Relieving platforms
  - Bulkheads and retaining walls
  - Seawalls and revetments
  - Gravity block walls
  - Paving in immediate vicinity of structure
  - Caisson, cofferdams and cellular structures
  - Floating structures
  - Mooring hardware and fender systems
  - Mooring buoy systems

NEW ASCE WATERFRONT FACILITIES INSPECTION & ASSESSMENT MANUAL | UI 2016 | 12

New Topics Introduced

- Extensive coverage of "Special Considerations" for specific structure and system types
  - Wave screens and attenuators
  - Waterfront security barriers
  - Cathodic protection systems
  - Marinas and small craft harbor components
  - Gangways
  - Boat ramps
  - Marine railways
  - Bullrails, ladders and safety features
  - Crane rails, trenching and cables
  - Waterfront utility systems

NEW ASCE WATERFRONT FACILITIES INSPECTION & ASSESSMENT MANUAL | UI 2016 | 13

Cutting-edge and Controversial Topics Introduced

- Guidance provided on "Significant Changes and Owner Responsibilities"
  - Significant changes include:
    - Reduction in design capacity due to damage or deterioration
    - Increased loads
      - Larger vessels
      - Increased sail or current area
      - Increased live loads
    - Upgrades that modify load paths
  - No "significant" deterioration or damage
    - Repair/rehabilitation may proceed normally

NEW ASCE WATERFRONT FACILITIES INSPECTION & ASSESSMENT MANUAL | UI 2016 | 14

Cutting-edge and Controversial Topics Introduced

- Guidance provided on "Significant Changes and Owner Responsibilities"
  - "Significant" deterioration or damage requires structural evaluation prior to repair or rehabilitation
    - Reduction in design capacity of primary members of 20% or more is considered potentially significant
    - Structures that are rated "Poor" or below are considered to exhibit potentially significant damage
    - Method of structural evaluation should be determined by a registered professional engineer
  - For upgrade projects where loads are "significantly" increased, performance of system should be ensured
    - "Significant" is when demand-capacity ratio is 10% or greater than without increased loads

Content of the Guidelines

1. Introduction

- Intent of Manual and target audience
- Importance of inspection over life cycle of asset
- Guidance on Owner responsibilities
- Terminology clarification
  - Preservation
  - Sustainment
  - Rehabilitation
  - Upgrade

Content of the Guidelines

2. Standards of Practice

- Introduction of the 8 inspection types
- Guidance on choosing the right inspection type based on project needs
- Guidance on inspection frequency
- Introduction to Service Life Modeling
- Minimum qualifications of inspection personnel
- Rating systems for both elements and overall systems
- Guidelines for follow-up actions

NEW ASCE WATERFRONT FACILITIES INSPECTION & ASSESSMENT MANUAL | UI 2016 | 17

Content of the Guidelines

3. Scope of Inspection

- Guidance on boundaries and limits
- Definitions of the three levels of inspection effort
- For each of the 8 inspection types:
  - Objectives
  - Methods of inspection and documentation
  - Guidance on evaluating, rating and recommending follow-up actions

NEW ASCE WATERFRONT FACILITIES INSPECTION & ASSESSMENT MANUAL | UI 2016 | 18

Content of the Guidelines

4. Service Life Modeling

- Guidance on when and how to conduct SLM as part of inspection & rehabilitation strategy for a project
- Guidance on field sampling and testing
- Guidance on laboratory testing & analysis
- Key modeling considerations
- How to find optimum solution for extending life of existing asset

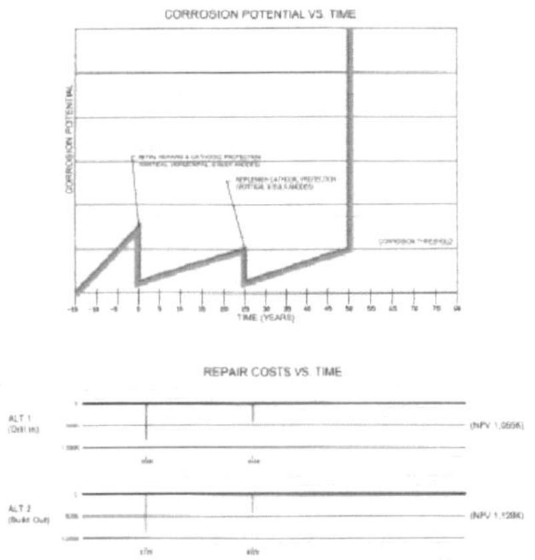

Content of the Guidelines

5. Documentation and Reporting

- Guidance on appropriate level of documentation and reporting
- Guidance on tailoring report content to project and client requirements

6. Administrative Considerations

- Guidance on contractual agreements
- Guidance on insurance considerations
  - Longshoreman's and Harbor Worker's Insurance
  - Jones Act Maritime Insurance
  - Professional Liability Insurance
  - Railroad Protective Insurance

Content of the Guidelines - Appendices

A. Special Considerations for Specific Structure Types and Systems

- Very Comprehensive!
- Detailed guidance for virtually every type of waterfront structure
- "What" to look for, not "how" to inspect
- Detailed guidance on inspection of utility systems
- Guidance on appurtenant systems and features

Content of the Guidelines - Appendices

B. Types and Causes of Defects/Deterioration

- Extensive guidance on defining defect types
- Insights on determining root cause of defects
- Materials and systems covered:
  - Concrete
  - Steel
  - Timber
  - Masonry
  - Composite materials
  - Coating systems
  - Load isolators and bearings
  - Undermining/scour

Content of the Guidelines - Appendices

C. Specialized Inspection Techniques

- Infrared thermography
- Ground penetrating radar
- Acoustic emission
- R-Meter testing
- Schmidt Hammer
- Impact echo testing
- Windsor Probe
- Half-cell corrosion testing
- Chloride ion testing
- Material sampling
- Ultrasonic testing
- Liquid dye penetrant
- Magnetic particle
- Structure monitoring systems
- Unknown foundation investigations
- Underwater acoustic imaging and channel bottom soundings
- Bacteria testing

Content of the Guidelines - Appendices

D. Inspection Nomenclature

- Guidance on standardized nomenclature for both components and defect types
- Guidance on numbering schemes
- Guidance on reporting schemes

E. Bibliography

- Comprehensive list of references

F. Glossary

- Compendium of definitions for waterfront facilities and inspections of same

What's Next?

PORTS 2016 Short Course

- June 2016 – New Orleans, LA

REHABILITATION MANUALS:

- Timber Waterfront Structures – Protection, Repair Analysis, and Rehabilitation
- Concrete Waterfront Structures - Protection, Repair Analysis, and Rehabilitation

NEW ASCE WATERFRONT FACILITIES
INSPECTION & ASSESSMENT MANUAL

QUESTIONS?

Ikaika Kincaid, PE
IKincaid@msimarinesolutions.com
Heath Pope, PE
HPope@moffattnichol.com

Demonstrating Impact: MATE ROV Competition Alumni Survey Preliminary Results

Jill Zande

Associate Director & Competition Coordinator, MATE Center

Chair, Marine Technology Society (MTS) - Monterey Bay section

MATE

marine technology SOCIETY

Opportunity runs deep™

BACKGROUND

Partnering with the MTS ROV Committee to...

- Address shortage of skilled individuals to support the marine industry
  - Reach more students & get them interested in the industry
  - Help students develop the skills to support marine activities

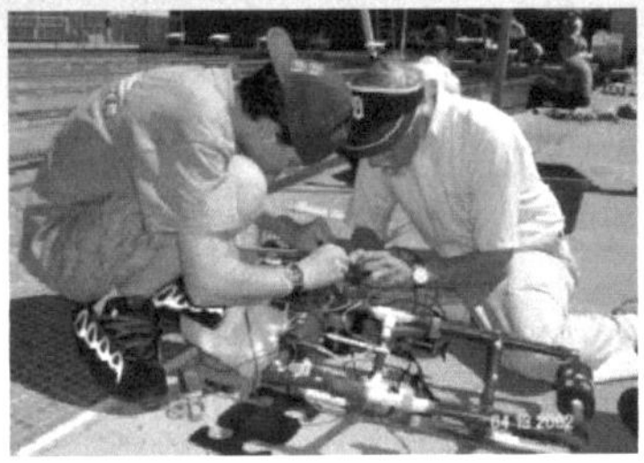

- Provide a venue for industry to become involved with education
  - Funds – to support student participation
  - Industry mentors – complement & expand students' & educators' knowledge and expertise

MATE

marine technology SOCIETY

Opportunity runs deep™

TO DATE...

- Has involved more than 13,000 students in grades 4-16 who work in teams to tackle missions based on the ocean STEM workplace
- Includes one international competition and a network of 26 (and growing!) regional contests
- Involves 100s of working professionals & organizations as mentors and sponsors
- Emphasizes entrepreneurship and includes underwater missions and technical reports, engineering presentations, and poster displays
- Helps students to develop technical as well as 21st century workplace skills (problem solving, critical thinking, teamwork, and more)
- Is encouraging students to pursue STEM degrees and jobs in the field
- Has been highlighted in 1 documentary, 1 Hollywood movie, and 1 book

MATE

2016 marks the 15th international competition!

marine technology SOCIETY
Opportunity runs deep

MATE

marine technology SOCIETY
Opportunity runs deep

IMPACT

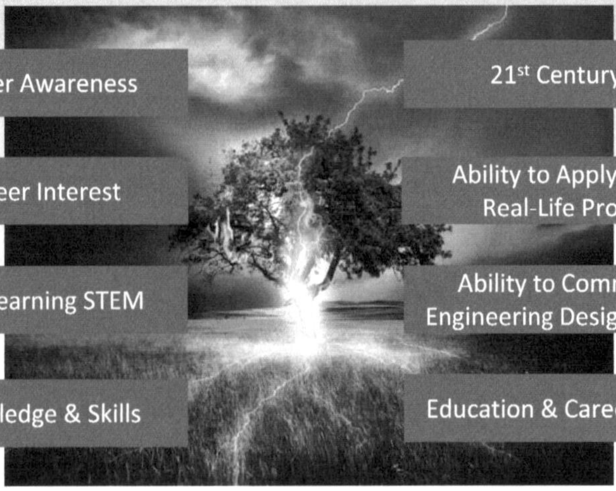

- STEM Career Awareness
- STEM Career Interest
- Interest in Learning STEM
- STEM Knowledge & Skills
- 21st Century Skills
- Ability to Apply STEM to Real-Life Problems
- Ability to Communicate Engineering Design & Process
- Education & Career Decisions

MATE

marine technology SOCIETY
Opportunity runs deep

SUMMARY RESULTS FOR 2015

STEM Careers

- After building their ROVs...
  - 77% more aware (N=1,420)
  - 77% more interested (N=1,418)

STEM Knowledge/Skills

- Because of their ROV projects...
  - 93% (N= 1,404) increased skills and knowledge in engineering
  - 90% (N= 1,388) increased skills and knowledge in technology
  - 87% (N= 1,393) better problem solvers
  - 83% (N= 1,398) better critical thinkers

Ability to communicate engineering design process

- 81% (N=1,412) stated that participating helped them learn how to communicate their engineering design to other people

Ability to apply STEM to real-world problems

- 86% (N=1,415) indicated that participating helped them learn to apply STEM to real world problems

MATE

marine technology SOCIETY
Opportunity runs deep

WHERE ARE THEY NOW?

MATE

marine technology SOCIETY

Opportunity runs deep

ALUMNI SURVEY

- **Launched in June 2015**
- **Targeted alumni over the age of 18**
- **432 respondents**
  - Response rate of 10.2%
- **Of those who responded:**
  - Length of competition "career" was 1-9 years with an average of 2.2 years
  - Participation years 2006-2015
- **Full circle**
  - 9% alumni became classroom/club mentors, 6% judges, 4% teachers leading teams

Education and Employment

- Highest level of education ranges from high school to doctorate
- Among the 236 current college and university students,
  - 85% are studying towards a STEM degree.
- Among the 220 alumni who earned a college degree,
  - 85% earned a degree in a STEM discipline.
- Among the 320 alumni currently employed,
  - 73% are currently working a STEM-related job.
  - 22% have worked in a job related to ROVs or other underwater technologies (14% currently).

Influence on Education/Career

"To what extent has participating in the MATE ROV Competition program influenced your educational or career path?" (N=432)

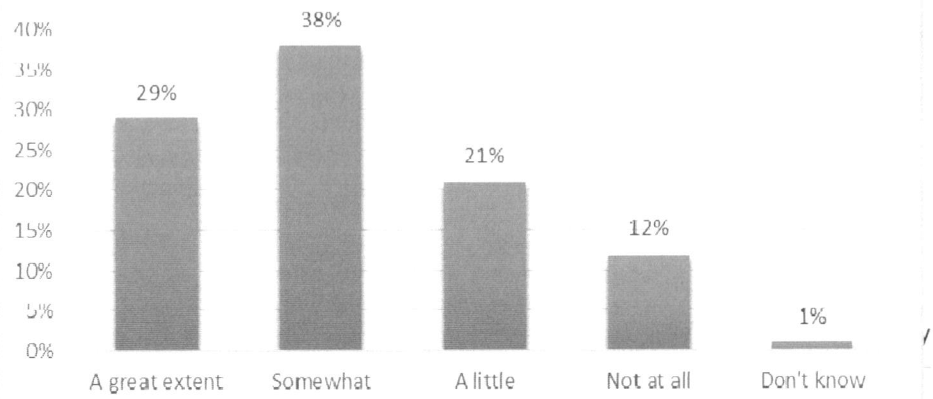

Alumni Quotes

"It showed me another path in life that I wouldn't have realized. It showed me a deeper love for sciences and has influenced me to become a processing engineer."

"The MATE ROV competition brought the realm of underwater robots into my life. Without it, I would have never worked for a company doing R&D for syntactic foams or R&D for sonar systems or even thought of applying for a job designing submarine systems."

Role in Attainment of Educational/Career Milestones

"Has participating in the MATE ROV Competition played a role in you attaining any of the following?"

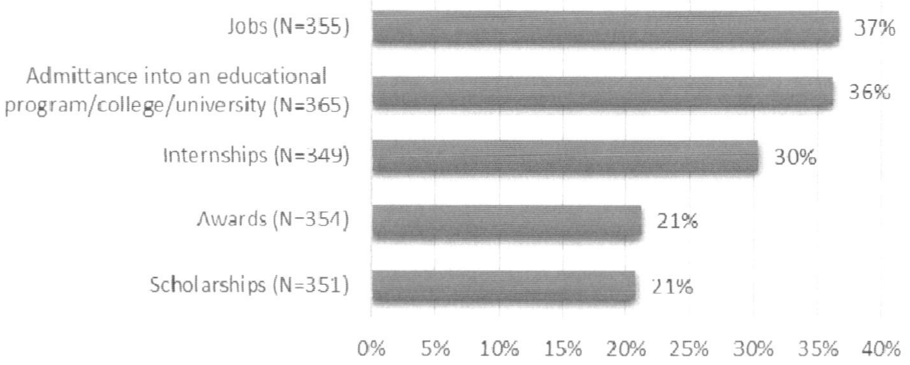

Employment

"I did a ton of electrical design as part of my involvement in the ROV team. It was as a result of that experience that I got hired for my co-op jobs, which led me to being hired full-time."

"I am currently an ROV technician and pilot for Oceaneering. The competition gave me exposure to the use of ROVs and allowed me to network with the right people, landing me a job."

Admittance into College/University

"In my acceptance letter into college, they referred to the MATE competition on my resume."

"George Fox University was very impressed with my work through MATE. They had never heard of such a program and were very excited about it."

Impact on STEM and 21st Century Workplace Skills

"To what extent did the ROV program help to strengthen your knowledge or skills in STEM?" (N= 426)

- ENGINEERING – 98%
- TECHNOLOGY – 97%
- SCIENCE – 91%
- MATH – 74%

"To what extent did the ROV program help strengthen your skills in the following areas?" (N= 440)

- Teamwork – 98%
- Leadership – 96%
- Critical thinking – 96%

"Did participating in the ROV program help you learn how to apply STEM to real-world problems?

- 91% said "yes"

WHERE ARE *WE* NOW?

- Planning another call to alumni
- Working with the WA State ERDC to create a comparison group of students
- Looking at ways to measure gains in knowledge and skills
- Focusing on the impact of multi-year participation on students' awareness/interest in STEM, stronger improvements in knowledge and skills, larger gains in 21st Century workplace skills, receipt of more awards and educational/career opportunities, etc.

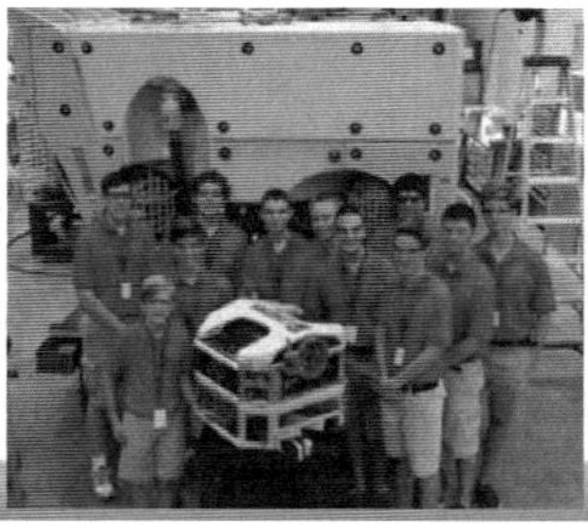

marine technology SOCIETY

Opportunity runs deep

Questions?

(Hope to see you at the 2016 international competition June 23-25 at NASA's Neutral Buoyancy Lab in Houston, Texas!)

Jill Zande
MATE Center
980 Fremont Street
Monterey, CA 93940
jzande@marinetech.org
(831) 646-3082
www.marinetech.org, www.mtsociety.org

MATE

First-year vs. Multi-year Participants

Statistically significant differences: Multi-year participants...

- Higher levels of awareness of and interest in STEM careers
- Greater gains in interest in taking STEM courses
- Stronger improvements in STEM knowledge and skills
- Larger increases in 21st Century skills
- Receipt of more awards, honors, and new educational/career opportunities

Data Sources

2015 Post-Competition Surveys
Students: N = 1,420
Teachers/Mentors: N = 304
Parents: N = 344

Sturgeon Research Project:

Students use an ROV to search for young of the year sturgeon in the St. Clair river.

Starting the Project/ Getting Contacted

- Instructor David Cummins (ACC) and Mark Gleason (GVSU) recommended us to Joe Krieger and Dave Brooks.
- Met over the summer with Joe and Dave.
- Excited to have a real world project.

In Need of Real World Projects

- Rather than being stuck in a classroom environment where we don't have as many opportunities to test and learn from what we have built. Our Advanced Underwater Robotics team would prefer working out in the field on what we've been building.
- We take advantage of almost every opportunity we are given.
- We were especially interested in Joe's project.

Joe's Work/ His Goal

- Joe Krieger is a student at U of M working towards getting his PhD.
- has been gathering data in the St. Clair river by using different methods of searching for "young of the year" sturgeon.
- Contacted us about using the ROV to scan the bottom of the river.

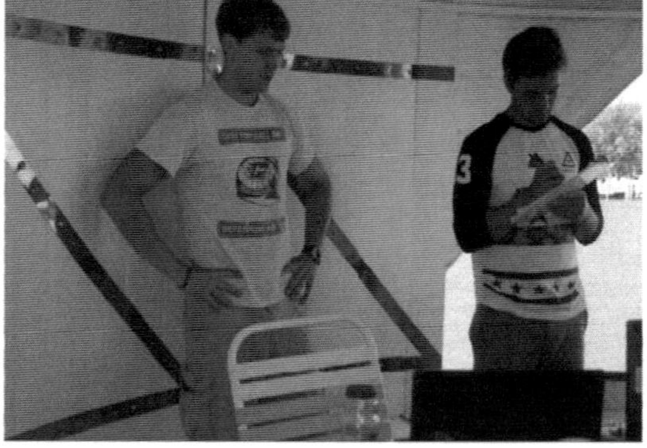

Collaborating With of U of M

- We worked with U of M because they provided us with the boat that we were on.
- We collaborated with Joe Krieger, a PhD student from U of M.

UNIVERSITY OF
MICHIGAN

Joe's Project Results

- Gave us an opportunity to work on science in the real world.
- Joe received the knowledge that there aren't sturgeon in that area

Joe's PhD Experience So Far

- Being able to help better inform better management of the species
- Has helped lead resource managers to better sampling/ monitoring efforts for young of year sturgeon

Being on the Nancy Kay

- Being on the Nancy Kay is a little more spacious than the boats In Palau.
- The view of the water was absolutely beautiful.
- While we were on the captain let us students have a feel of what it's like behind the wheel.

Our Class Experience

Our Personal Experience

- The sturgeon project was our first robotics trip.
- Learning hands on contributed towards are classwork.
- We had a fun learning experience.

Film Team experience

- Being on the film team.
- Film team member Jake was involved in this project more than other projects.
- Film Team member Sam he experienced more than usual with things on the technical side.

Gaining Palau Experience

- Being on the boat gave our team the feel of being in Palau.
- Lack of space.

SENSE IT

Transitioning from research to retail with an ocean sensing education program

Liesl Hotaling, Eidos Education

Acknowledgments

- Dr. Rustam Stolkin, University of Birmingham, UK
- Dr. Susan Lowes and Ms. Peiyi Chen, Columbia University, Teachers College
- Teachers and students participating in SENSE IT
- National Science Foundation ITEST program
- National Science Foundation ICorp-L program

SENSE IT

NSF

SENSE IT

SENSE IT integrates STEM skills into a robust interdisciplinary curricula and teacher development effort by teaching students to build, test, deploy and interpret their own water quality sensors.

SENSE IT

SENSE IT Description of Activities

SENSE IT is comprised of four modules:

1) Sensor development
(temperature, conductivity, turbidity, depth)
2) Sensor deployment and data gathering
3) Water Quality and Environmental Science
4) Sharing Data - Wireless sensor networks

The modules are currently implemented with middle and high school students and integrated into Environmental Science, Chemistry, General Science, Physics, Integrated Algebra, and/or Pre-engineering courses.

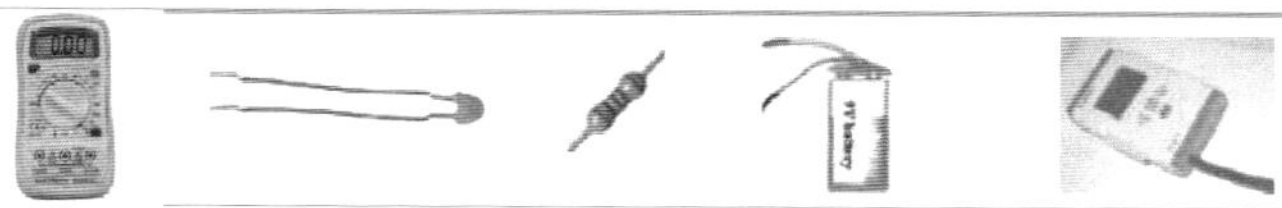

SENSE IT Professional Development

SENSE IT has an associated professional development program in order to provide teachers with:

- thorough instruction on the curricula;
- equipment necessary to implement the modules with students; and
- support to strengthen teacher content knowledge of sensors and relevant STEM concepts necessary to implement the SENSE IT activities.

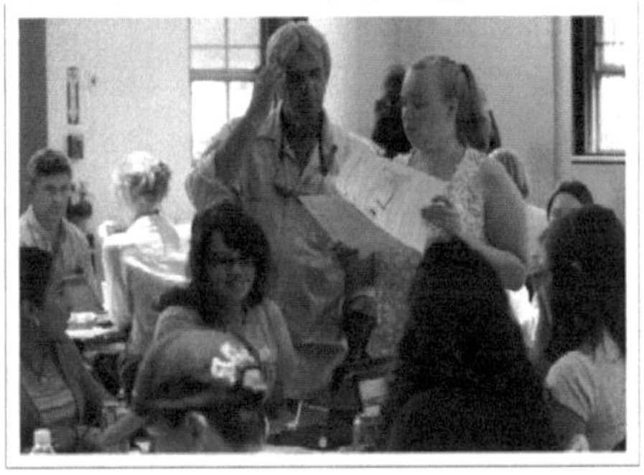

SENSE IT Evaluation

The evaluation of SENSE IT is assessing student increases in:

- technological proficiency;
- learning of specific concepts and skills;
- awareness of IT and science careers (including understanding of the fundamental role of mathematics in many STEM careers); and
- engagement in STEM courses in general, and changes in career aspirations.

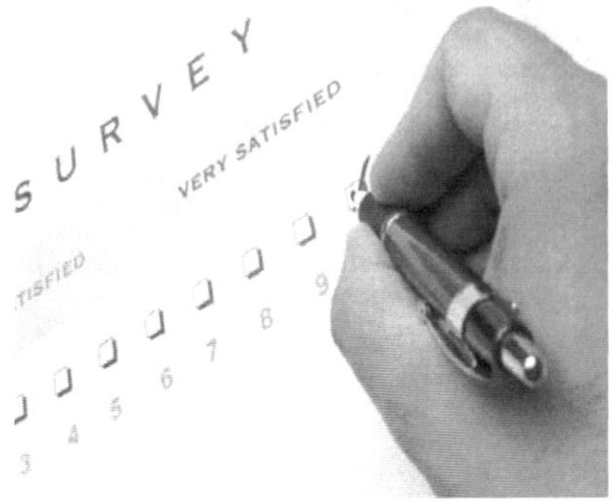

SENSE IT Evaluation

SENSE IT assessments

Technological proficiency, concepts and skills:

- Algebra – pre/post
- Electricity – pre/post

(5 questions each with confidence Likert-type scale from 1 - 5)

- Water Quality – post test only
- Sensor Pile Sort – pre/post

Awareness of STEM careers, STEM engagement/aspirations:

- Pre/post STEM interest attitudinal survey

SENSE IT Evaluation

Results indicate that the teachers were positive about the benefits of the SENSE IT curriculum for their students, both in terms of how it engaged them and in terms of what they had learned.

High school students liked the hands-on aspects of the project best, while middle school students liked building and working in groups.

SENSE IT Evaluation

And a very important result was the positive shift in student attitudes toward considering studying science and math and in math and science careers.

There was a statistically significant increase in the percent of students who said they might like to major in engineering or science in college, as well as those who might like to be scientists or mathematicians.

SENSE IT Next Steps

- Develop a Virtual Academy
- Develop additional sensors
- Develop additional platforms (Arduino, Raspberry Pi, PicoBoard)
- Create materials for the college level classroom
- Export SENSE IT to other countries
- Partner with existing programs and projects
- Transition to retail

SENSE IT Next Steps

Transition to retail

Taking lessons learned from 6 years of SENSE IT in classrooms around the country, we took the next step by exploring commercializing the program by creating SENSE IT kits.

Eidos received an ICorp-L grant from NSF for initial exploration of the concept.

Over 120 interviews were conducted with educators all over the country with positive reception.

We are exploring the pathways to produce kits and create retail partnerships.

Please let me know if you would like to place an order ☺

Sample Results

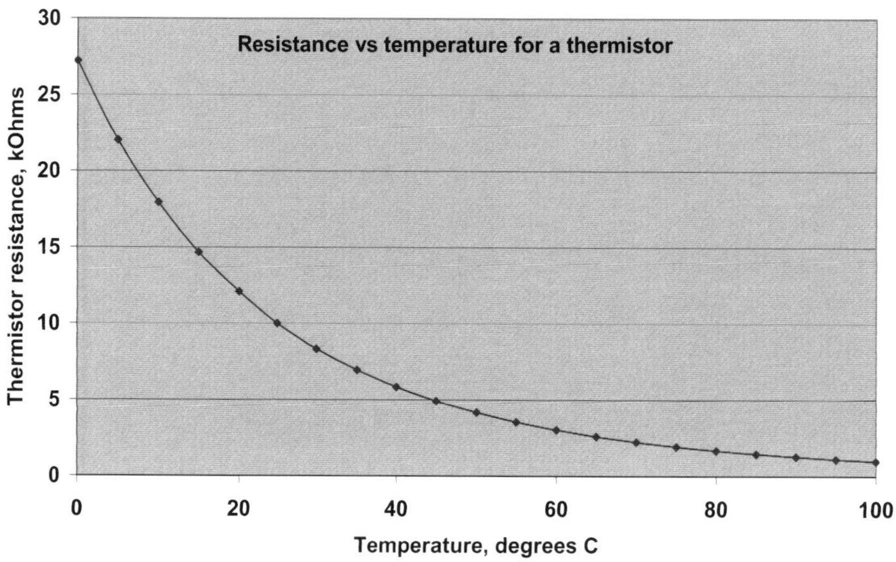

Conclusion

SENSE IT promotes inquiry based learning strategies through challenging hands-on projects. It also promotes student awareness of the interdisciplinary nature of modern engineering and the interdependence of diverse areas of science and math.

SENSE IT

SENSE IT

http://senseit.org

Thank You!   Questions?

Liesl Hotaling
Eidos Education
liesl@eidoseducation.org

EXPLORING SHIPWRECK ALLEY

UNDERWATER INTERVENTIONS, FEBRUARY 2016

MADI HOWARD & FAITH WHITT

MATE UNDERWATER ROV COMPETITION

- We started in underwater robotics by being on a MATE competition team.
- We placed very high, and decided to expand our career.

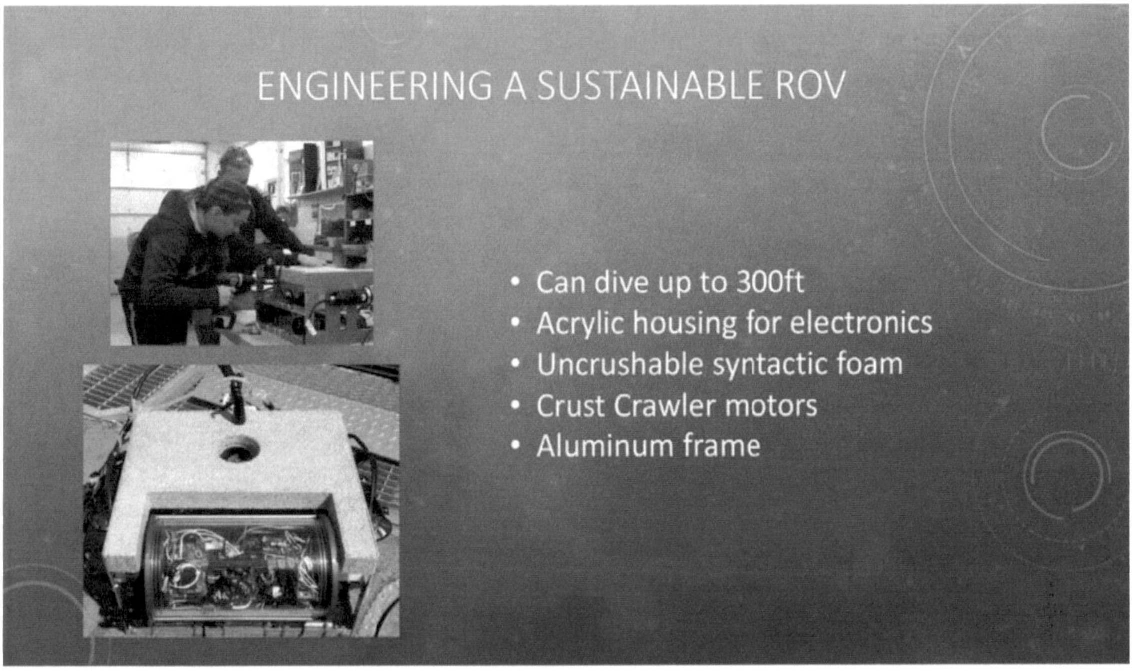
ENGINEERING A SUSTAINABLE ROV
• Can dive up to 300ft
• Acrylic housing for electronics
• Uncrushable syntactic foam
• Crust Crawler motors
• Aluminum frame

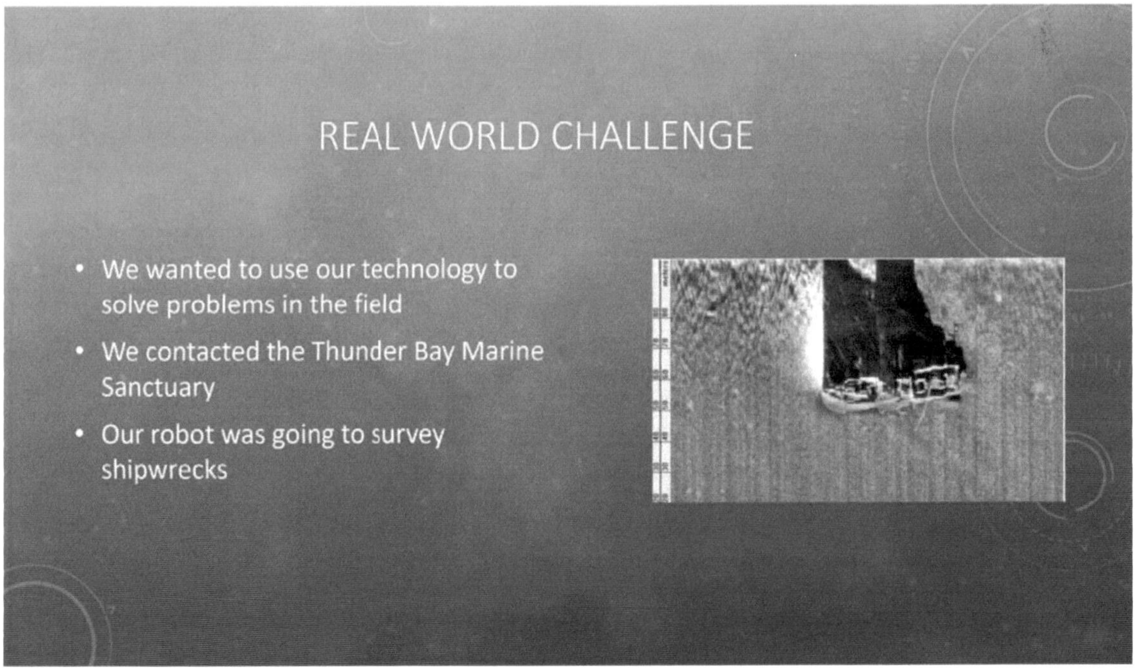
REAL WORLD CHALLENGE
• We wanted to use our technology to solve problems in the field
• We contacted the Thunder Bay Marine Sanctuary
• Our robot was going to survey shipwrecks

JOURNEY TO ALPENA

- Packed all of our equipment
- Fit everything in one truck
- Drove 5 hrs. to Alpena

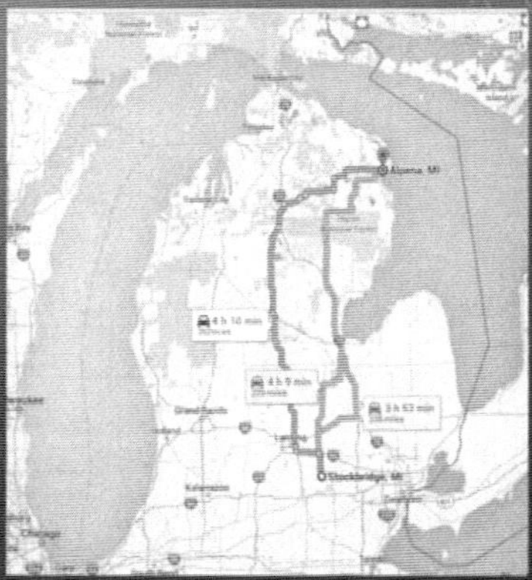

WORKING WITH THUNDER BAY NATIONAL MARINE SANCTUARY

- We worked and fixed our ROV in their warehouse
- Accommodations were in the scientist quarters in the shipwreck museum

NEW YORK NEWS CREW
• They wanted to shadow us on our trip to make a short TV series
• It was focused on the extraordinary learning experiences we get

SURVEYING AND DOCUMENTING SHIPWRECKS
• The Oscar T. Flint
• The Montana
• The Grecian
• E.B. Allen

UNPLANNED PROBLEMS- 1ST DAY IN ALPENA

- Signal issues from the microcontroller to the computer
- Adjusted buoyancy
- Spent 12 hrs. fixing the robot

GETTING PREPPED FOR THE BOAT

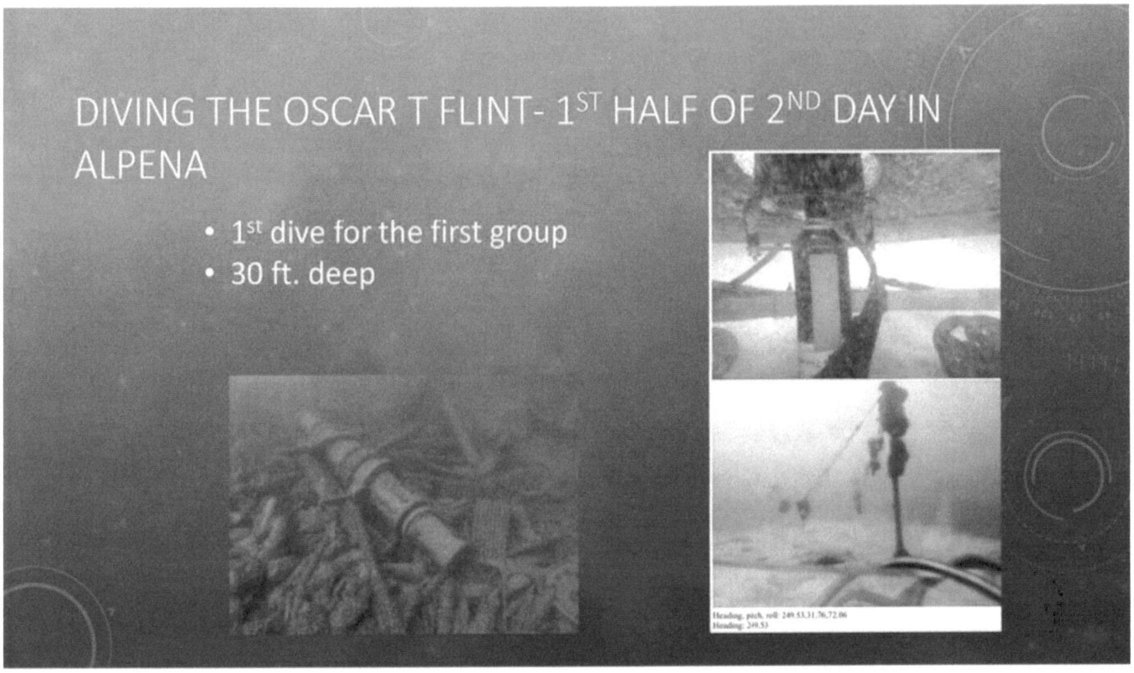
DIVING THE OSCAR T FLINT- 1ST HALF OF 2ND DAY IN ALPENA
• 1st dive for the first group
• 30 ft. deep

DIVING THE MONTANA- 2ND HALF OF 2ND DAY IN ALPENA
• Down 63 ft.
• First dive for group 2

VISITING ALPENA COMMUNITY COLLEGE
OCEANEERING
ALPENA COMMUNITY COLLEGE

WEATHER PERMITING- 3RD DAY

CHANNEL 10 INTERVIEW

- 5 am TV interview via skype
- When we went on the air, connection cut out.
- We had to stop and cancel our interview

DIVING THE GRECIAN AND E.B ALLEN - 4TH DAY IN ALPENA

- Depths of approximately 100 ft.
- Had issues with the ROV

MICHIGAN BUSINESS NETWORK INTERVIEW

- Radio interview in Grayling
- Karl Klimek did an interview about our experiences

MICHIGAN BUSINESS
NETWORK.COM

EXPLORING SHIPWRECK ALLEY

GAINED EXPERIENCES

- Problem solving
- Teamwork
- Real world job days
- How to work on little sleep
- How to be prepared for different conditions.

Stockbridge InvenTeam
Mid Grant Technical Review

JANUARY 13, 2016

Lemelson-MIT Grant

- Lemelson-MIT InvenTeams™ are comprised of high school students, educators, and mentors that receive up to $10,000 each to invent technological solutions to real-world problems of their own choosing.
- The Stockbridge InvenTeam was awarded a Lemelson-MIT grant on October 14th, 2015.
- With this grant we are able to accomplish everything we need to make a successful product besides travel fare.

Lemelson-MIT InvenTeams

- An InvenTeam is "A team created to inspire young people to solve real-world problems through invention."
- As an InvenTeam we are trying to solve an environmental problem in the Great Lakes.

A Challenge From The President

In 2013 President Remengesau challenged the team to find a way to use their technology to study and preserve the ocean, and coral reefs around Palau.

Invent Like a Girl:

- Experience in the MATE ROV Competition leads to Impactful Invention.
- We are also presenting at Underwater Interventions in New Orleans

Our First Prototype

A Go Pro Camera attached to an external battery.

The Great Lakes Fishery Commission

- The GLFC is a shared organization of the United States and Canada.
- It is worth more than $7 billion dollars annually.
- They protect the Great Lakes environment and support the fishing industry.

Sea Lampreys Great Lakes History

- Sea lampreys invaded the upper Great Lakes in the 1920s
- They had caused a serve loss in stocks of Great Lakes Fish by the1950s

Sea Lamprey Information

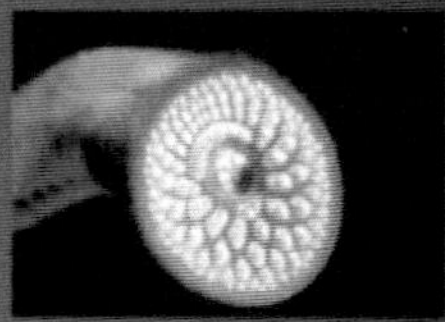

- Sea Lampreys (Petromyzon marinus) are a parasitic fish that originates in the Atlantic Ocean. They are invasive to the Great Lakes.
- Sea Lampreys resemble eels and have a large sucking disk and razor sharp teeth and tongue.
- They kill fish by attaching on the surface of fish, and sucking blood and other body fluids out of them.
- sea lampreys act as predators, with each individual capable of killing up to 40 pounds (18 kilograms) of fish

Life Cycle of the Sea Lamprey

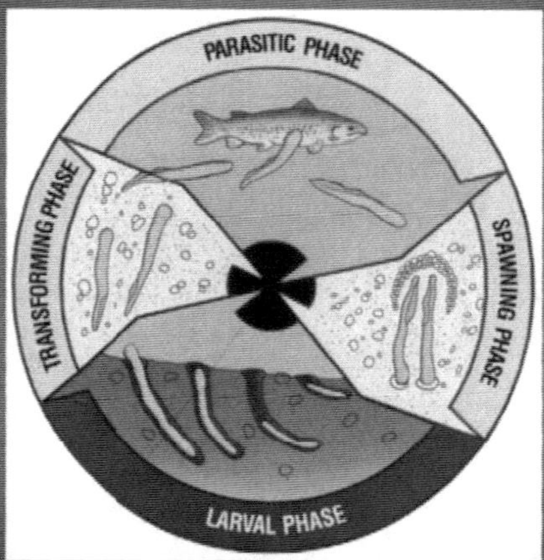

Sea Lamprey Control

- Lampricides
- Barriers
- Traps
- Pheromones and alarm cues

Lampricides

- TFM- The primary method used to kill sea lampreys in a liquid form that doesn't harm most other organisms before they develop lethal mouths and begin migrating.
- Bayluscide- a less costly treatment in granular form that is released at the surface of slow-moving or stationary waters. This method is used where TFM cannot be.
- About 200 Great Lakes tributaries and larval "hot spots" are regularly treated with lampricides to kill larvae before they have a chance to prey on fish.

St. Clair River (Huron-Erie Corridor)

- The number of adult sea lampreys and markings on fish species in Lake Erie have been exceedingly above targets set by the lake committee.
- This may be due in part to an untreated larval sea lamprey population in the St. Clair River.
- The current estimate of larval sea lamprey in the St. Clair River is 1.5 million. A population of this size warrants treatment.

Problem

Researchers need a way to observe the behavior of larval sea lamprey during a lampricide treatment

Current Method

- Individual researches report observations from their location.
- May or may not have photographic/video recording capabilities.

Partnership with the Great Lakes Fisheries Commission

We will be using our cameras to assist the GLFC in monitoring Sea Lamprey behavior after the application of Lampricide. In addition, our camera will be measuring water quality data.

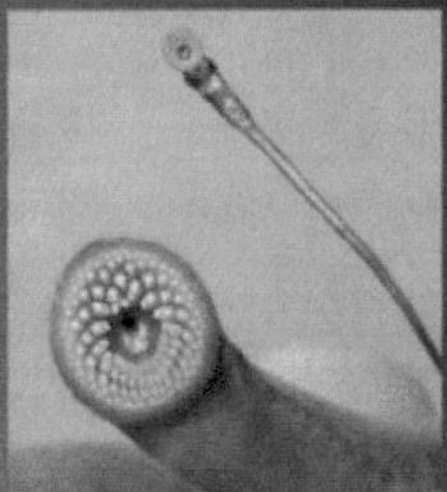

Why Our Invention is Better

- There currently isn't a method being used to monitor the sea lampreys reaction to the lampricide other than a chance encounter while studying a certain area being treated.
- We are creating a sustainable and productive method for monitoring sea lampreys reaction to lampricide that doesn't currently exist.

Project Description V-3

- The Stockbridge High School InvenTeam's proposed invention is an accessory for a GoPro style sports/action camera that turns the Camera into a low cost, remote underwater camera and water quality monitoring system capable of shooting time-lapse photography and/or video for up to 5 days at a time in remote locations without having to change batteries or SD cards.
- At the heart of our project is the unique base and mounting system which also houses a sensor package capable of collecting water quality data and logging the data on an SD card. Image, GPS and water quality data which has been logged on the SD card can then be used by scientists to study the effect of lampricide on reducing numbers of invasive Sea Lamprey in the Great Lakes.
- This invention will be used by marine biologists, teachers, researchers and citizen scientist to capture photographic and water quality data in aquatic environments over extended periods of time.
- "This project will gather important observations on the behavior of larval sea lampreys during a lampricide treatment, which could improve the efficiency of lamprey control, particularly in large systems like the St. Clair River."

Current Prototype

Our system consists of a water tight camera housing with battery separator plates to hold all of the electrical components in place.

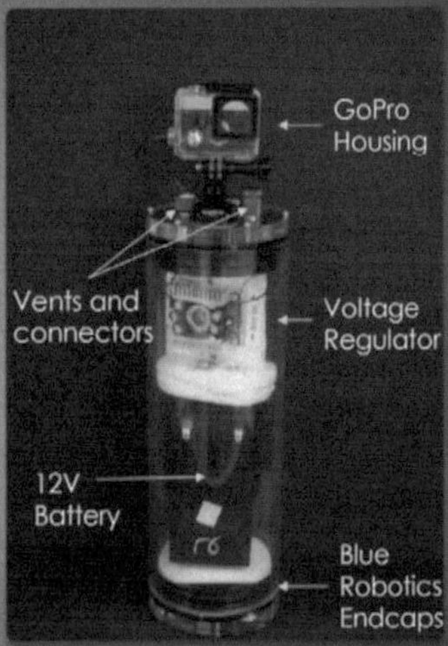

Quality Tested

- Camera tested to a depth of 30 ft
- Runs for 24 hrs. on single battery while shooting time-lapse photography
- Can run a total of 48 hrs. on the double battery design
- View of vision has been determined

Next steps in Engineering

We are currently incorporating sensors into our camera system that will measure water quality data to assist scientists studies.

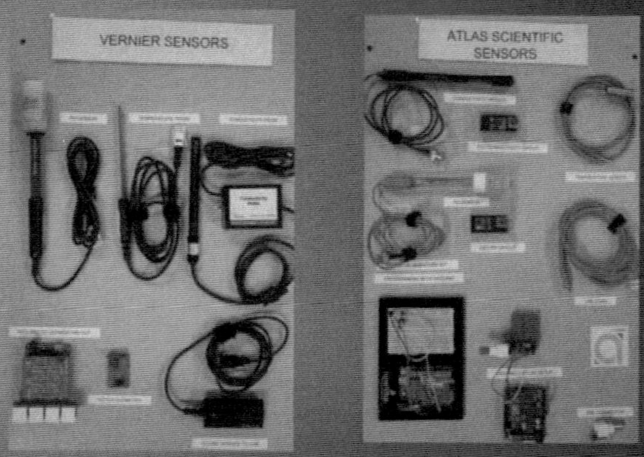

Field work

- May 31, 2016 to June 6, 2016
- Middle Channel of the St. Clair River
- 19 Cameras
- Collecting video, time lapse photography and sensor data
- 3 test plots 25m X 20m each in 8' to 10' of water
- Project partners
  - The Great Lakes Fishery Commission
  - The U.S. Fish and Wildlife Service
  - The U.S. Geological Service
  - Oceans Canada

Sustainable Design

- Rechargeable batteries
- We used recycled PVC while prototyping
- Recycled HDPE plastic for battery holders
- Presentation materials made from recycled cabinets

Budget

Month	Projected Spending	Actual Spending
Oct	0%	5%
Nov	25%	22%
Dec	45%	30%
Jan	60%	30%
Feb	70%	30%
Mar	75%	30%
Apr	85%	30%
May	95%	30%
Jun	100%	30%

*report percent spend of grant amount less the educator fellowship

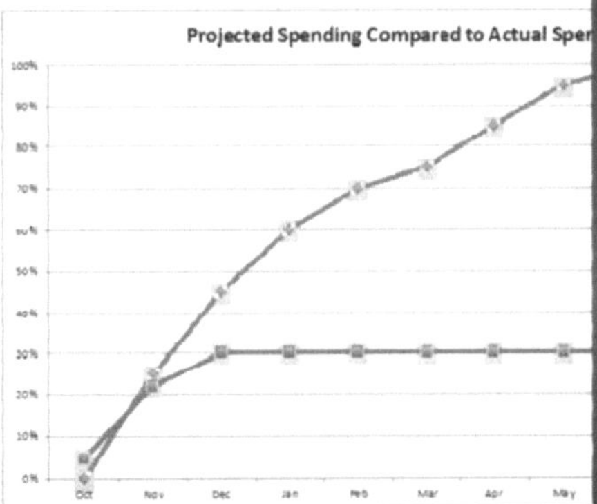

Follow Our Progress

To raise awareness and funds, we have a lot of media outreach. We go to various events, and run social media pages plus a blog.

Stockbridge Robotics on Facebook: www.facebook.com/StockbridgeRobotics

Stockbridge Robotics on Go Fund Me: www.gofundme.com/SHS-InvenTeam

Stockbridge InvenTeam on Twitter: @SHSInvenTeam

The Help We Need

- Financially, we need funds to travel to Boston in June to attend Lemelson- MIT EurekaFest. There, we will showcase our final product.
- On the technical side, we could use assistance in programing sensors and for you to give feedback on our design.

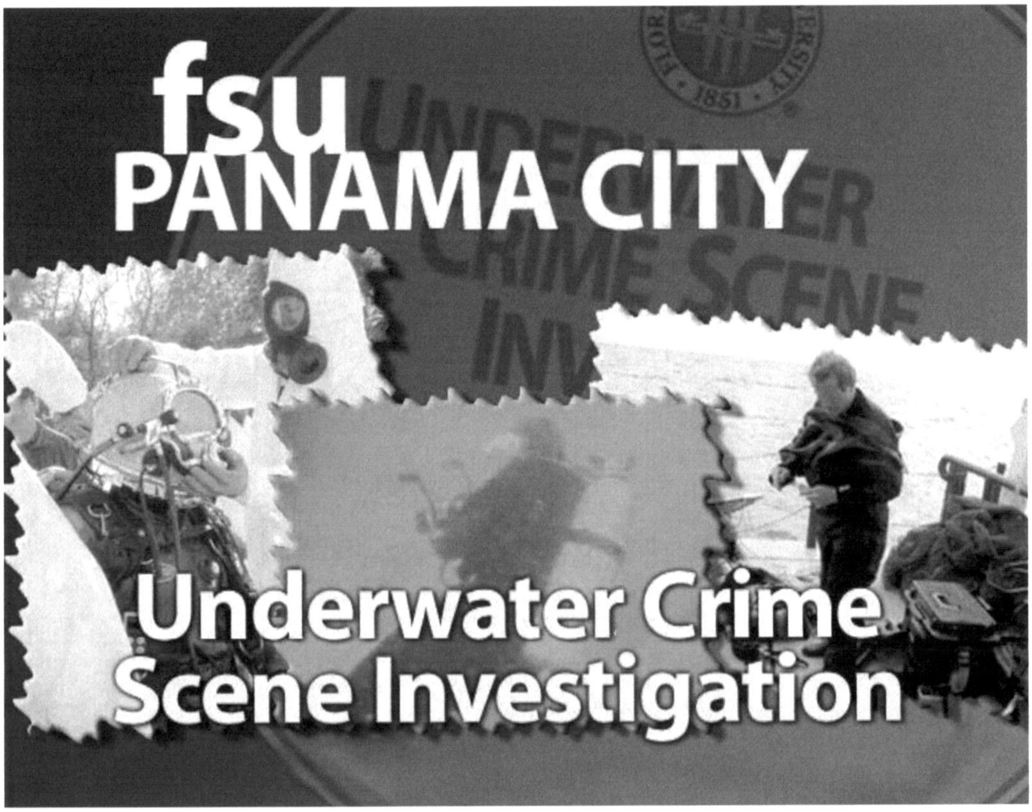

2015 UCSI

UCSI Process
Evidence Collection

Michael A. Zinszer

UCSI Process

CONTENTS

- Effects From Demographics
- Introduction of a New Concept
- Legal Content
- Collection of Evidence
- Advantages of Technology

UCSI Process

According to the U.S. Department of Commerce:

Florida has 3,331 miles of tidal coastline.

This area does not include

- Swamps
- Rivers
- Lakes
- Aquifers

That make up a large portion of the interior of the state

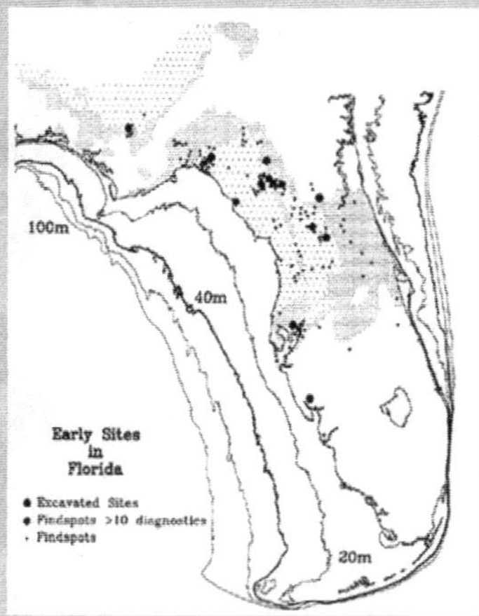

UCSI Process

Many of the individuals who are involved in criminal activities tend to use an aquatic environment as a safe haven to hide the evidence of their crimes. Evidence when handled properly can be used to convict the perpetrators of:

- Insurance Fraud
- Theft
- Burglary
- Murder

Just as critical is the proper collection of evidence when Investigating Boating, Diving and Drowning accidents that often become part of excessive payoffs by insurance companies or civil litigation involved in Law Suits.

UCSI Process

Bringing Forensic Science underwater is a relatively new concept that has been largely ignored by investigation authorities.

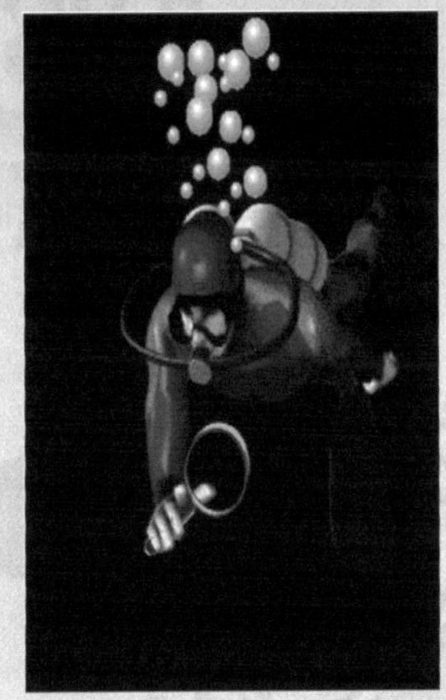

Like many of its predecessors that are now common place within the field, technology and general acceptance is slower than the need for change.

Examples Include:

- Finger printing, introduced 1904 general acceptance 1915.
- Firearms rifling comparisons, introduced during the 1920's general acceptance 1930's
- DNA testing, developed during the 1950's general acceptance 1980's

UCSI Process

Theory of Search

- What is the probability the evidence is in a particular area ?
  - An Investigative Question
- What is the probability the evidence is there but we missed it?
  - A Scientific Question

UCSI Process

Evidence: Location - Time

- Where in the world is the crime scene?
- Where i the crime scene is the evidence?
- When did the incident happen?
- When did the evidence get to the crime scene?

UCSI Process

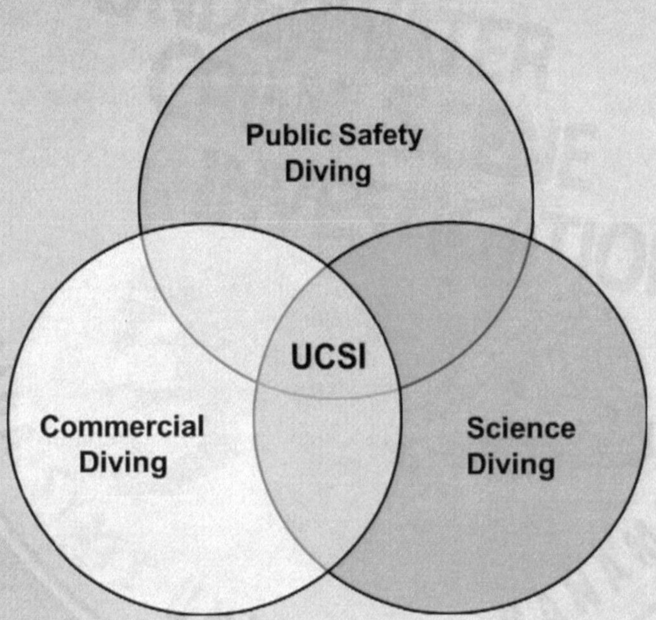

UCSI Process

All crime and accident scenes, whether persons, vehicles, or places, should be treated with the same attention to procedures.

Production of Physical Evidence:

- Accidents and crime are the phenomena studied in forensic science.

- Physical evidence is the tangible results of a perturbation in the environment.

From physical changes we can infer what physically happened, when it occurred, where it occurred, and what substances, objects, or persons were involved.

UCSI Process

A Crime scene is any physical location having a connection with a crime or potential crime. An accident scene is any physical location having a connection with an accident.

Forensic Science

- **Observations** (directly pertains to the scene being investigated)
- **Questions** (which are posed to, or by, the forensic scientists)
- **Examinations** (to answer questions)
- **Findings** (answers from an examination)
- **Evidence** (findings presented in court)

UCSI Process

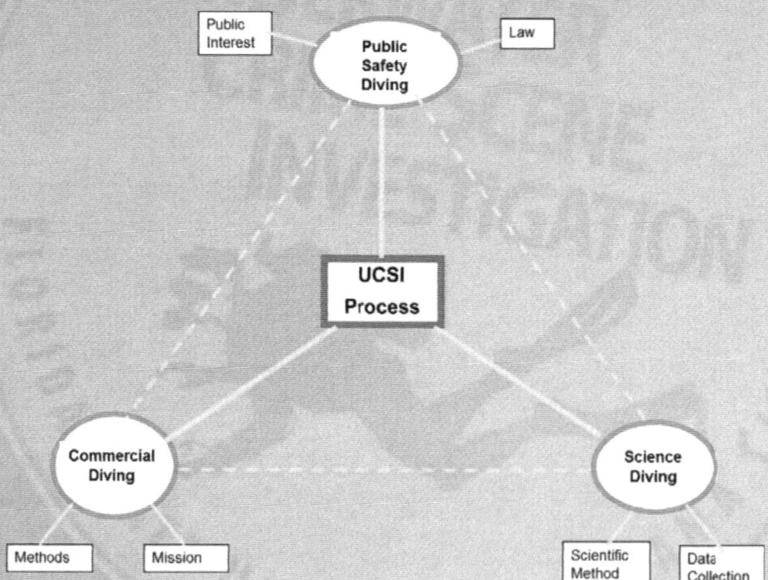

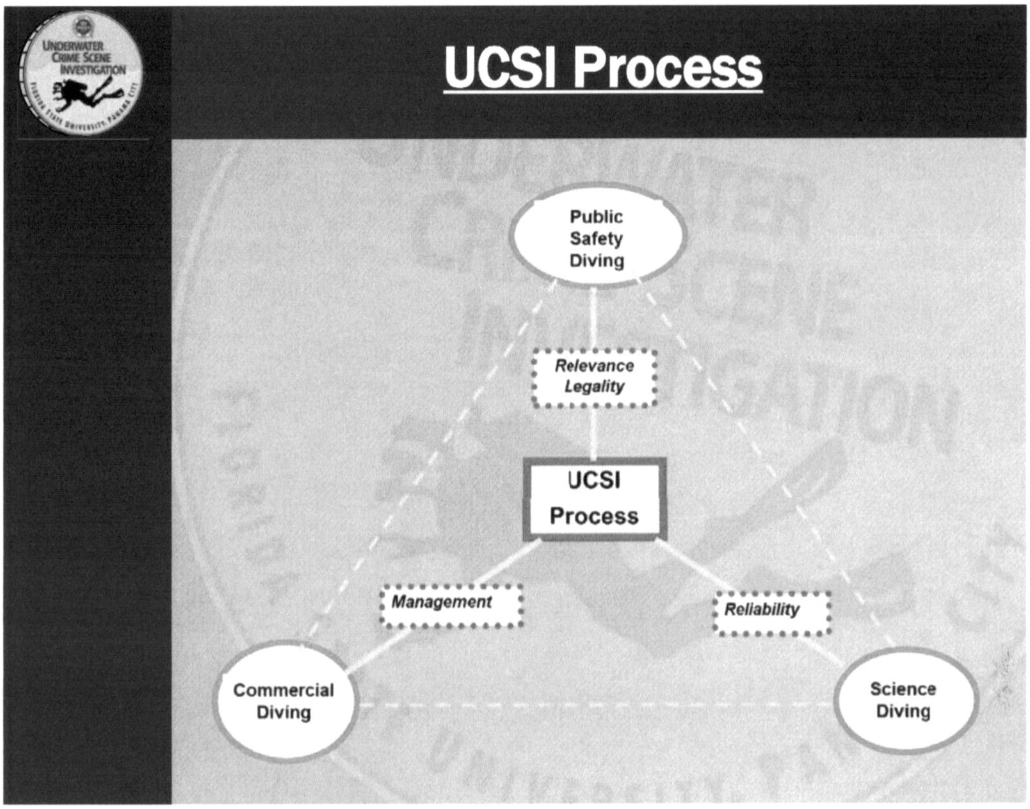
UNDERWATER CRIME SCENE INVESTIGATION
UCSI Process
Public Safety Diving
Relevance
Legality
UCSI Process
Management
Reliability
Commercial Diving
Science Diving

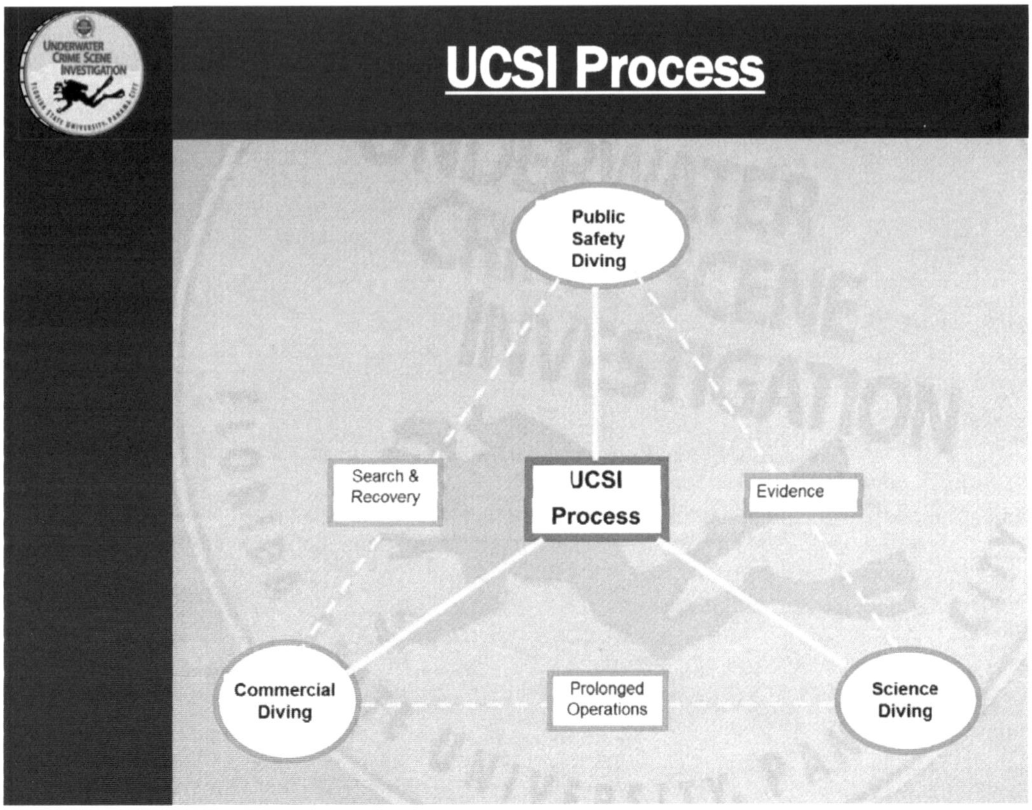
UNDERWATER CRIME SCENE INVESTIGATION
UCSI Process
Public Safety Diving
Search & Recovery
UCSI Process
Evidence
Commercial Diving
Prolonged Operations
Science Diving

UCSI Process

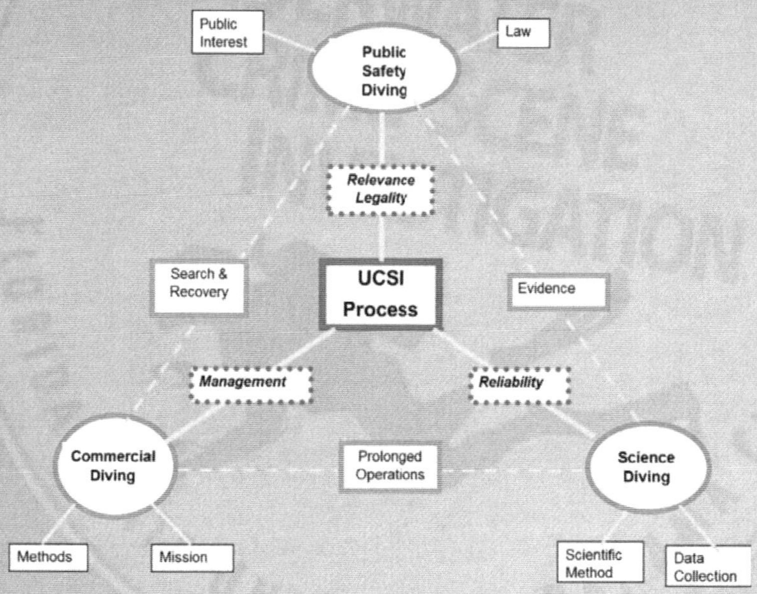

UCSI Process

Scientist

+

Investigator

+

Diver

UCSI Process

Evidential rules that apply on land also apply in an aquatic environment

Frye v. United States, 54 App.D.C.46, 293p. 1013 (D.C Cir. 1923)

- Evidence based on a novel scientific theory is inherently unreliable and inadmissible in a legal proceeding unless the theory has been adequately tested and accepted by the relevant scientific community.

- When a scientific principle or discovery crosses the line between the experimental and demonstrable stages is difficult to define.

- The underlying theory for this rule is that a courtroom is not a laboratory, and, as such, it is not the place to conduct scientific experiments.

UCSI Process

The United States Supreme Court in Daubert v. Merrell Dow Pharmaceuticals (92-102), 509 U.S. 579(1923)

- Eschewed the *Frye* ruling of "General Acceptance" in favor of a rule of **reliability** wherein the trial court must assess the scientific validity of the theory in issue based on various criteria including general acceptance.

- *Daubert* now extended to be applicable to all expert's testimony, even if the testimony is based on an expert's personal experience rather than on scientific knowledge.

- The Florida Supreme Court rejected the *Daubert* rule in favor of continued use of *Frye*. The Frye test requires the court to peruse disparate sources-e.g., expert testimony, scientific, and legal publications, and judicial opinions

UCSI Process

Forensic Science Process

Location — Relevance

Measurement — Reliability

Collection — Authenticity

Interpretation

Safety

UCSI Process

The Goal for Crime Scene Investigation, whether on the land, sea or Underwater, is to collect data that will aid a jury in making their decision about an issue being disputed in a court of law.

The tasks required to meet that goal commonly include:

- **Locating and recognizing** potential evidence
- **Measuring and documenting** the scene and its evidence
- **Collecting and packaging** evidence being recovered
- **Examining and interpreting**, in place, evidence that cannot be recovered

These tasks must be accomplished using reproducible, scientific techniques while maintaining a chain of custody and adhering to the other legal constraints necessary to present evidence into court.

UCSI Process

Scientist + Investigator

Reliability	Relevance
Conducting valid, reproducible, Objective examinations	Applying Inferences from the data obtained to the Elements of Proof
In Order to: Obtain and interpret data	In Order to: Solve and adjudicate Crimes

UCSI Process

Incorporating Scientific Methodology To Recover Evidence

Evidence can only be admitted into court if it is considered:

- **Relevant** (directly pertains to the case at hand)
- **Reliable** (the collection of evidence is reproducible and scientific)
- **Authentic** (evidence can be traced to the scene, chain of custody)

These legal requirements mandate that investigators incorporate verifiable and reproducible procedures during evidence recovery while assuring that a chain of custody is properly maintained

UCSI Process

Types of Physical Changes

Positional Change	*Macro Change*	*Micro Change*
Of Item or Object move to site change location at site change orientation at site move from site	**Impressions in Surface** Static Sliding	**Compositional Transformation** decompose (explode, metabolize) synthesis (burn, rust)
Of Contaminants to/from Item With contact (Pattern imprint by object) Without contact, (pattern of movement)	**Separation of Item** break tear cut saw disperse (shatter & scatter)	**Temperature Changes** (engine cooling, body cooling) **Phase Change** (ice melting, water evaporating, blood drying)
		Component Separtation (pyrophoric carbon, plasma separation)

Hathaway Bridge Accident

UCSI Process

New Tools

- To "see" underwater
- To "measure" underwater
- To "document" underwater
- To "survive" underwater

UCSI Process

Needed Technology for The Layman Investigator

Mapping and Imaging are fundamental concerns in forensic Science

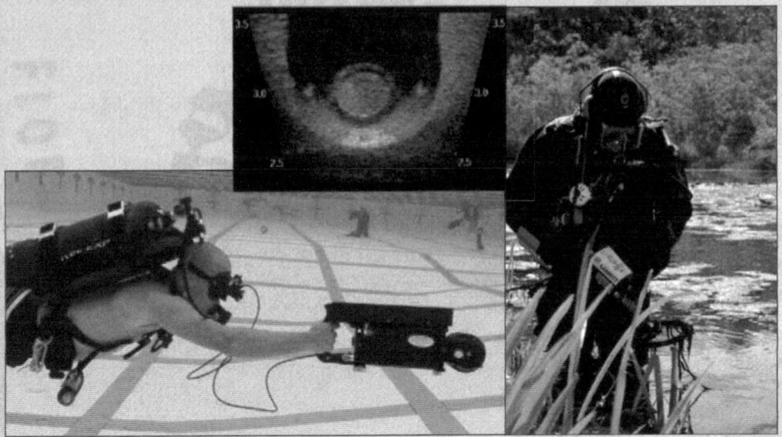

UCSI Process

Knowing Your Environment

- Environmental HAZARDS
- Tides / Currents
- Zero Visability
- Location of Site

UCSI Process

Knowing Your Environment

Holloway Investigation

UCSI Process

UCSI Process

UCSI Process

UCSI Program

What make UCSI unique; is understanding that there is a difference between Public Safety Diver and Underwater Investigator. A Public Safety Diver recovers objects, while the Underwater Investigator collects evidence.

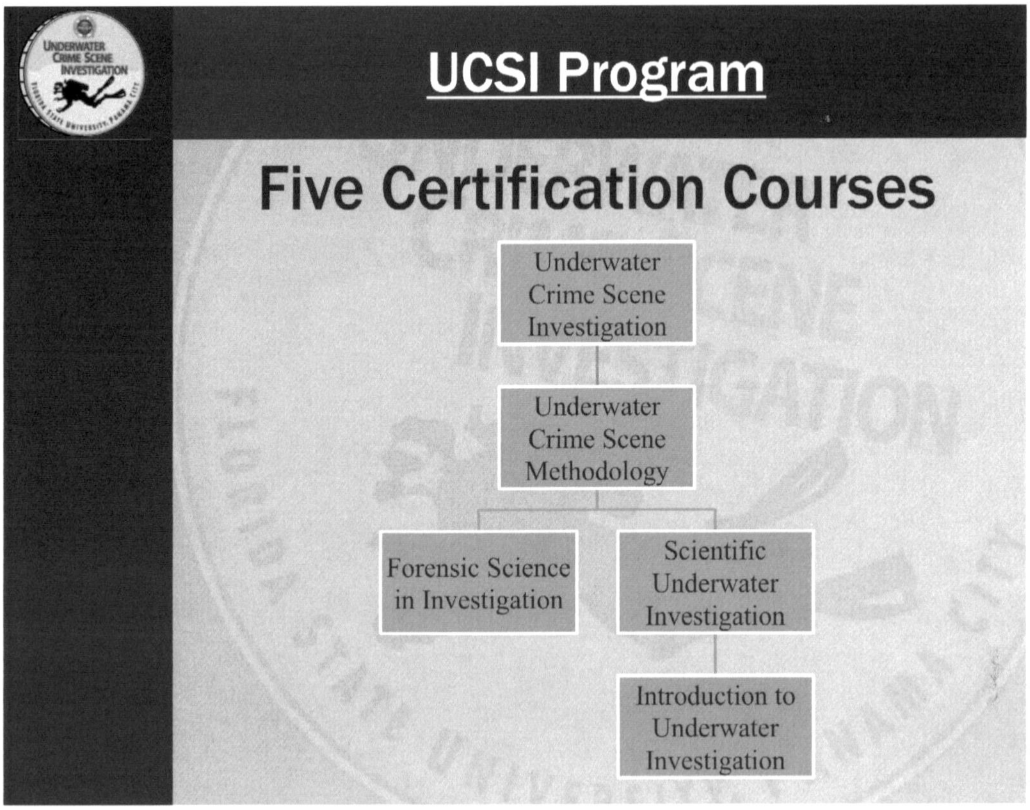
UCSI Program
Five Certification Courses
Underwater Crime Scene Investigation
Underwater Crime Scene Methodology
Forensic Science in Investigation
Scientific Underwater Investigation
Introduction to Underwater Investigation

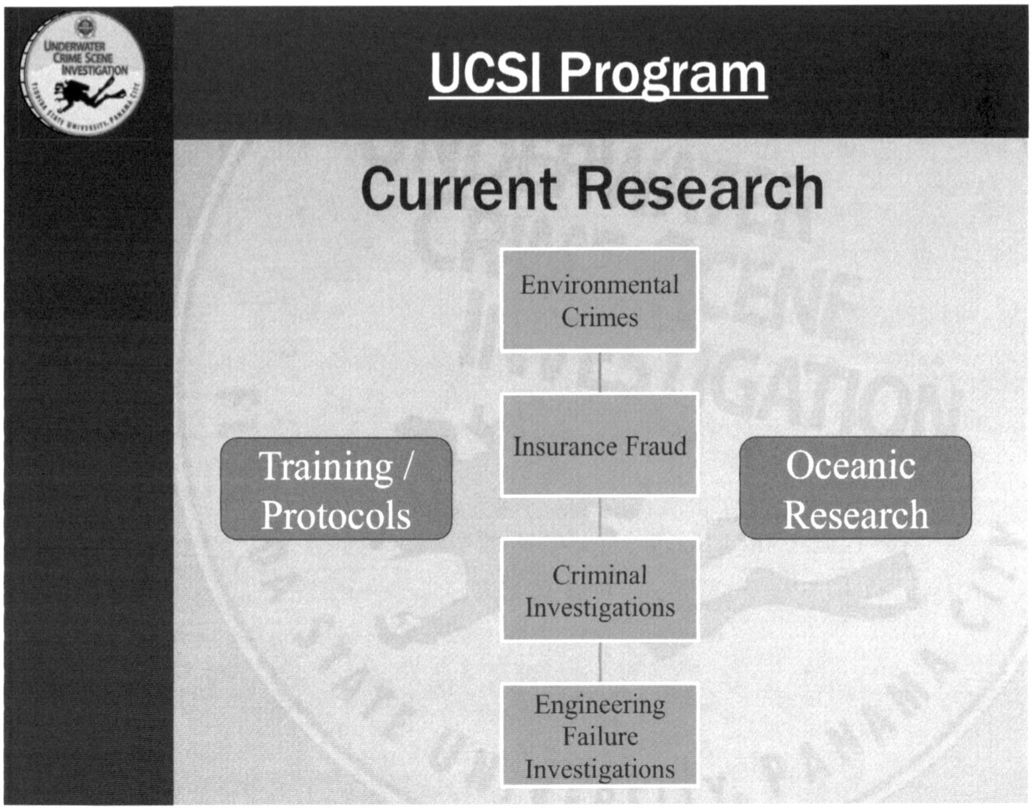
UCSI Program
Current Research
Environmental Crimes
Insurance Fraud
Criminal Investigations
Engineering Failure Investigations
Training / Protocols
Oceanic Research

UCSI Process

Questions

Design and Development of the ROV Xotic for the 2015 MATE International ROV Competition

TECXOTIC

Brissa Romero*
Marco Uribe
Enrique Carvajal
Oliver Santos
Maïa Renaud
Hilda Rojas
Karla Aguirre
Michelle Zenteno
Victor González
Gianpiero Trane

Tecnológico de Monterrey

TECXOTIC Tecnológico de Monterrey

Contents

- Introduction
- Block Diagram
- Power System
- Propulsion System
- Control System
- Structural Frame
- Tools
- Vision System
- Sensors
- Results
- Conclusion
- References

Introduction

- Nowadays, the ROVs take an important role in our society in the development of marine industry and research.
- Because of the versatility of the use of ROVs , the MATE Center organizes an annual themed competition where young people are invited to participate in the design and construction of an ROV to solve different problems.
- The Mexican team TecXotic designed the ROV Xotic in order to accomplish the tasks of the international competition. It was divided in subsystems, these systems where:
  - Power system
  - Propulsion system
  - Control system
  - Structural frame
  - Tools
  - Vision system
  - Sensors

Block diagram

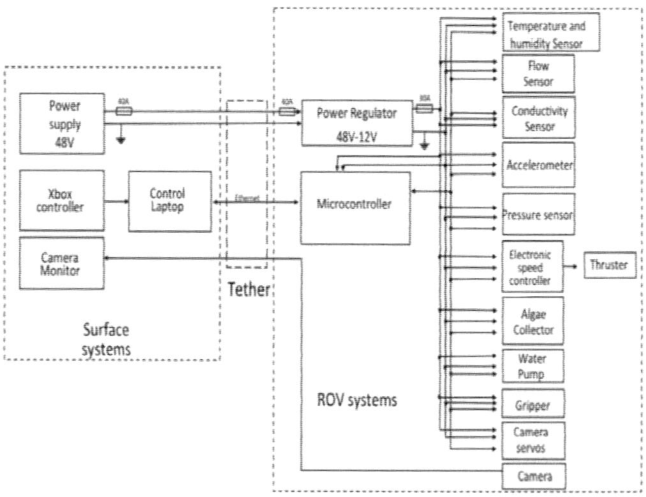

Power System

- To fulfill the competition power requirements of 48 v DC and 40 A, a power DC to DC converter was mounted inside the electronic control box.
- The DC to DC converter gives 12 volts maintaining the 40 amperes stable to power the ESCs and the Brushless motors.
- The digital control systems and sensors work with 5 V at no more than 5 A.
  - A property electronic implementation was required to regulate the input voltage to 5 V up to 5 A, to power microcontrollers, sensors and the GoPro underwater camera.

Propulsion System

- Xotic is driven by eight NTM Prop Drive Series 28 - 30A 800kv located at 45° in order to move the ROV in the 3D.
- Each motor receive a PWM signal (Pulse-Width Modulated) given by a Turnigy TrackStar 25A 1/18th Scale Brushless Car ESC.
- Two different mounting bases were designed, modeled and manufactured in Nylamid to fix each motor to the main structure.

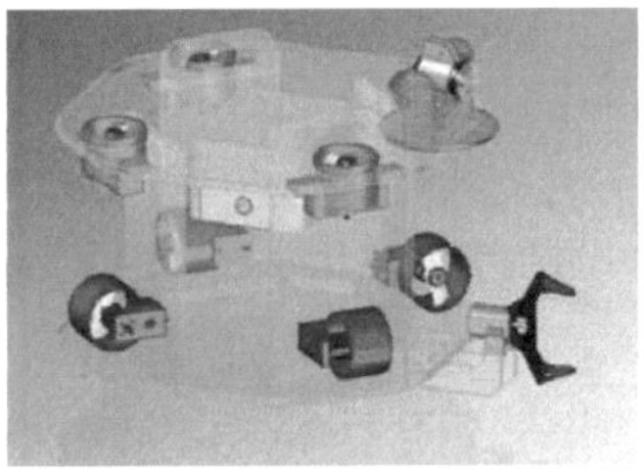

TECXOTIC Tecnológico de Monterrey

Displacements

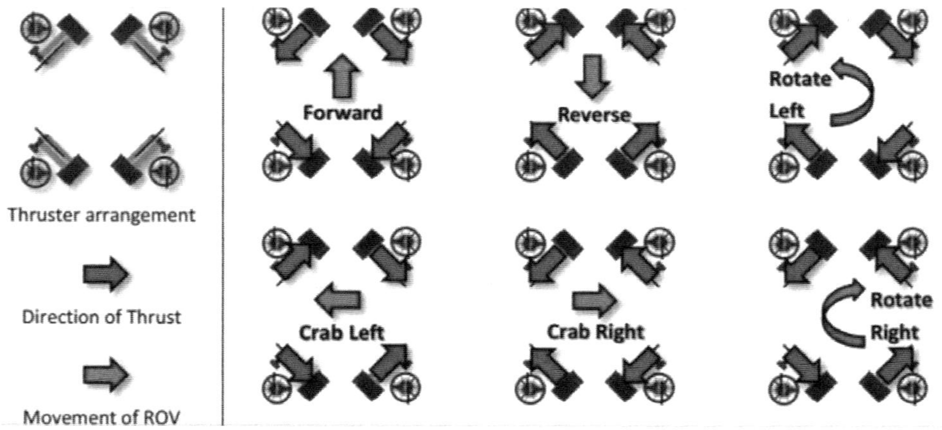

TECXOTIC Tecnológico de Monterrey

Control System

- X-box control is used to control the ROV outside the water connected through an Ethernet cable to the computer, where a program in Processing is used to process the data.
- Inside the ROV, the system is controlled by an Arduino Mega ADK that receives data from the Ethernet cable to the ESC.
- The control allows the ROV to move in every direction at the 3D plane.

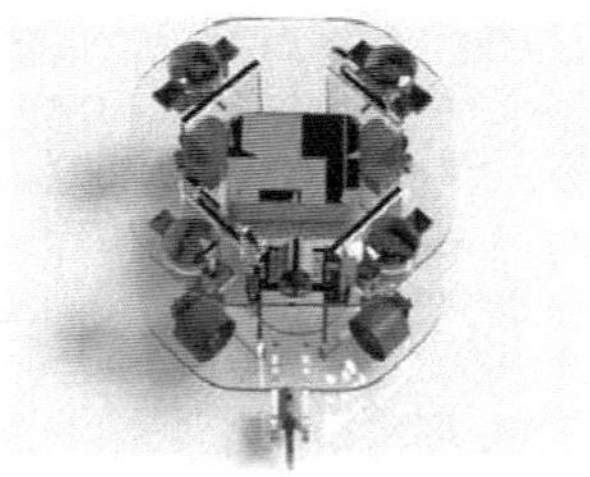

Structural Frame

- The structure was designed to allow the installation of exchangeable tools according to the challenge, also considering hydrodynamics, buoyancy and stability.
- The material chosen for the main structure was 9 mm thick acrylic due to its zero permeability, which is also a resistant material.
- A waterproof electronic control housing was adapted with a multipurpose industrial box made by Schneider Electric.

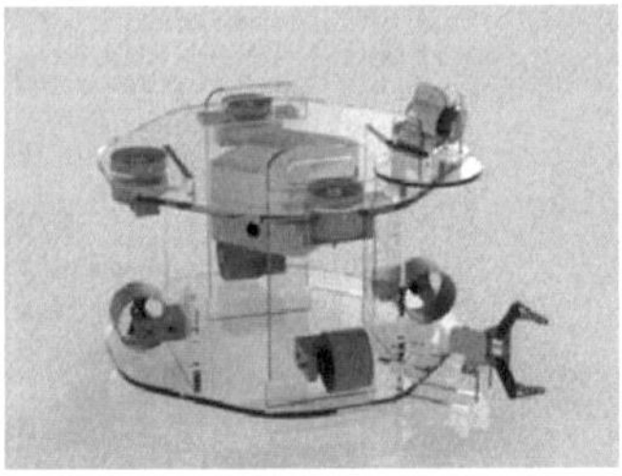

Tools

To facilitate the accomplishment of the tasks and the distribution of the tools on the main structural frame, it was decided to design exchangeable tools.

- **Algae Collector**:

An algae collector by suction was designed to suck out ping pong balls from a plastic layer. A DC motor and PVC pipes were used.

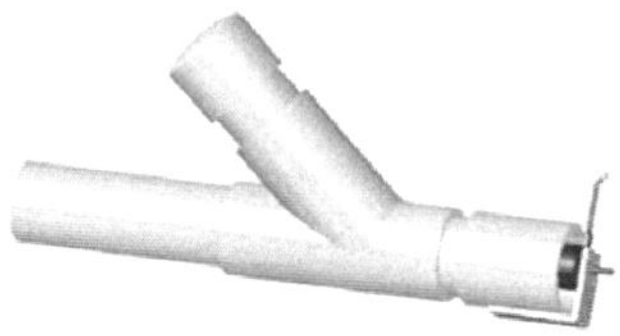

Algae collector CAD model.

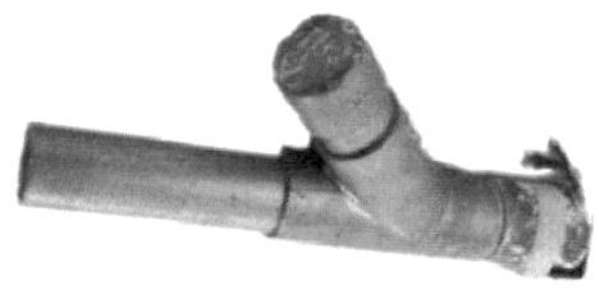

Algae collector real model.

Tools

➢ **Pump:**

A pump was designed to pump water through pipelines. It was made of a PVC pipe and a DC motor. It also has two 3D-printed supports for its fixation to the structural frame.

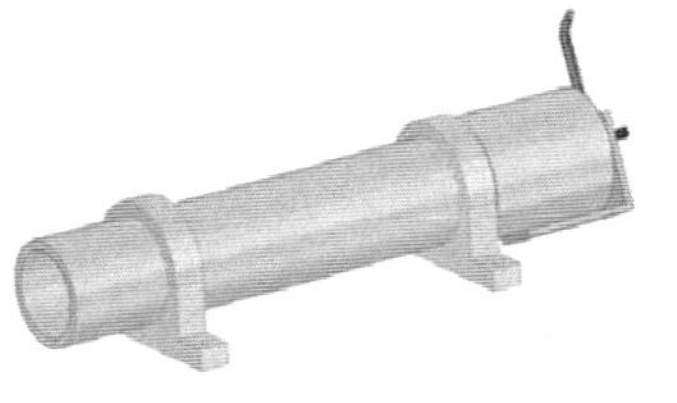

Pump CAD model.

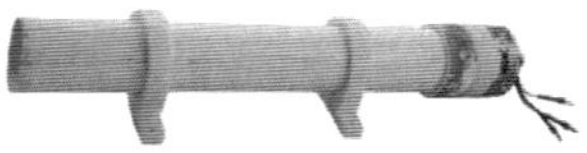

Pump real model.

Tools

➢ **Gripper:**

Two types of grippers were used for tasks in which were necessary to manipulate objects.

- The first gripper that was used is a commercial gripper made of plastic, with a micro gear motor for the closing mechanism, and acrylic claws that were adapted for a better grip.
- The final gripper was CAD designed and CNC machined. It has three degrees of freedom: rotation and vertical tilt.

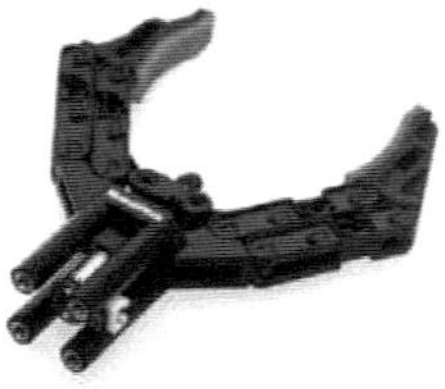

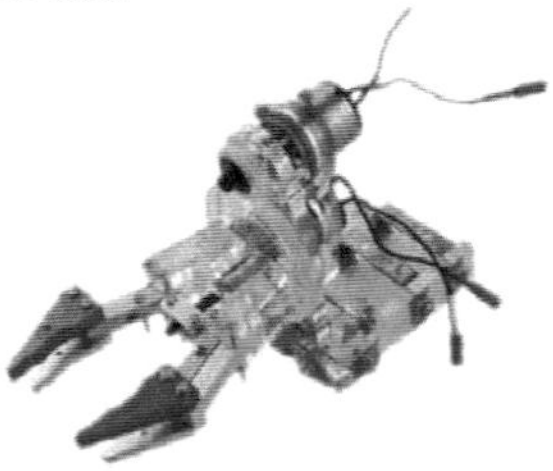

Vision System

- A GoPro camera was used with an under because of its graphics quality and wide vision angle.
- In order to transmit the video to the surface, we used a video out cable and another cable for powering.
- The video is transmitted in a Standard Definition (SD).
- A mechanism with was designed to increase vision field, it has 3 DOF to allow angular and X-Y plane movement.

Sensors

- **Voltage sensor**
  - This sensor consists in a circuit with two plates that will be able to measure voltage touching the common ground and the anode.
  - This sensor is connected to the Arduino Mega ADK in an analog pin and the data will be shown in the interface of the pilot.
- **Temperature and humidity**
  - A DTH11 was used inside the waterproof electronic control housing for safety in order to know if water is leaking through our electronic box or if our system is overheating.

DTH11 Sensor.

TECXOTIC Tecnológico de Monterrey

Sensors

Image Processing Measuring System

- The system works with the camera of the ROV; it was programmed using the Vision Acquisition Module in LabVIEW.
- A target is selected, then the inclination degree is calculated with an analogic accelerometer, the result is the distance from the ROV to the target.
- Once the distance is obtained, a pixel/cm scale is calculated.
- With the target image, the amount of pixels is obtained and the application returns the real size.

TECXOTIC Tecnológico de Monterrey

Results

- TecXotic team participated at the Regional MATE Competition helded at the NBL in Houston, TX where after three attempts an extra opportunity was given to us to send a video with the ROV completing the tasks. With this evidence the MATE Center give us the acceptance to participate at the 2015 MATE International ROV Competition in Newfoundland and Labrador, Canada.
- At the final competition in Newfoundland and Labrador Canada, the TecXotic team faced very well some of the challenges, including the security inspection and the sales presentation.
- The development of the ROV, from the design until the team work and participation gives the opportunity to win the Judges' Choice Award 2015.

Conclusion

- As TecXotic team, we are proud of our participation at the 2015 MATE International ROV Competition.
- The achievements fill us with satisfaction and encourage us to continue working.
- Now we know the improvements that must be made to the prototype in order to have a better development at the competition.
- The experience that this competition give us is unique and amazing.

References

1. S. Jin, J. Kim, J. Kim, and T. Seo, "Hovering underwater robotic platform with four tilting thrusters," in Advanced Intelligent Mechatronics (AIM), 2014 IEEE/ASME International Conference on, 2014, pp. 1547-1551.
2. W. M. Bessa, M. S. Dutra, and E. Kreuzer, "Depth control of remotely operated underwater vehicles using an adaptive fuzzy sliding mode controller," Robotics and Autonomous Systems, vol. 56, pp. 670-677, 2008.
3. H.-H. Chen, W.-N. Chuang, and C.-C. Wang, "Vision-based line detection for underwater inspection of breakwater construction using an ROV," Ocean Engineering, vol. 109, pp. 20-33, 2015.
4. H.-T. Choi, J. Choi, Y. Lee, Y. S. Moon, and D. H. Kim, "New concepts for smart ROV to increase efficiency and productivity," in Underwater Technology (UT), 2015 IEEE, 2015, pp. 1-4.
5. S. Bhattacharyya and H. H. Asada, "Control of a compact, tetherless ROV for in-contact inspection of complex underwater structures," in Intelligent Robots and Systems (IROS 2014), 2014 IEEE/RSJ International Conference on, 2014, pp. 2265-2272.
6. V. N. Kuhn, P. L. J. Drews Jr, S. C. P. Gomes, M. A. B. Cunha, and S. S. da Costa Botelho, "Automatic control of a ROV for inspection of underwater structures using a low-cost sensing," Journal of the Brazilian Society of Mechanical Sciences and Engineering, vol. 37, pp. 361-374, 2013.
7. T. Salgado-Jimenez, J. Gonzalez-Lopez, J. Pedraza-Ortega, L. García-Valdovinos, L. Martínez-Soto, and P. Resendiz-Gonzalez, "Design of ROVs for the Mexican power and oil industries," in Applied Robotics for the Power Industry (CARPI), 2010 1st International Conference on, 2010, pp. 1-8.
8. T. Salgado-Jimenez, J. Gonzalez-Lopez, L. Martinez-Soto, E. Olguin-Lopez, P. Resendiz-Gonzalez, and M. Bandala-Sanchez, "Deep water ROV design for the Mexican oil industry," in OCEANS 2010 IEEE-Sydney, 2010, pp. 1-6.

Thank you for your attention.

Questions

Tecnológico de Monterrey

Abstract

The Marine Advanced Technology Education (MATE) is an association dedicated to improve and support marine technical education and research. Remote Operated Vehicles (ROV) have been used to explore and interact with the marine environment. MATE uses the development of this technology to involve students in science, technology and engineering. At the MATE International ROV Competition, each year, students from all over the world design proprietary ROVs, to accomplish determined tasks. This 2015 the theme was Science and Industry in the Arctic.

This paper describes the design and technical integration of an ROV developed by the mexican team TecXotic, to participate at the 2015 MATE International ROV Competition. The main tasks to accomplish were pipelines visual inspection and repair, offshore oilfield maintenance and science under the ice. According to this, the design included navigation, vision, taking samples and sensing marine environment. The prototype implemented is composed of an acrylic based octagonal structure, 8 thrusters to displace in all directions, an X-Box controller with a graphical interface to operate the ROV, a power management system, a GoPro camera underwater vision system, 2 interchangeable grippers with 1 and 3 degrees of freedom and multiple interchangeable tools. In addition, an ethernet communication system was used to transmit and receive data from the ROV to the control system in the surface programed in Processing.

The overall design and implementation of this project was done by the team, no pieces or structures were purchased or outsourced manufactured. The prototype worked mostly properly and the outcome was successful: TecXotic was the only team that got the pass to the international competition at Houston TX. The project development from the design to the final competition got the acceptance of the judges, leading TecXotic to the Judges Choice Award at the 2015 MATE International ROV Competition.

The Development of an Autonomous Underwater Vehicle with Locomotion Inspired by the Thunnus Thynnus

Wayne Stegner

Introduction/Background - AUVs

- AUV - Autonomous Underwater Vehicle
- "A robotic device that is driven through the water by a propulsion system, controlled and piloted by an onboard computer, and maneuverable in three dimensions" (Alt, 2003)

Introduction/Background - AUVs (*cont.*)

- First AUV - 1960's
- Continuously developing
- Purposes
  - Search for wreckage
  - Underwater Exploration/Experimentation
  - Studying marine life

("Special purpose underwater research vehicle (SPURV)," n.d.)

Introduction/Background - Fish

Forces acting upon a fish:

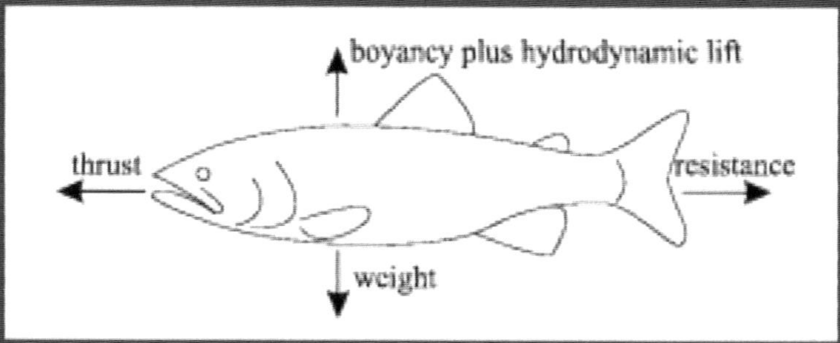

(Sfakiotakis et al., 1999)

Introduction/Background - Fish (*cont.*)

Basic anatomy:

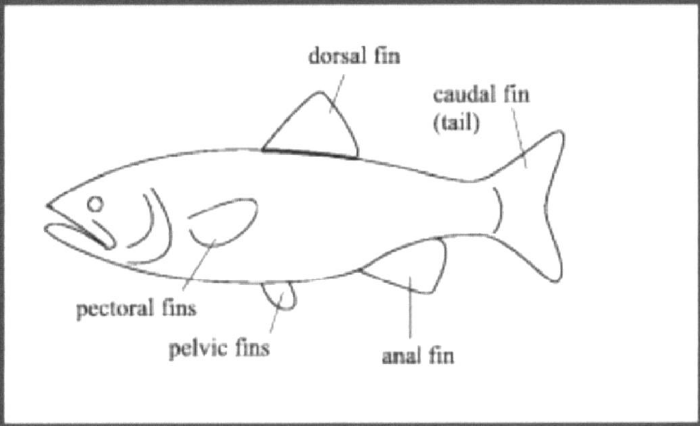

(Sfakiotakis et al., 1999)

Introduction/Background - Existing Robotic Fish

- MIT
  - Pneumatically actuated
  - Moves as fast as real fish
  - Expensive

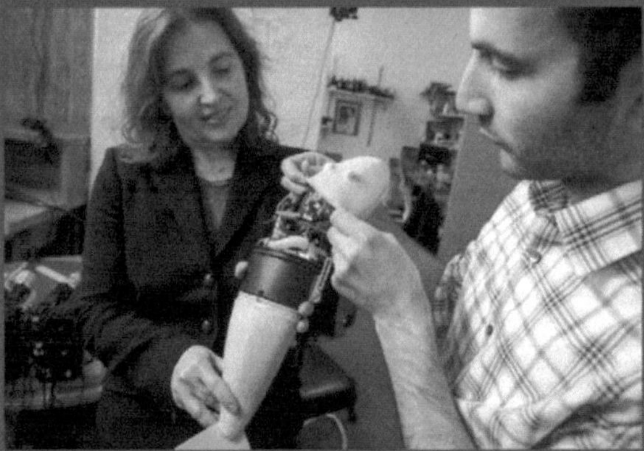

(Dredge, 2014)

Introduction/Background - Existing Robotic Fish (*cont.*)

- DIY Arduino Project
  - Inexpensive
  - Low functioning

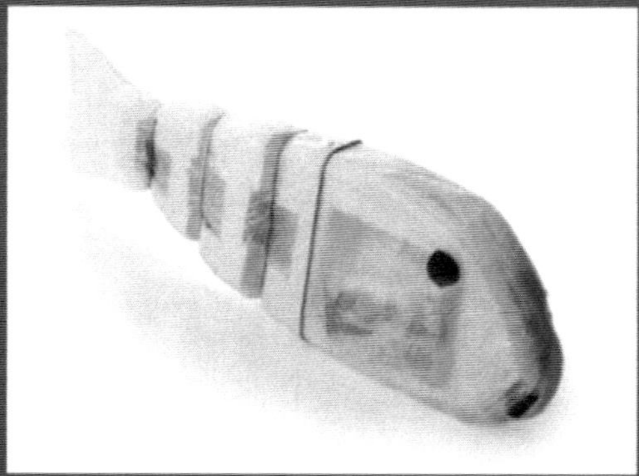

(Landoni, 2013)

Purpose

- Construct a robotic fish combining functionality and cost effectiveness
- Less invasive alternative to traditional AUVs
- Potential applications:
  - Ocean exploration
  - Searching for underwater wreckage
  - Covert military surveillance

Methods

- Based on the Atlantic Bluefin tuna
  - Most efficient mode of swimming

("Why some chefs just can't quit serving bluefin tuna," n.d.)

Methods - Frame Design

- Frame design
  - 3 body segments
    - main body
    - 2 tail segments
  - Dimensioned in Autodesk Inventor

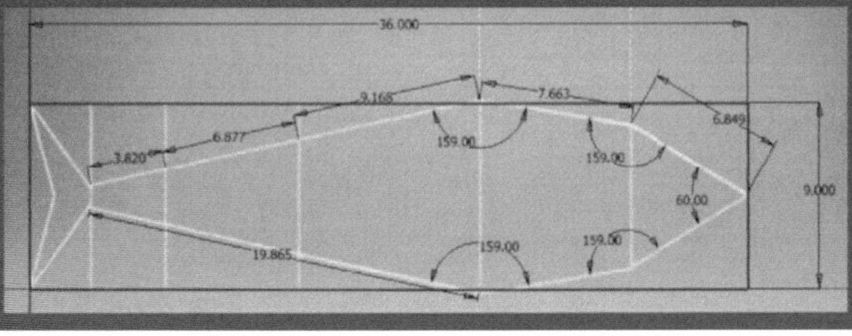

Methods - Custom PVC Joints

- Needed specific angles
    - 60 degrees
    - 159 degrees
- 3D Printing
- First design
    - Tight fit, like manufactured joints
    - Too much stress
    - Plastic layers split

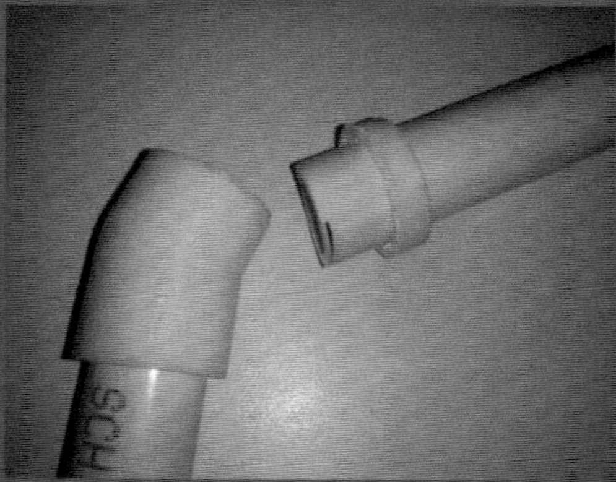

Methods - Custom PVC Joints (*cont.*)

- Second design
    - Looser fit
    - Added screw holes
    - Did not break

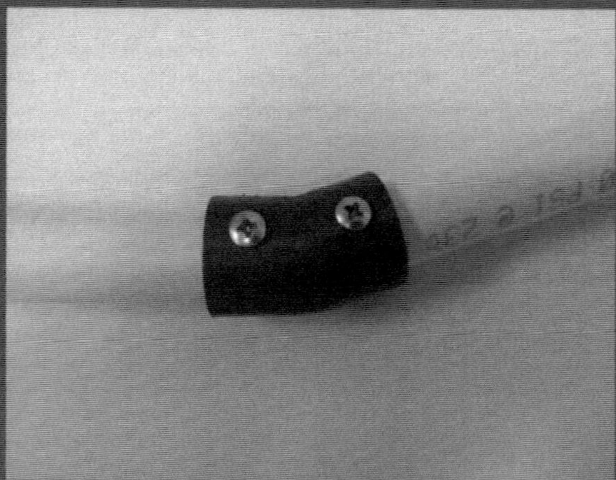

Methods - Custom PVC Joints (*cont.*)

- Test different layering
  - Horizontal vs vertical
  - Strength

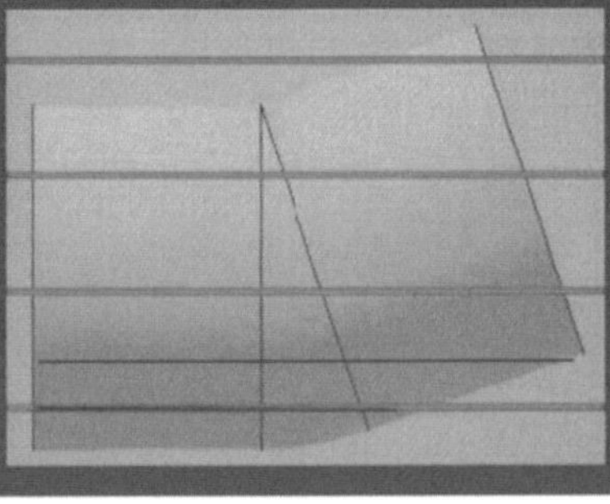

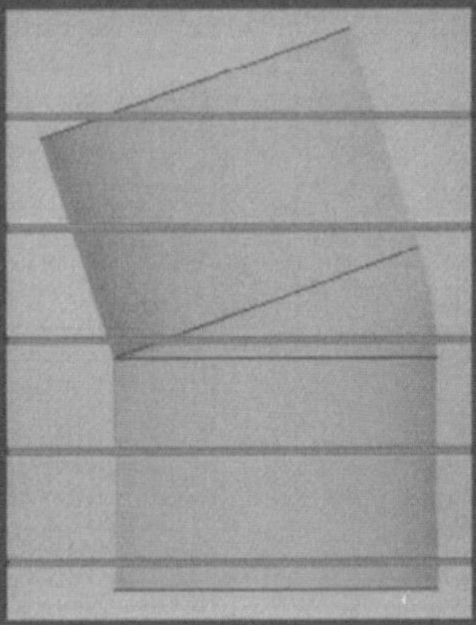

Methods - Custom PVC Joints (*cont.*)

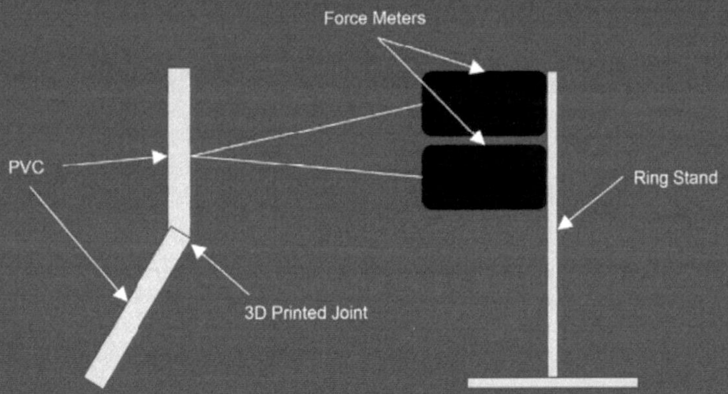

Methods - Custom PVC Joints (*cont.*)

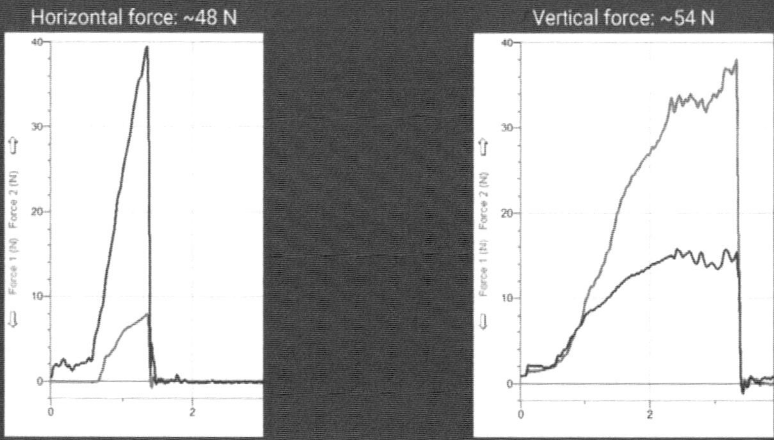

Methods - Custom PVC Joints (*cont.*)

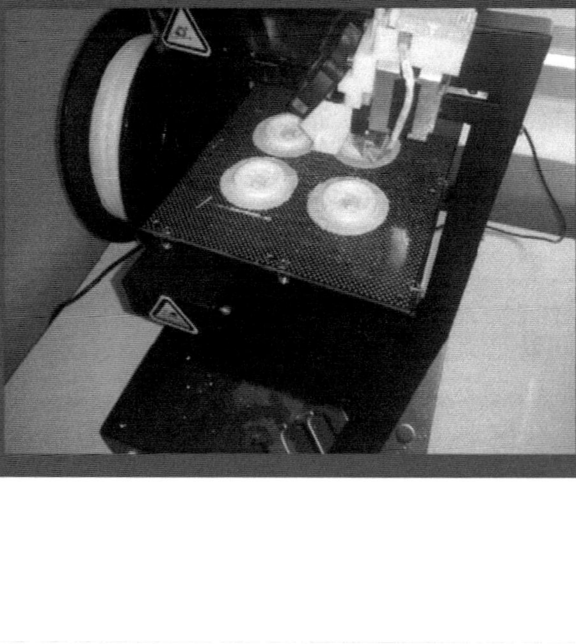

Methods - Motors

- Stepper motors
  - Allows precise control
  - Not enough torque
  - Skips steps

Methods - Motors (*cont.*)

- Servo motors
  - Allow precise control
  - "Smart" - know position

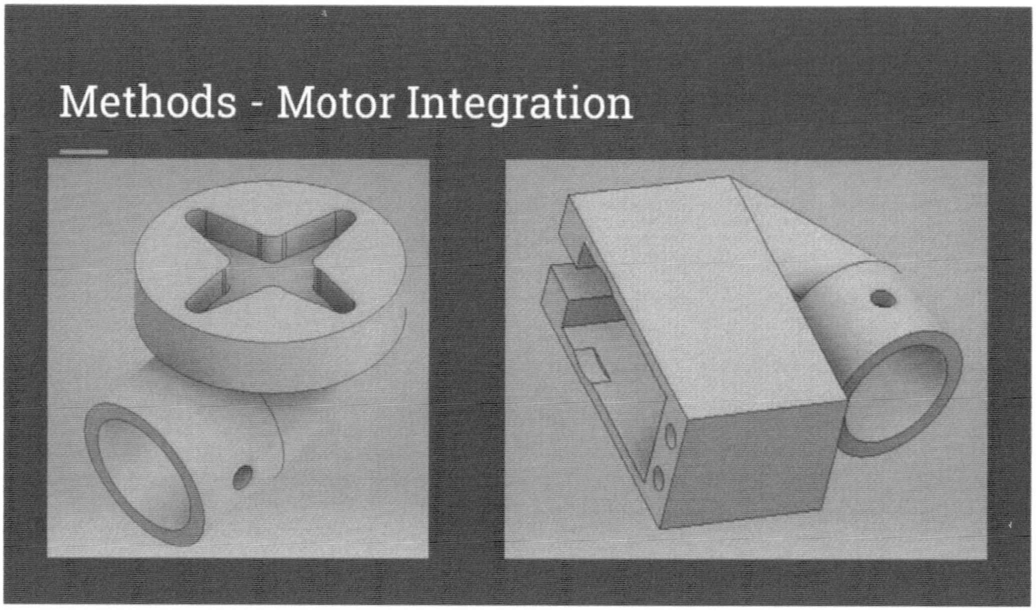
Methods - Motor Integration

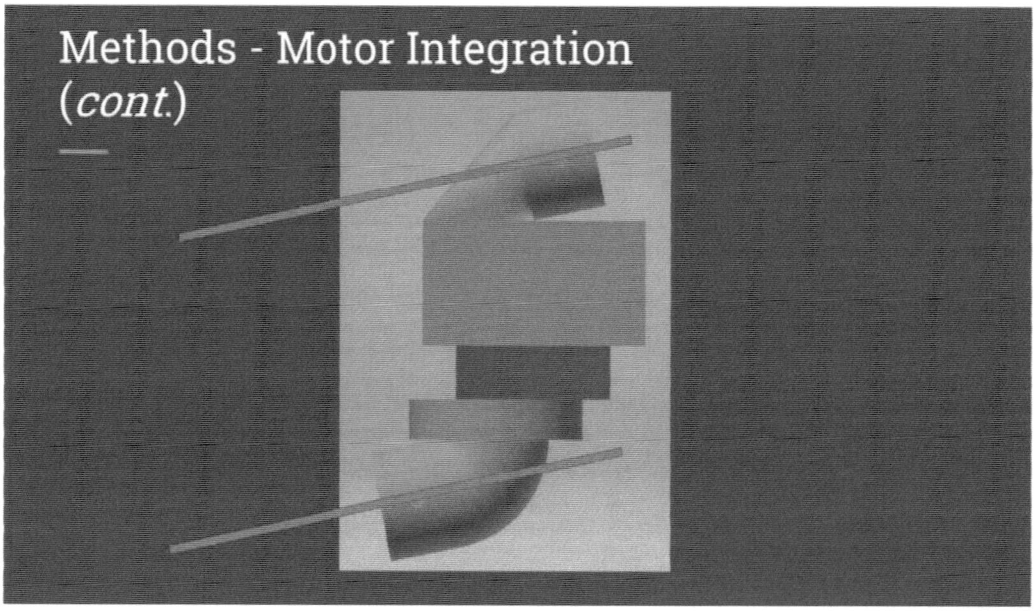
Methods - Motor Integration (*cont.*)

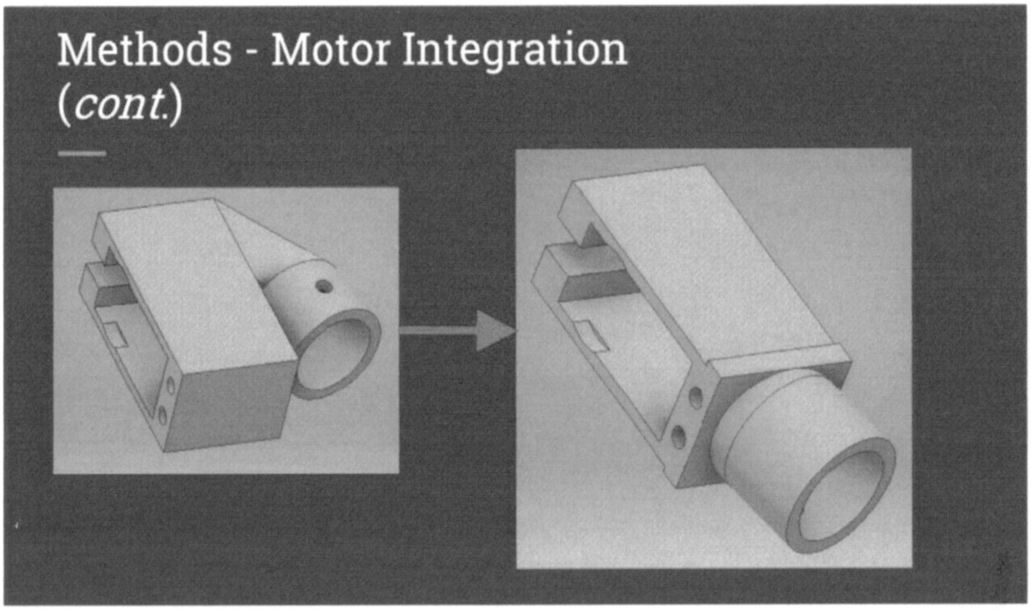
Methods - Motor Integration
(*cont.*)

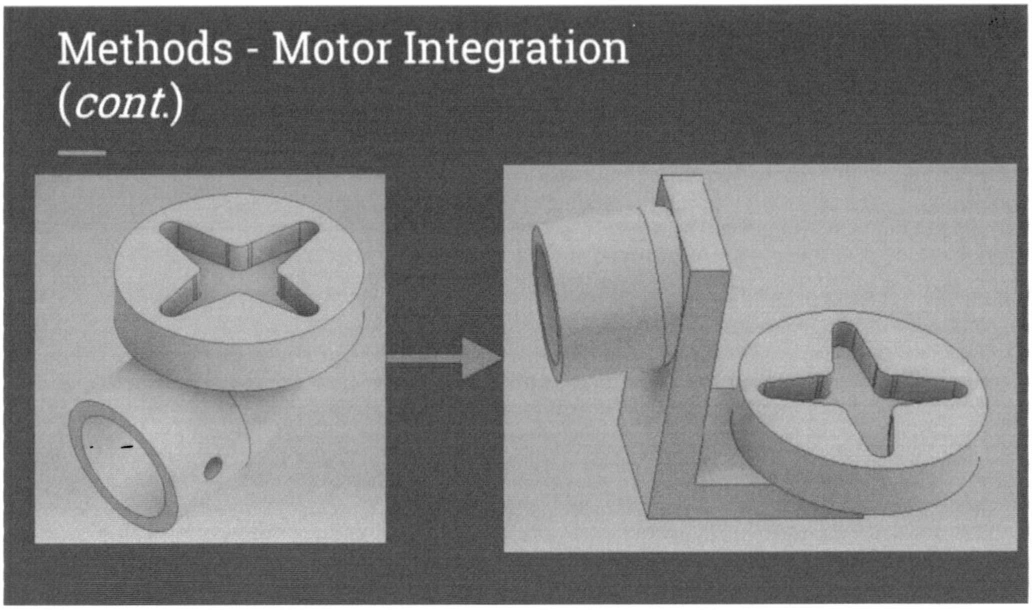
Methods - Motor Integration
(*cont.*)

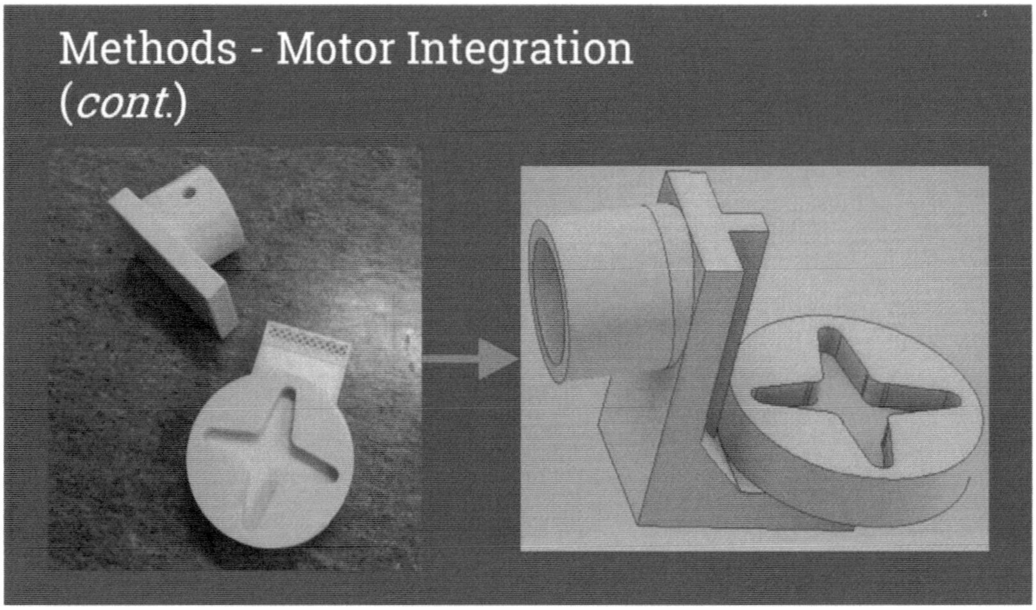
Methods - Motor Integration
(*cont.*)

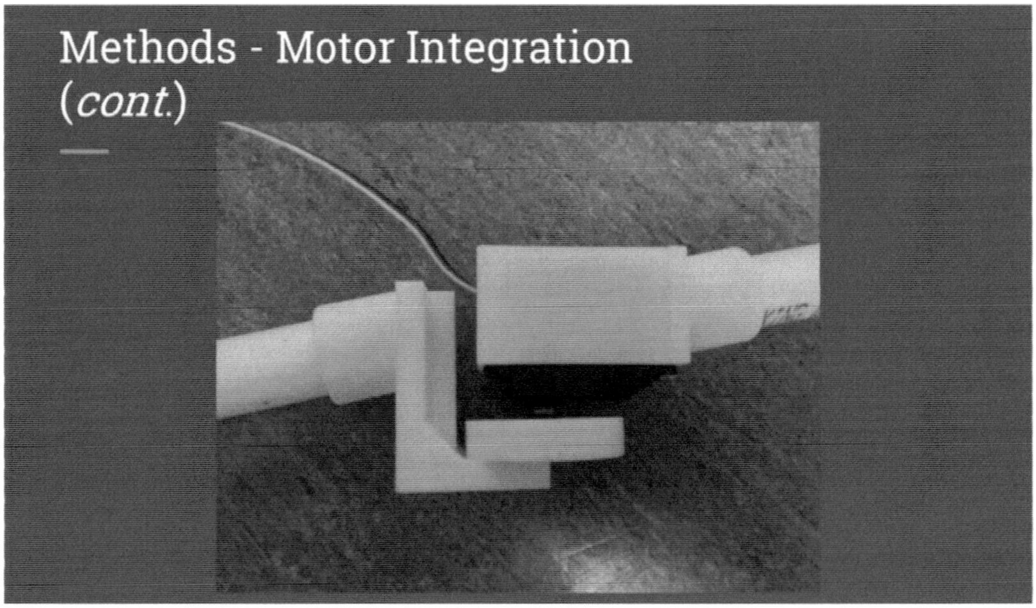
Methods - Motor Integration
(*cont.*)

Methods - Bottom Rotatable Joints

Methods - Bottom Rotatable Joints (*cont.*)

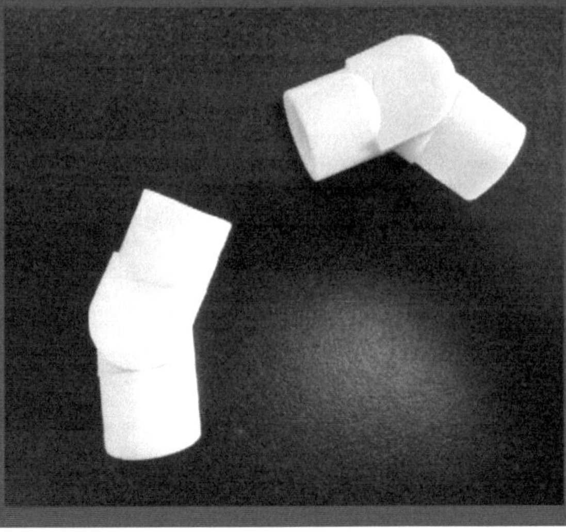

Methods

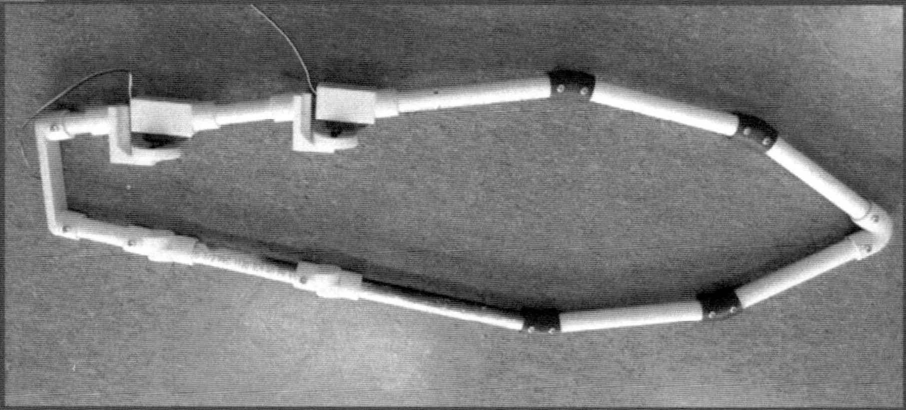

Methods - Electronics

- Arduino Mega
  - Programmed in C/C++
  - I/O pins
    - Controls motors
    - Can read sensors
- Powered by 2200mah battery

("Arduino Mega," n.d.)

Methods - Waterproofing Electronics

Methods - Programming

- PWM - Pulse Width Modulation
  - Tells servos the position
- The program
  - Moves 2 servos
  - Oscillate tail segments

```
int servoPin1 = 3;
int servoPin2 = 5;
int pwmMin = 130;
int pwmMax = 240;
int pwmMid = (pwmMin+pwmMax)/2;
int pause = 1;
int maxRotAngle = 55;
int posComp - pwmMid:
void setup() {
  pinMode(servoPin1,OUTPUT);
  pinMode(servoPin2, OUTPUT);
  Serial.begin(9600);
}
void loop() {
  for(int i = -maxRotAngle; i<maxRotAngle; i++){
    toPos(i);
    delay(pause);
  }
  delay(10C0);
  for(int i = maxRotAngle; i>-maxRotAngle; i--){
    toPos(i);
    delay(pause);
  }
}
void toPos(int x){
  analogWrite(servoPin1, x + posComp);
  analogWrite(servoPin2, x + posComp);
}
```

Results/Analysis – Video

Results/Analysis (*cont.*)

- Good movement
  - Mimics Atlantic Bluefin Tuna
  - Joints and Servo Adapters integrate well
- Motor failure
  - Plastic threads stripped
  - Servo horn falls off

Conclusion

- Stronger gears required
  - Metal gear servos
- Shows promise to emulate the Atlantic Bluefin Tuna
- No method of propulsion yet

Recommendations

- Implement metal gear servos
- Construct caudal fin for propulsion
  - 3D printed
    - More precise
    - More time
  - Wire modeled
    - Less time
    - Less precise

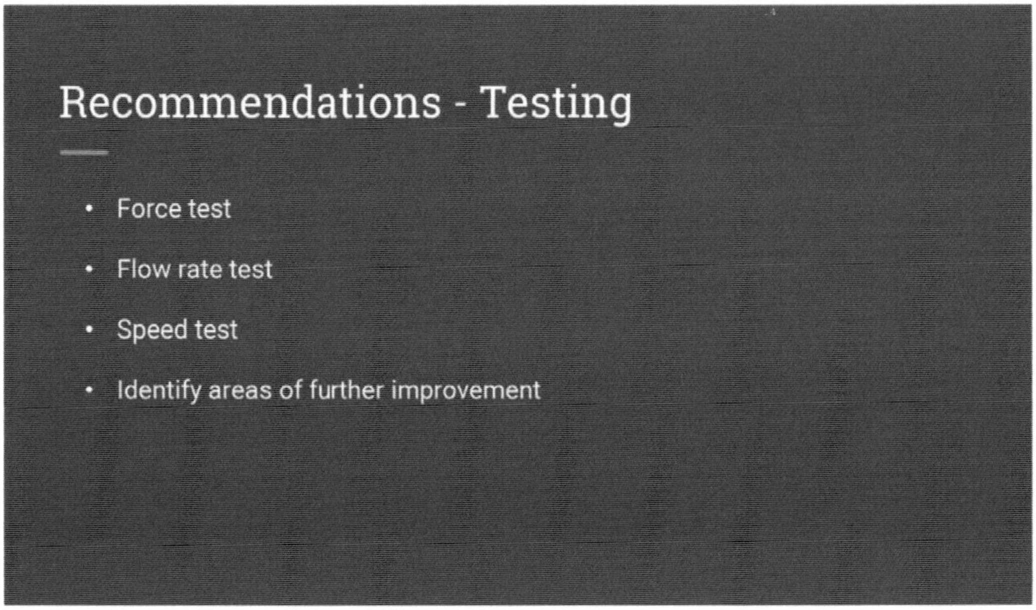
Recommendations - Testing
Force test
Flow rate test
Speed test
Identify areas of further improvement

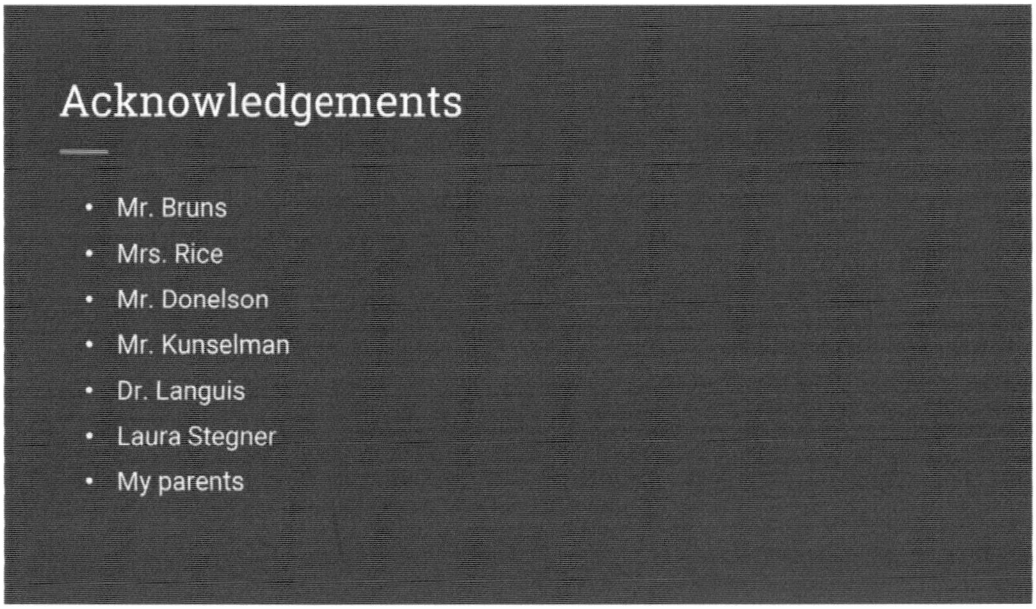
Acknowledgements
Mr. Bruns
Mrs. Rice
Mr. Donelson
Mr. Kunselman
Dr. Languis
Laura Stegner
My parents

References

Alt, C. V. (2003). Autonomous underwater vehicles. doi:10.5772/923

Arduino Mega. (n.d.). Retrieved December 6, 2015, from https://www.arduino.cc/en/Main/arduinoBoardMega

Autonomous underwater vehicle ABE. (n.d.). Retrieved September 27, 2015, from https://www.whoi.edu/main/ABE

Benyus, J. (n.d.). Biomimicry Explained. Retrieved September 25, 2015, from http://www.cbid.gatech.edu/biomimicry_defined.html

Biomimetics. (n.d.). Retrieved September 28, 2015, from https://en.wikipedia.org/wiki/Biomimetics

Dredge, S. (2014, March 14). Fish'n'microchips: MIT's soft robotic fish swims 'like the real thing' Retrieved July 27, 2015, from http://www.theguardian.com/technology/2014/mar/14/mit-soft-fish-robot-swimming

Fish anatomy. (2010). Retrieved September 22, 2015, from http://www.versaquatics.com/fishanatomy.htm

Graham, J. B., & Dickson, K. A. (2000). The evolution of thunniform locomotion and heat conservation in scombrid fishes: New insights based on the morphology of allothunnus fallai [Abstract]. *Zoological Journal of the Linnean Society, 129*(4). doi:10.1006/zjls.1998.0211

Hirata, K., & Kawai, S. (2001, November 15). Hydrodynamic performance of stream-lined body. Retrieved September 25, 2015, from https://www.nmri.go.jp/eng/khirata/fish/experiment/upf2001/body_e.html

ISE Theseus. (n.d.). Retrieved September 27, 2015, from http://www.ise.bc.ca/theseus.html

Landoni, B. (2013, February 26). Robofish: Create your robot fish with Arduino. Retrieved July 28, 2015, from http://www.open-electronics.org/robofish-create-your-robot-fish-with-arduino/

References (*cont.*)

Partridge, B. L., & Pitcher, T. J. (1980). The sensory basis of fish schools: Relative roles of lateral line and vision [Abstract]. *Journal of Comparative Physiology, 135*(4), 315-325. doi:10.1007/BF00657647

Pitcher, T. J., Partridge, B. L., & Wardle, C. S. (1976, November 26). A blind fish can school [Editorial]. *Science, 194*(4268), 963-965. Retrieved September 23, 2015, from http://www.sciencemag.org/content/194/4268/963.short

PWM. (n.d.). Retrieved December 6, 2015, from https://www.arduino.cc/en/Tutorial/PWM

REMUS. (n.d.). Retrieved September 27, 2015, from http://www.whoi.edu/main/remus

Sfakiotakis, M., Lane, D. M., & Davis, J. C. (1999). Review of fish swimming modes for aquatic locomotion. *IEEE JOURNAL OF OCEANIC ENGINEERING, 24*(2), 237-252. doi:10.1109/48.757275

Special purpose underwater research vehcle (SPURV). (n.d.). Retrieved December 4, 2015, from http://www.navaldrones.com/SPURV.html

Stepper motor. (n.d.). Retrieved December 6, 2015, from http://home.roboticlab.eu/en/examples/motor/stepper

Tu, X. (1996, January). Pectoral Fin Motor Controllers. Retrieved September 23, 2015, from http://www.dgp.toronto.edu/people/tu/thesis/node61.html

What is Arduino? (n.d.). Retrieved September 28, 2015, from https://www.arduino.cc/en/Guide/Introduction

Why some chefs just can't quit serving bluefin tuna. (n.d.). Retrieved December 5, 2015, from http://www.npr.org/sections/thesalt/2015/01/07/375366742/why-some-chefs-just-cant-quit-serving-bluefin-tuna

References (*cont.*)

Yu, J., Tan, M., & Zhang, J. (2010). Fish-inspired swimming simulation and robotic implementation. *Fish-Inspired Swimming Simulation and Robotic Implementation.* Retrieved July 23, 2015, from http://www.researchgate.net/profile/Junzhi_Yu/publication/224232560_Fish-Inspired_Swimming_Simulation_and_Robotic_Implementation/links/0c9605216a9d76fd1d000000.pdf

Zhou, J., & Clark, C. M. (2006). Autonomcus fish tracking by ROV using monocular camera [Abstract]. doi:10.1109/CRV.2006.16

Questions?

Cost Analysis

- Current:
  - Arduino Mega - $40
  - Waterproof Servo Motors (2) - $60
  - Waterproof Ultrasonic Sensor - $15
  - 1/2" PVC pipe - $0.75
  - 3" PVC Pipe and Caps - $10
  - 6" Printer Cable - $5
  - Battery - $10
  - ABS Plastic Filament - $30
- Total: $170.75

Further Improvements

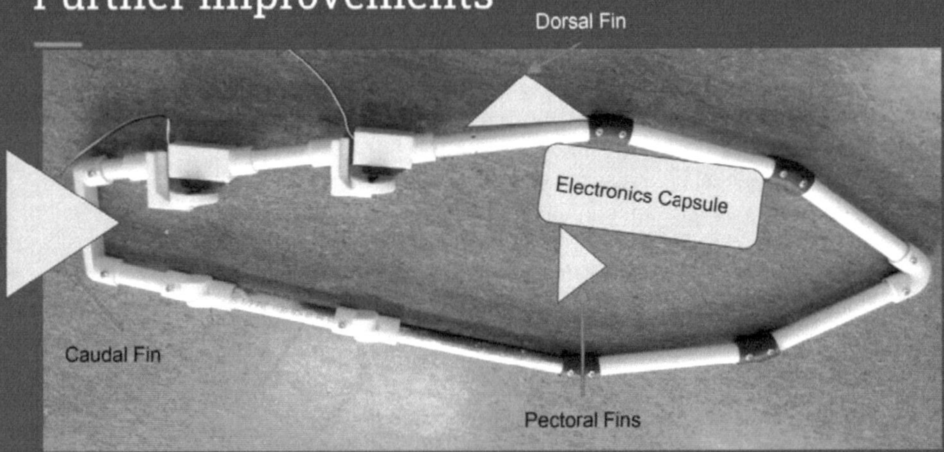

Modes of Locomotion

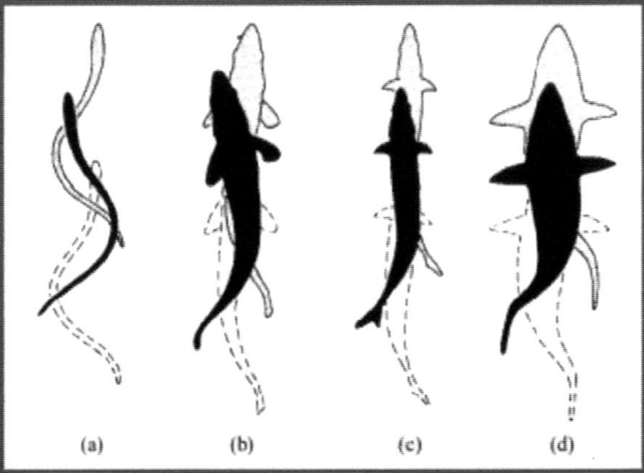

(Sfakiotakis et al., 1999)

PWM

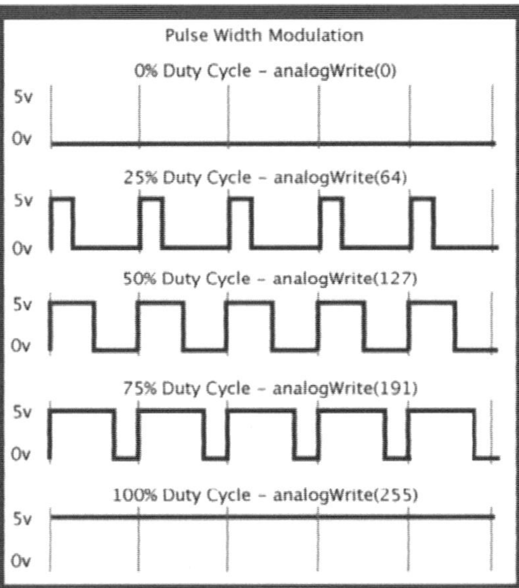

("PWM," n.d.)

Stepper Motors

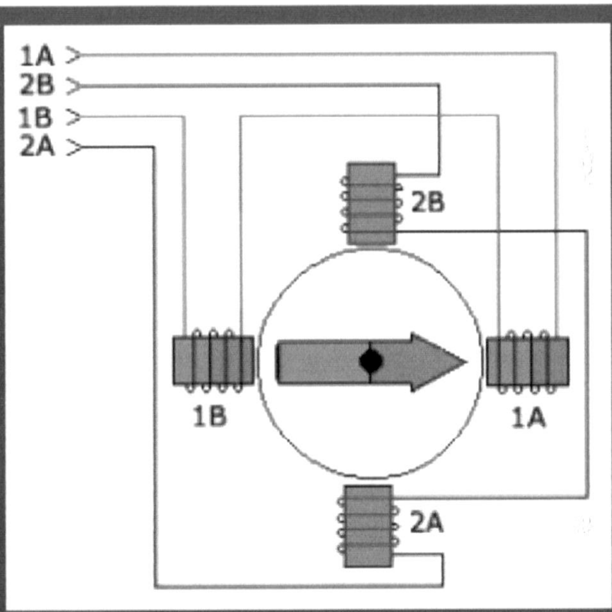

("Stepper motor," n.d.)

MTS marine technology society

UNDERWATER INTERVENTION 2016

2016 MTS MUV PROGRAM

UI2016, New Orleans, LA USA
February 23-25, 2016

CALL for PAPERS

Dear Friends and Colleagues,

I would like to take this opportunity to invite you to join us at UI2016 and be a presenter for the Manned Underwater Vehicles program. The success of the MTS MUV Program over the past 12 years comes from the global participation from members across all sectors (commercial, private, tourism, research and military) and strives to stand as a professional representative group from the world of human occupied vehicles.

It is the time of year where we need to start organizing the program with your ideas for papers and presentations. All you need to do at this stage is to decide if you wish to present at the conference and submit an idea. A Title and a few words on the subject will be a great start.

Deadlines

ABSTRACTS	Oct. 20, 2015
FINAL PAPERS/PRESENTATIONS	Jan. 12, 2016

Some of the theme ideas planned for 2016 include:

1) New Vehicles & Technologies
2) Submersible Operations Updates
3) Advanced Underwater Engineering
4) MUV Operations – Consensus Standards
5) Regulatory updates

Participation at the UI conference offers a unique opportunity to share your work, projects, accomplishments, new ideas and stories with professionals in the global Underwater Intervention industry and the Manned Submersible community specifically.

The biggest benefit of participating as a Speaker is that the entire conference registration, show floor and all technical sessions, is complimentary. It also offers a unique opportunity to discuss face-to-face with professionals around the world under one roof, for three days.

If you have any questions or general theme ideas, please do not hesitate to contact me at the coordinates below. Even if it is to suggest topics that you wish someone would present. I look forward to hearing from you and working with you to make UI2016 another great conference. If in doubt or have any questions, don't hesitate to call me.

Sincerely,

William Kohnen
Track Chair, UI2015 MUV Program
MTS MUV Committee
Tel: (951) 323-5377
Email: wkohnen@hydrospacegroup.com

UI 2016 CONFERENCE DETAILS

VISIT: www.underwaterintervention.com

Technical Program Tracks

1. Work Class ROVs
2. Commercial Diving
3. Manned Submersibles
4. Instruments and Sensors
5. Marine Education/Academia
6. Observation Class ROVs
7. Ocean Research and Marine Archaeology Technology
8. Deepwater Field Development
9. AUV Technology
10. Advanced Underwater Engineering
11. Import/Export Compliance *or How to Avoid Federal Prison*
12. Subsea Survey
13. ASV Technology

WHO ATTENDS?

- Oil & Gas industry professionals / Operators
- Information and technologies futurists
- Senior International and Regional underwater operators
- Underwater Operations Service and equipment suppliers
- Ocean Engineering and construction companies
- Commercial Diving and ROV Contractors
- Underwater Operations Consultants
- Engineers, Project managers, Purchasing Agents
- Law Enforcement Dive Teams
- Fire Department Dive Teams
- Public Safety and Rescue Teams
- Military Specialists and Purchasing Agents

Information for Authors

The main body of your abstract should be between 150 and 300 words. Please include a brief biography for publication in your abstract submittal. You must provide a valid, working, e-mail. Deadline is October 20, 2015.

As authors are accepted, they will be featured in web postings, social media announcements, UI Mobile App, and UI Newsletter. Program Scheduling will begin as presentations are approved and accepted. If your presentation is accepted, you will be automatically registered for the conference and will receive a confirmation letter by e-mail.

(The term "papers" is used to include formal research papers, PowerPoint presentations, Case Studies, Short Film with Presenter Narration, etc.)

Deadline for Final Papers/Presentations is January 12, 2016 Papers submitted after this date will not be listed in the Conference Guide but can be added to the UI Mobile App, the Conference Proceedings and posted to the web site post-conference.

1. Final selection of papers will be determined by the Underwater Intervention Technical Program Committee. Papers will be evaluated on the basis of the abstract submitted. The papers should be in English, completely original, and address underwater operations issues as outlined in the conference focus areas (Tracks). Papers and presentations should avoid any commercialism and overt marketing.
2. The following criteria will be observed by the Underwater Intervention Technical Program Committee in the selection of papers:
    a. The paper should be of interest for underwater operations applications.
    b. The paper should emphasize innovative technologies including an analysis of research.
    c. The paper should address case studies and/or technical interest for indicated topics.
3. Twenty-minutes are programmed to present a paper (presentation in English). A ten-minute discussion will follow each presentation. If your presentation requires a different format, please indicate such requirements at the time of submission.
4. Authors of papers selected for the UI program will be notified by the Track Leader.
5. A manuscript or technical PowerPoint® presentation will be required for each paper selected. Manuscripts should be provided to the Track Leader by email in Word® and PDF format. Copyright of papers and presentations belongs to Underwater Intervention.
6. Maximum length of paper shall be 15 typewritten pages, including references. Bibliography tables should not exceed 6 pages.
7. Full instructions on preparation of manuscripts and presentations will be sent to authors of selected papers. Complete manuscripts must be provided. Slides, sketches, graphs or photographs to illustrate the PowerPoint® presentation shall be provided in a PowerPoint document.
8. Complimentary conference registration will be provided only for the author who makes the presentation. Underwater Intervention assumes no obligation for expenses incurred by authors for travel, lodging, food, or other expenses.

To submit an abstract, go to:

https://meetings.mtsociety.org/eventmanager/abstractmanager/author_login.asp?customercode=MTS&eventcode=WPA. You will need to create an account for yourself. Please record the login and password for future use.

General enquiries may be directed to **UITech@underwaterintervention.com**. This is a committee mailbox and the appropriate Track Leader will respond.

AUTHOR INDEX

Underwater Intervention
5206 Cypress Creek Parkway
Suite 202
Houston, Texas 77069
USA

ISBN 978-1-5108-2297-9
9 781510 822979

CURRAN ASSOCIATES INC.
proceedings.com

Association of Reptilian and Amphibian Veterinarians Conference 2017 (ARAV 2017)

Dallas, Texas, USA
23 – 28 September 2017

ARAV
ASSOCIATION OF REPTILIAN
AND AMPHIBIAN VETERINARIANS